INFRARED HEATING
FOR FOOD AND
AGRICULTURAL PROCESSING

Contemporary Food Engineering

Series Editor

Professor Da-Wen Sun, Director

Food Refrigeration & Computerized Food Technology
National University of Ireland, Dublin
(University College Dublin)
Dublin, Ireland
http://www.ucd.ie/sun/

Infrared Heating for Food and Agricultural Processing, *edited by Zhongli Pan and Griffiths Gregory Atungulu* (2010)

Mathematical Modeling of Food Processing, *edited by Mohammed M. Farid* (2009)

Engineering Aspects of Milk and Dairy Products, *edited by Jane Sélia dos Reis Coimbra and José A. Teixeira* (2009)

Innovation in Food Engineering: New Techniques and Products, *edited by Maria Laura Passos and Claudio P. Ribeiro* (2009)

Processing Effects on Safety and Quality of Foods, *edited by Enrique Ortega-Rivas* (2009)

Engineering Aspects of Thermal Food Processing, *edited by Ricardo Simpson* (2009)

Ultraviolet Light in Food Technology: Principles and Applications, *Tatiana N. Koutchma, Larry J. Forney, and Carmen I. Moraru* (2009)

Advances in Deep-Fat Frying of Foods, *edited by Serpil Sahin and Servet Gülüm Sumnu* (2009)

Extracting Bioactive Compounds for Food Products: Theory and Applications, *edited by M. Angela A. Meireles* (2009)

Advances in Food Dehydration, *edited by Cristina Ratti* (2009)

Optimization in Food Engineering, *edited by Ferruh Erdoğdu* (2009)

Optical Monitoring of Fresh and Processed Agricultural Crops, *edited by Manuela Zude* (2009)

Food Engineering Aspects of Baking Sweet Goods, *edited by Servet Gülüm Sumnu and Serpil Sahin* (2008)

Computational Fluid Dynamics in Food Processing, *edited by Da-Wen Sun* (2007)

Contemporary Food
Engineering Series
Da-Wen Sun, Series Editor

INFRARED HEATING
FOR FOOD AND
AGRICULTURAL PROCESSING

Edited by
Zhongli Pan
Griffiths Gregory Atungulu

CRC Press
Taylor & Francis Group
Boca Raton London New York

CRC Press is an imprint of the
Taylor & Francis Group, an **informa** business

CRC Press
Taylor & Francis Group
6000 Broken Sound Parkway NW, Suite 300
Boca Raton, FL 33487-2742

First issued in paperback 2019

ISBN-13: 978-1-4200-9097-0 (hbk)
ISBN-13: 978-0-367-38378-7 (pbk)

This book contains information obtained from authentic and highly regarded sources. Reasonable efforts have been made to publish reliable data and information, but the author and publisher cannot assume responsibility for the validity of all materials or the consequences of their use. The authors and publishers have attempted to trace the copyright holders of all material reproduced in this publication and apologize to copyright holders if permission to publish in this form has not been obtained. If any copyright material has not been acknowledged please write and let us know so we may rectify in any future reprint.

Library of Congress Cataloging-in-Publication Data

Infrared heating for food and agricultural processing / editors, Zhongli Pan and Griffiths Gregory Atungulu.
    p. cm.
    "A CRC title."
    Includes bibliographical references and index.
    ISBN 978-1-4200-9097-0 (alk. paper)
    1. Radiation preservation of food. 2. Infrared heating. 3. Agricultural processing. 4. Farm produce--Radiation preservation. 5. Infrared radiation--Industrial applications. I. Pan, Zhongli. II. Atungulu, Griffiths Gregory.

TP371.8.I5623 2010
664'.0288--dc22
                                         2010006333

Visit the Taylor & Francis Web site at
http://www.taylorandfrancis.com

and the CRC Press Web site at
http://www.crcpress.com

Contents

Series Preface...vii
Series Editor...ix
Preface...xi
Editors... xiii
Contributors xv

Chapter 1   Fundamentals and Theory of Infrared Radiation 1

            *Soojin Jun, Kathiravan Krishnamurthy, Joseph Irudayaraj, and
            Ali Demirci*

Chapter 2   Infrared Radiative Properties of Food Materials.............................. 19

            Griffiths Gregory Atungulu and Zhongli Pan

Chapter 3   Heat and Mass Transfer Modeling of Infrared Radiation for
            Heating 41

            Fumihiko Tanaka and Toshitaka Uchino

Chapter 4   Emitters and Infrared Heating System Design 57

            Ipsita Das and S.K. Das

Chapter 5   Infrared Drying...89

            Caleb Nindo and Gikuru Mwithiga

Chapter 6   Combined Infrared and Hot Air Drying 101

            Habib Kocabiyik

Chapter 7   Combined Infrared Radiation and Freeze-Drying........................... 117

            Griffiths Gregory Atungulu and Zhongli Pan

Chapter 8   Vacuum Infrared Drying.. 141

            Chatchai Nimmol and Sakamon Devahastin

Chapter 9   Infrared Dry Blanching..169

Zhongli Pan and Griffiths Gregory Atungulu

Chapter 10   Infrared Baking and Roasting...203

Servet Gülüm Sumnu and Semin Ozge Ozkoc

Chapter 11   Infrared Radiation Heating for Food Safety Improvement..............225

Kathiravan Krishnamurthy, Soojin Jun, Joseph Irudayaraj, and Ali Demirci

Chapter 12   Industrial Applications of Infrared Radiation Heating and Economic Benefits in Food and Agricultural Processing...............237

Belgin S. Erdoğdu, İbrahim H. Ekiz, Ferruh Erdoğdu, Griffiths Gregory Atungulu, and Zhongli Pan

Index..275

Series Preface

Food engineering is the multidisciplinary field of applied physical sciences combined with the knowledge of product properties. Food engineers provide the technological knowledge transfer essential to the cost-effective production and commercialization of food products and services. In particular, food engineers develop and design processes and equipment in order to convert raw agricultural materials and ingredients into safe, convenient, and nutritious consumer food products. However, food engineering topics are continuously undergoing changes to meet diverse consumer demands, and the subject is being rapidly developed to reflect market needs.

In the development of food engineering, one of the many challenges is to employ modern tools and knowledge, such as computational materials science and nanotechnology, to develop new products and processes. Simultaneously, improving quality, safety, and security remain critical issues in the study of food engineering. New packaging materials and techniques are being developed to provide more protection to foods, and novel preservation technologies are emerging to enhance food security and defense. Additionally, process control and automation regularly appear among the top priorities identified in food engineering. Advanced monitoring and control systems are developed to facilitate automation and flexible food manufacturing. Furthermore, energy savings and minimization of environmental problems continue to be important issues in food engineering, and significant progress is being made in waste management, efficient utilization of energy, and reduction of effluents and emissions in food production.

The Contemporary Food Engineering book series, which consists of edited books, attempts to address some of the recent developments in food engineering. Advances in classical unit operations in engineering related to food manufacturing are covered as well as such topics as progress in the transport and storage of liquid and solid foods; heating, chilling, and freezing of foods; mass transfer in foods; chemical and biochemical aspects of food engineering and the use of kinetic analysis; dehydration, thermal processing, nonthermal processing, extrusion, liquid food concentration, membrane processes, and applications of membranes in food processing; shelf-life, electronic indicators in inventory management, and sustainable technologies in food processing; and packaging, cleaning, and sanitation. These books are aimed at professional food scientists, academics researching food engineering problems, and graduate-level students.

The editors of these books are leading engineers and scientists from all parts of the world. All of them were asked to present their books in such a manner as to address the market needs and pinpoint the cutting-edge technologies in food engineering. Furthermore, all contributions are written by internationally renowned experts who have both academic and professional credentials. All authors have attempted to provide critical, comprehensive, and readily accessible information on the art and science of a relevant topic in each chapter, with reference lists for further information. Therefore, each book can serve as an essential reference source to students and researchers in universities and research institutions.

Da-Wen Sun
Series Editor

Series Editor

Born in southern China, Professor Da-Wen Sun is a world authority in food engineering research and education; he is a member of the Royal Irish Academy, which is the highest academic honor in Ireland. His main research activities include cooling, drying, and refrigeration processes and systems; quality and safety of food products; bioprocess simulation and optimization; and computer vision technology. Especially, his innovative studies on vacuum cooling of cooked meats, pizza quality inspection by computer vision, and edible films for shelf-life extension of fruits and vegetables have been widely reported in national and international media. The results of his work have been published in over 200 peer-reviewed journal papers and more than 200 conference papers.

Sun received his B.Sc. honors (first class), his M.Sc. in mechanical engineering, and his Ph.D. in chemical engineering in China before working in various universities in Europe. He became the first Chinese national to be permanently employed in an Irish university when he was appointed as college lecturer at the National University of Ireland, Dublin (University College Dublin) in 1995, and was then continuously promoted in the shortest possible time to senior lecturer, associate professor, and full professor. He is currently the professor of food and biosystems engineering and the director of the Food Refrigeration and Computerized Food Technology Research Group at University College Dublin.

Sun has contributed significantly to the field of food engineering as a leading educator in this field. He has trained many Ph.D. students who have made their own contributions to the industry and academia. He has also regularly given lectures on advances in food engineering in international academic institutions and delivered keynote speeches at international conferences. As a recognized authority in food engineering, he has been conferred adjunct/visiting/consulting professorships from 10 top universities in China including Zhejiang University, Shanghai Jiaotong University, Harbin Institute of Technology, China Agricultural University, South China University of Technology, and Jiangnan University. In recognition of his significant contribution to food engineering worldwide and for his outstanding leadership in this field, the International Commission of Agricultural Engineering (CIGR) awarded him the CIGR Merit Award in 2000 and again in 2006. The Institution of Mechanical Engineers (IMechE) based in the United Kingdom named him Food Engineer of the Year 2004. In 2008, he was awarded the CIGR Recognition Award in honor of his distinguished achievements in the top 1% of agricultural engineering scientists in the world. In 2010, he received the CIGR Fellow Award for his sustained outstanding contributions worldwide, the title of Fellow is the highest honor in CIGR.

Sun is a fellow of the Institution of Agricultural Engineers and a Fellow of Engineers Ireland (the Institution of Engineers of Ireland). He has received numerous awards for teaching and research excellence, including the President's Research Fellowship and the President's Research Award of University College Dublin on two occasions. He is the CIGR Incoming President 2011–2012, and a member of the CIGR Presidium CIGR Executive Board; the editor-in-chief of *Food and Bioprocess Technology—An International Journal* (Springer); the former editor of the *Journal of Food Engineering* (Elsevier); and an editorial board member for the *Journal of Food Engineering* (Elsevier), the *Journal of Food Process Engineering* (Blackwell), *Sensing and Instrumentation for Food Quality and Safety* (Springer), and *Czech Journal of Food Sciences*. He is also a chartered engineer.

Preface

One of the primary objectives of the food industry is to transform raw agricultural materials into foods suitable for human consumption with ensured safety and reduced production cost. The processing methods have become sophisticated and diverse in response to the growing demand for food quality, nutrition, and safety. Consumers' expectations for convenience, variety, adequate shelf-life and caloric content, reasonable cost, and environmental soundness have equally demanded advancement and modification to existing food processing techniques and adoption of more novel processing technologies. This book is the most comprehensive and ambitious undertaking we are aware of that documents novel engineering approaches and applications of infrared radiation (IR) heating to process food and agricultural products by prominent food processing researchers in response to the current consumers' needs.

IR heating was first used in the 1930s for automotive curing applications and rapidly became a widely applied technology in the manufacturing industry. Contrarily, a similar pace in the development of IR technologies for processing foods and agricultural products was not achieved due to a slower progress in research addressing the heterogeneous agricultural field landscape. The major obstacle to the adoption of the IR technology in agricultural and food industries was lack of understanding of the technology. But with the change in events which currently demand that the agricultural sector adopt energy efficient, less water-intensive and environmentally friendly technologies, the application of IR technologies has resurfaced as an alternative to most processing technologies in the agricultural sector with attractive merits such as uniform heating, high heat transfer rate, reduced processing time and energy consumption, and improved product quality and safety. Presently, the merits of IR application for food processing are evidenced by scattered literature on the subject in different research journals in the food processing area. The most recent single book related to IR heating for food processing was published in the 1960s. In the last two decades, researchers have made significant progress in understanding the mechanism of the IR heating of food products and interactions between the IR and food components. The design and efficiency of infrared emitters have also been improved, such as the use of selective wavelength emitters. This book brings to researchers and professionals updated knowledge and novel applications, as well as the potential of IR heating technology.

A number of important elements in this book stand to distinguish it from others in the field. It is the only book in a single volume solely dealing with IR heating for food and agricultural processing. Great emphasis is given to novel applications, and fundamental information is included where necessary to make the book comprehensible to readers who have limited process engineering background. Thermal processing of foods and agricultural products using IR is discussed and, where applicable, review case studies are incorporated to address specific industrial concerns. The volume contains 12 chapters covering fundamentals, emitters, infrared heating system design, drying, blanching, baking, thawing, pest management, food safety improvement, and industrial and economic benefits. The development of combination treatment of IR and other technologies such as combined infrared and freeze-drying, vacuum infrared drying, and IR blanching are expected to receive a lot of attention because of their novel applications captured in this book. The last chapter documents the economic benefits of IR technology. This book contains a significant database of research references on IR application for food and agricultural processing and is expected to be of great value to food process engineers, food processing companies, education and research institutions, and quality control and safety managers in food processing and food manufacturing operations.

Editors

Zhongli Pan is a research engineer in the Processed Foods Research Unit, Western Region Research Center, U.S. Department of Agriculture Agricultural Research Service, and associate adjunct professor in the Department of Biological and Agricultural Engineering, University of California–Davis. Dr. Pan received his B.S. degree in agricultural engineering from Northeast Agricultural University, China, and M.S. and Ph.D. degrees in food engineering from the University of Illinois at Urbana–Champaign and University of California–Davis, respectively. His research interests include the development of novel food processing technologies for improving the values of agricultural products and their components; characterization of the physical, chemical, and rheological properties of agricultural and food products; and modeling and optimization of food processing for improved food quality and ensured food safety. He is the author or coauthor of over 150 scientific and popular articles on food processing technologies. He is a leader in infrared research for food processing with extensive knowledge and industrial experience and is a recipient of the Presidential Early Career Award for Scientists and Engineers in 2007 and the Bring Charm to the World Award in 2008 due to the high impact of his research in food processing and safety. He has served in various leadership positions in several professional societies including the American Society of Agricultural and Biological Engineers and Association of Overseas Chinese Agricultural, Biological and Food Engineers. He currently also serves as the vice chairman of the *International Journal of Agricultural and Biological Engineering*.

Griffiths Gregory Atungulu was born in western Kenya. He received his B.S. degree in agricultural engineering from Jomo Kenyatta University of Agriculture and Technology (JKUAT). He obtained his M.S. and Ph.D. degrees in agricultural engineering (food processing and safety engineering major) from Iwate University (2001) and the United Graduate School of Agricultural Sciences of Iwate University (2004) in Japan, respectively. He served as a faculty member at JKUAT and taught courses related to food process engineering on topics encompassing development of novel food processing technologies, modeling, and analysis of physical and biological processes for food processing and food safety, improving the values of agricultural products and their components. For several years he worked in Iwate and Kyushu universities in Japan, until in 2008 he joined the University of California–Davis. He is currently engaged in cutting-edge engineering research on novel thermal and non-thermal techniques for food processing and safety, including applications of infrared radiation and pulsed electric fields.

Contributors

Dr. Ipsita Das
Department of Electrical Engineering
Indian Institute of Technology–Bombay
Mumbai, Maharashtra, India
ipsita.das@iitb.ac.in

Dr. S.K. Das
Department of Agricultural and Food
    Engineering
Indian Institute of Technology–
    Kharagpur
Kharagpur, West Bengal, India
skd@agfe.iitkgp.ernet.in

Dr. Ali Demirci
Department of Agricultural and
    Biological Engineering
Pennsylvania State University
University Park, Pennsylvania, USA
DEMIRCI@psu.edu

Dr. Sakamon Devahastin
Department of Food Engineering
King Mongkut's University of
    Technology–Thonburi
Tungkru, Bangkok, Thailand
sakamon.dev@kmutt.ac.th

Dr. İbrahim H. Ekiz
Department of Food Engineering
University of Mersin
Mersin, Turkey
hiekiz@mersin.edu.tr

Dr. Belgin S. Erdoğdu
Department of Food Engineering
University of Mersin
Mersin, Turkey

Dr. Ferruh Erdoğdu
Department of Food Engineering
University of Mersin
Mersin, Turkey
ferruherdogdu@yahoo.com,
    ferruherdogdu@mersin.edu.tr

Dr. Joseph Irudayaraj
Department of Agricultural and
    Biological Engineering
Purdue University
West Lafayette, Indiana, USA
josephi@purdue.edu

Dr. Soojin Jun
Department of Human Nutrition, Food
    and Animal Science
University of Hawaii
Honolulu, Hawaii, USA
soojin@hawaii.edu

Dr. Habib Kocabiyik
Agricultural Faculty
Department of Agricultural Machinery
Canakkale Onsekiz Mart University
Cankkale, Turkey
kocabiyikh@comu.edu.tr

Dr. Kathiravan Krishnamurthy
National Center for Food Safety and
    Technology
Summit-Argo, Illinois, USA
and
Department of Chemical and Biological
    Engineering
Illinois Institute of Technology
Chicago, Illinois, USA
kathiravan.rps@gmail.com

Dr. Gikuru Mwithiga
School of Bioresource Engineering and
   Environmental Hydrology
University of Kwazulu Natal
Scottsville, South Africa
gikurum@yahoo.com

Dr. Chatchai Nimmol
Department of Materials Handling
   Engineering
Faculty of Engineering
King Mongkut's University of
   Technology–North Bangkok
Thailand
ccnimmol@gmail.com

Dr. Caleb Nindo
School of Food Science
University of Idaho
Moscow, Idaho, USA
cnindo@uidaho.edu

Semin Ozge Ozkoc
Vocational School of Ihsaniye
Kocaeli University
Kocaeli, Turkey
seminozgeozkoc@gmail.com

Dr. Servet Gülüm Sumnu
Department of Food Engineering
Middle East Technical University
Ankara, Turkey
gulum@metu.edu.tr

Dr. Fumihiko Tanaka
Department of Bio-Production
   Environmental Sciences
Faculty of Agriculture
Kyushu University
Higashi-ku, Fukuoka, Japan
fumit@bpes.kyushu-u.ac.jp

Toshitaka Uchino
Department of Bio-Production
   Environmental Sciences
Faculty of Agriculture
Kyushu University
Higashi-ku, Fukuoka, Japan
toshiu@bpes.kyushu-u.ac.jp

1 Fundamentals and Theory of Infrared Radiation

Soojin Jun, Kathiravan Krishnamurthy,
Joseph Irudayaraj, and Ali Demirci

CONTENTS

1.1 Introduction 1
1.2 Basic Laws of Radiative Heat Transfer.. 2
    1.2.1 Planck's Law... 3
    1.2.2 Wien's Displacement Law 3
    1.2.3 Stefan–Boltzmann's Law 5
1.3 Extinction of Radiation.. 5
1.4 View Factor and Energy Balance 7
1.5 Physiological Effects of IR Radiation.. 10
1.6 Spectral Selectivity for IR Heating.. 13
1.7 Conclusion 16
Nomenclature.. 16
References.. 17

1.1 INTRODUCTION

Infrared (IR) radiation is one of the oldest ways to heat-treat foods. A traditional drying method for food products by exposure to intensive sunlight, it was aimed at reducing water activity and allowing longer periods of storage with minimal packaging requirements. It is known that IR radiation has some advantages over convective heating. Heat transfer coefficients are high, the process time is short, and the cost of energy is low. Since air is transparent to IR radiation, the process can be done at ambient air temperature. Equipment can be compact and automated with a high degree of control over process parameters (Nowak and Lewicki, 2004). Similar to other electromagnetic waves such as microwaves and radio frequencies, IR rays attain their unique radiative characteristics. Two key radiative aspects of interest for designing the IR heater are its spectral distribution and energy intensity. The spectral region of IR radiation can be controlled by the use of appropriate optical filters and the surface temperature of its heating elements. According to Jun and Irudayaraj (2004), the differential energy absorption of protein among several key components in the food complex can be found when the IR ray emits light in the narrow spectral region between 6 and 11 μm. Also, the radiation properties of food materials vary with decreasing water content; consequently, its reflectivity increases and the

1

absorptivity decreases. From the food engineer's standpoint, it is very important to fully understand the above optic-thermal phenomena associated with IR and food products. This chapter reviews and presents a theoretical basis for IR heat processing of food materials and the interaction of IR radiation with food components.

1.2 BASIC LAWS OF RADIATIVE HEAT TRANSFER

IR radiation is the part of the electromagnetic spectrum that is predominantly responsible for the heating effect of the sun, as shown in Figure 1.1 (Modest, 1993). IR radiation can be divided into three different categories: near-IR (NIR), mid-IR radiation (MIR), and far-IR radiation (FIR; Table 1.1; Sakai and Hanzawa, 1994). Since IR radiation is an electromagnetic wave, it has both a spectral and directional dependence. Spectral dependence of IR heating needs to be considered because energy coming out of an emitter is composed of different wavelengths, and the fraction of the radiation in each band is dependent on a number of factors such as the temperature of the emitter, emissivity of the lamp, etc. Radiation phenomena become more complicated because the amount of radiation that is incident on any surface does not have only a spectral dependence but also a directional dependence.

The wavelength at which the maximum radiation occurs is determined by the temperature of the heater. This relationship is described by the basic laws for blackbody radiation, such as Planck's law, Wien's displacement law, and Stefan–Boltzmann's law (Sakai and Hanzawa, 1994; Dangerskog and Österström, 1979).

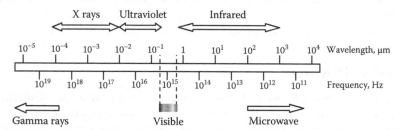

FIGURE 1.1   Electromagnetic wave spectrum.

TABLE 1.1
Classes of Infrared Radiation

Class	Spectral Range
Near-infrared (NIR)	0.75–1.4 μm
Mid-infrared (MIR)	1.4–3 μm
Far-infrared (FIR)	3–1000 μm

Source: Sakai, N., and T. Hanzawa. 1994. *Trends in Food Science and Technology* 5: 357–362. With permission.

1.2.1 PLANCK'S LAW

Planck's law presents the spectral distribution of radiation from a blackbody source that emits 100% IR radiation at a given single temperature (Modest, 1993).

IR sources are made up of thousands of *point sources* at different temperatures. By combining the point sources, an entire spectral distribution for specific regions can be obtained. Hence, an approximation of the spectral distribution using an average surface temperature and emissivity value can be used to characterize the IR radiation.

Max Planck (1901) reported the spectral blackbody emissive power distribution, now commonly known as Planck's law, for a black surface bounded by a transparent medium with refractive index n, as

$$E_{b\lambda}(T,\lambda) = \frac{2\pi hc_0^2}{n^2\lambda^5\left[e^{hc_0/n\lambda kT} - 1\right]} \tag{1.1}$$

where k is known as Boltzmann's constant (1.3806×10^{-23} J/K), and n is the refractive index of the medium. By definition, the refractive index of a vacuum is $n = 1$. For most gases, the refractive index is very close to unity. λ is the wavelength (μm), T is the source temperature (K), c_0 is the speed of light (km/s), and h is Planck's constant (6.626×10^{-34} J·s).

Figure 1.2(a) shows Planck's curve based on Equation 1.1 for a number of blackbody temperatures. Overall, the level of emissive power rises with an increase of temperature, while the wavelength of the corresponding maximum emissive power shifts toward shorter wavelengths. The total amount of IR emissive power within a specific region considered can be estimated by integration of Planck's law at a given temperature with respect to the wavelength.

Planck's law can be applied to estimate the total amount of radiative heat flux when a specific surface temperature of the heating element is known. An energy balance, to assess the amount of energy emitted from the IR source, proportionally directed through a conveying chamber known as a *waveguide* to the surface of the food materials at the receiving end is known as *view factor*. Hence, the actual amount of heat flux absorbed by food can be estimated by calculating the total emissive power and view factors from the source to the target.

1.2.2 WIEN'S DISPLACEMENT LAW

Wien's displacement law gives the wavelength (denoted as peak wavelength) where the spectral distribution of radiation emitted by a blackbody reaches a maximum emissive power. The maximum of the curves (Figure 1.2) can be determined by differentiating Equation 1.1:

$$\frac{d}{d(n\lambda T)}\left(\frac{E_{b\lambda}}{n^3\lambda^5}\right) = 0 \tag{1.2}$$

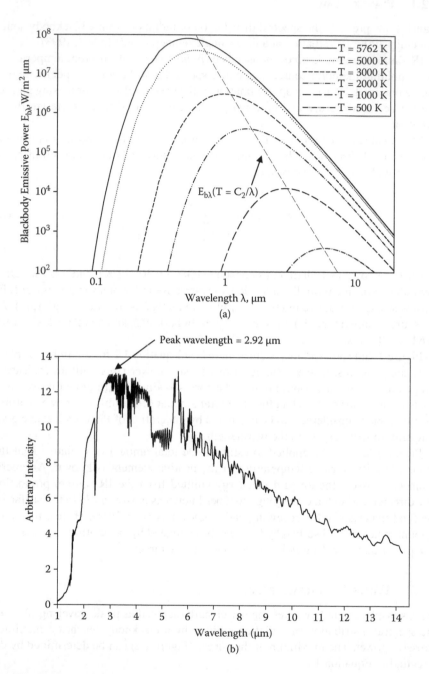

FIGURE 1.2 (a) Blackbody emissive power spectrum and (b) measured emissive power spectrum of IR heating elements.

Source temperatures of IR lamps needed for a desired spectral distribution can be estimated by (Modest, 1993)

$$\lambda_{max} = \frac{2898}{T} \tag{1.3}$$

where T is the source temperature and λ_{max} is the peak wavelength. If the source temperature is known, the peak wavelength can be derived from Equation 1.3. The dotted line in Figure 1.2(a) demonstrates the relationship between the source temperature and the peak wavelength. As an example, the emissive power spectrum of the original IR source with unknown surface temperature can be measured and recorded using the Fourier transform IR (FTIR) spectrometer (Figure 1.2(b)). Based on the plot and Equation 1.3, a peak wavelength of 2.92 μm and correspondent IR source temperature of 720°C are obtained.

1.2.3   STEFAN–BOLTZMANN'S LAW

Stefan–Boltzmann's law gives the total power radiated at a specific temperature from an IR source. The entire amount of heat flux estimated using this law should be consistent with integration of the spectral amount of heat flux estimated using Planck's law given in Sakai and Hanzawa (1994):

$$E_h(T) = \int_0^\infty E_{b\lambda}(T,\lambda)d\lambda = n^2 T^4 \int_0^\infty \frac{C_1 d(n\lambda T)}{(n\lambda T)^5 \left[e^{C_2/(n\lambda T)} - 1 \right]} = n^2 \sigma T^4 \tag{1.4}$$

where $C_1 = 2\pi hc_0^2 = 3.7419 \times 10^{-16}$ Wm^2, $C_2 = hc_0/k = 14{,}388$ μmK, and σ is known as the Stefan–Boltzmann constant (5.670×10^{-8} W/m^{2}K^4). Stefan–Boltzmann's law is available for prompt estimation of the total amount of heat flux at a given source temperature.

1.3   EXTINCTION OF RADIATION

The mechanisms to explain the attenuation of electromagnetic radiation as it propagates through a medium are absorption and scattering. Converting the radiation to some other forms of energy (or some other spectral distribution) is called *absorption phenomena*, whereas scattering mechanisms redirect the radiant energy from its original direction of propagation due to the combined effect of reflection, refraction, and diffraction. The sum of the mechanisms of attenuation of electromagnetic radiation as it passes through a medium (absorption plus scattering) is generally called *extinction* of radiation (Sandu, 1986; Modest, 1993).

When the extinguishing material is agglomerated into particles, separated by regions of different transmissivities (such as emulsions and dispersions), or when variations occur in the density of the sample (as in capillary-porous bodies or in bodies subject to a temperature or moisture gradient or in solid bodies that contain a liquid, free phase inside), Beer's law should be formally adjusted for nonhomogeneous systems using

$$H_\lambda = H_{\lambda 0} \exp\left(-\sigma_\lambda^* u\right) \tag{1.5}$$

where H_λ is the transmitted spectral irradiance (W/m²·μm), $H_{\lambda 0}$ is the incident spectral irradiance (W/m²·μm), u is the mass of absorbing medium per unit area (kg/m²), and σ_λ^* is the spectral extinction coefficient (m²/kg).

Beer's law states that the amount of light absorbed by a solution varies exponentially with the concentration of the solution and the length of the light path in the solution. The spectral extinction coefficient, σ_λ^* (m²/kg), for a nonhomogeneous system is a complex function of the chemical composition of the radiated medium, the physiochemical state of the radiated medium, and the physiochemical parameters defining the radiated medium (density, porosity, diameter of particles, water content, etc.).

In radiative heating, an energy balance can be defined in relation to the extinction of radiation by a physical body. Assuming that this body is an infinite slab of given physicochemical composition and absorbed energy is the total radiation converted into heat inside the body, the entire process of extinction can be defined in terms of reflection, absorption, and transmission of radiation. As illustrated in Figure 1.3, the three fundamental radiative properties are reflectivity (ρ) as the ratio of reflected part of incoming radiation to the total incoming radiation, absorptivity (α) as the ratio of absorbed part of incoming radiation to the total incoming radiation, and transmissivity (τ) as the ratio of transmitted part of incoming radiation to the total incoming radiation.

Under these terms, the energy balance leads to the well-known relation given by

$$\rho + \alpha + \tau = 1 \tag{1.6}$$

Understanding the extinction of radiation is crucial because most IR heat transfer models count on the amount of local heat flux imparted to the food material in relation to the penetration depth.

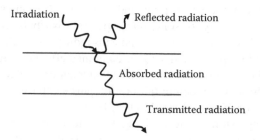

FIGURE 1.3 Extinction of radiation (absorption, transmission, and reflection).

1.4 VIEW FACTOR AND ENERGY BALANCE

To get the energy balance on a certain surface element, one needs to determine how much energy leaves one arbitrary surface element and travels toward other surfaces. The geometric relations governing this process for the participating surfaces are known as view factors (Modest, 1993). For example, Figure 1.4 shows a simplified schematic of the IR heating system, where boundary 1 stands for an assembly of FIR source lamps, boundary 2 for a cone-shaped waveguide to keep IR radiation from dispersing out into the air, boundary 3 for an opening outlet, and boundary 4 for food surface. Under the condition that the IR lamps (boundary 1) and the waveguide (boundary 2) are gray and diffuse reflectors, the total heat flux absorbed by food sample can be calculated using the energy balance equation for each surface (Modest, 1993), given as

$$\frac{q_i}{\varepsilon_i} - \sum_{j=1}^{3}\left(\frac{1}{\varepsilon_j}-1\right)F_{i-j}q_j = E_{bi}(T_i) - \sum_{j=1}^{3}F_{i-j}E_{bj}(T_j), \quad i=1,2,3 \tag{1.7}$$

$$A_1 q_1 + A_3 q_3 = 0 \quad (q_2 = 0, \quad \text{for insulation})$$

where q is the heat flux (W/m²) absorbed or emitted by each boundary (Figure 1.4), F is the view factor, and ε is the emissivity of each boundary. Each view factor is calculated as given by

$$F_{1-3} = \frac{1}{2}\left\{X - \sqrt{X^2 - 4\left(\frac{R_3}{R_1}\right)^2}\right\}, \quad R_1 = \frac{r_1}{a}, \quad R_3 = \frac{r_3}{a}, \quad X = 1 + \frac{1 + R_3^2}{R_1^2} \tag{1.8}$$

(Disk to a parallel coaxial disk with different radius)

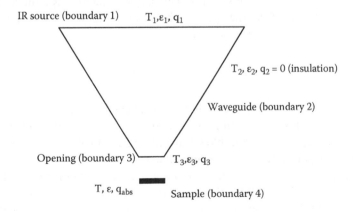

IR source (boundary 1) T_1, ε_1, q_1

$T_2, \varepsilon_2, q_2 = 0$ (insulation)

Waveguide (boundary 2)

Opening (boundary 3) T_3, ε_3, q_3

T, ε, q_{abs} Sample (boundary 4)

FIGURE 1.4 Heating chamber and its constituting surfaces.

where r_1 = radius of side 1 (cm), r_3 = radius of side 3 (cm), and a = height of wave-guide (cm).

$$F_{3-4} = \frac{1}{2}\left\{X - \sqrt{X^2 - 4}\right\}, \quad X = 1 + \frac{a^2 + r^2}{r^2} \quad (1.9)$$

(Disk to a parallel coaxial disk with same radius)

where r = radius of side 3 and side 4 (cm), and a = distance between side 3 and side 4 (cm). Since the summation relationship for an enclosure is unity

$$\left(\sum_{j=1}^{3} F_{1-j} = 1\right), \quad (1.10)$$

$$F_{1-2} = 1 - F_{1-1} - F_{1-3} \ (F_{1-1} = 0)$$

$$= 1 - F_{1-3} \quad (1.11)$$

$$F_{2-1} = \frac{A_1}{A_2} F_{1-2} \text{ (Rules of reciprocity)}$$

$$F_{2-3} = \frac{A_2}{A_3} F_{3-2} = \frac{A_2}{A_3}(1 - F_{3-1}) = \frac{A_2}{A_3}\left(1 - \frac{A_1}{A_3} F_{1-3}\right) \quad (F_{3-3} = 0) \quad (1.12)$$

$$F_{2-2} = 1 - F_{2-1} - F_{2-3} \quad (1.13)$$

The amount of heat flux absorbed by food surface, q_{abs}, is dependent upon the spectral absorptivity (α) of food, the spectral distribution of filtered IR radiation, and the view factor as obtained from the opening (boundary 3) to the food surface (boundary 4). This relationship can be expressed as

$$q_{abs} = F_{3-4} \cdot \alpha(\lambda) \cdot \tau(\lambda) \cdot q_3 \quad (1.14)$$

where τ is the filter transmissivity, which is a function of the wavelength (λ). The incoming IR heat flux transmitting the boundary 3, q_3 is spectral dependent and, hence, food absorptivity (α) and filter transmissivity (τ) can be combined into an integral form of q_3 with respect to the wavelength.

An energy balance, to assess the amount of energy emitted from the IR source, and proportionally directed through a conveying chamber known as a waveguide to the surface of the food materials at the receiving end was estimated by calculating appropriate view factors. Equation 1.7 for each boundary can be solved simultaneously as follows:

For $i = 1$:

$$\frac{q_1}{\varepsilon_1} - \left(\frac{1}{\varepsilon_1}-1\right)F_{1-1}q_1 - \left(\frac{1}{\varepsilon_2}-1\right)F_{1-2}q_2 - \left(\frac{1}{\varepsilon_3}-1\right)F_{1-3}q_3 =$$

$$(1-F_{1-1})E_{b1}(T_1) - F_{1-2}E_{b2}(T_2) - F_{1-3}E_{b3}(T_3) \tag{1.15}$$

For $i = 2$:

$$\frac{q_2}{\varepsilon_2} - \left(\frac{1}{\varepsilon_1}-1\right)F_{2-1}q_1 - \left(\frac{1}{\varepsilon_2}-1\right)F_{2-2}q_2 - \left(\frac{1}{\varepsilon_3}-1\right)F_{2-3}q_3 =$$

$$-F_{2-1}E_{b1}(T_1) + (1-F_{2-2})E_{b2}(T_2) - F_{2-3}E_{b3}(T_3) \tag{1.16}$$

For $i = 3$:

$$\frac{q_3}{\varepsilon_3} - \left(\frac{1}{\varepsilon_1}-1\right)F_{3-1}q_1 - \left(\frac{1}{\varepsilon_2}-1\right)F_{3-2}q_2 - \left(\frac{1}{\varepsilon_3}-1\right)F_{3-3}q_3 =$$

$$-F_{3-1}E_{b1}(T_1) - F_{3-2}E_{b2}(T_2) + (1-F_{3-3})E_{b3}(T_3) \tag{1.17}$$

where $F_{1-1} = 0$, $q_2 = 0$ (insulated), and $\varepsilon_3 = 0$ (opening). Hence, Equations 1.15 and 1.16 can be simplified as follows:

$$\frac{q_1}{\varepsilon_1} = E_{b1}(T_1) - F_{1-2}E_{b2}(T_2) - F_{1-3}E_{b3}(T_3) \tag{1.18}$$

$$-\left(\frac{1}{\varepsilon_1}-1\right)F_{2-1}q_1 = -F_{2-1}E_{b1}(T_1) + (1-F_{2-2})E_{b2}(T_2) - F_{2-3}E_{b3}(T_3) \tag{1.19}$$

If the term, $E_{b2}(T_2)$ is eliminated from the above two equations,

$$q_1 = \frac{1}{\dfrac{1-F_{2-2}}{\varepsilon_1} - \left(\dfrac{1}{\varepsilon_1}-1\right)F_{2-1}F_{1-2}}$$

$$\left[(1-F_{2-2}-F_{2-1}F_{1-2})E_{b1}(T_1) - (F_{1-3}-F_{1-3}F_{2-2}+F_{2-3}F_{1-2})E_{b3}(T_3)\right] \tag{1.20}$$

Since boundary 2 is insulated ($q_2 = 0$),

$$A_1 q_1 + A_3 q_3 = 0$$

$$q_3 = -\frac{A_1}{A_3}q_1 \tag{1.21}$$

From Equation 1.12

$$q_{abs} = F_{3-4} \cdot \alpha(\lambda) \cdot \tau(\lambda) \cdot q_3$$

$$= -F_{3-4} \cdot \frac{A_1}{A_3} \alpha(\lambda) \cdot \tau(\lambda) \cdot q_1$$

$$= -F_{3-4} \cdot \frac{A_1}{A_3} \int_0^\infty \alpha(\lambda) \cdot \tau(\lambda) \cdot \frac{1}{\frac{1-F_{2-2}}{\varepsilon_1} - \left(\frac{1}{\varepsilon_1} - 1\right)F_{2-1}F_{1-2}} \cdot \tag{1.22}$$

$$\left[(1 - F_{2-2} - F_{2-1}F_{1-2})E_{b1}(T_1, \lambda) - (F_{1-3} - F_{1-3}F_{2-2} + F_{2-3}F_{1-2})E_{b3}(T_3, \lambda)\right]d\lambda$$

This approach can be extended to the case studies for other IR heaters with different structural schemes.

1.5 PHYSIOLOGICAL EFFECTS OF IR RADIATION

The effect of IR radiation on the optical and physical properties of food materials is important for the design of IR heating system and thermal process optimization of food components.

Generally, foods are complex mixtures of different biochemical macromolecules, biological polymers, inorganic salts, and water. The IR spectra of such mixtures originate with the mechanical vibrations of molecules or particular molecular aggregates within a very complex phenomenon of reciprocal overlapping (Halford, 1957). The strongest absorption is often localizable within the so-called group frequencies, which are generated by the vibrations of these molecular aggregates of molecular structural groups. The influence of the molecular environment on the IR spectra of complex biochemicals analyzed indicated that there were two groups of parameters acting on the group frequencies: (1) intramolecular environmental parameters and (2) extramolecular environmental parameters. The first group has interactions due to the chemical bonds characterizing the given biochemical molecule itself, and the second group has well-known hydrogen bonds (as important forms of interactions with the extramolecular environment).

When radiant electromagnetic energy impinges upon a food surface, it may induce changes in the electronic, vibrational, and rotational states of atoms and molecules. The type of mechanisms for energy absorption determined by the wavelength range of the incident radiative energy are (Decareau, 1985) (1) changes in the electronic state corresponding to the wavelength range 0.2–0.7 μm (ultraviolet and visible rays), (2) changes in the vibrational state corresponding to wavelength range 2.5–100 μm (FIR), and (3) changes in the rotational state corresponding to wavelengths above 100 μm (microwaves). In general, the substances absorb FIR energy most efficiently through the mechanism of changes in molecular vibrational state, which can lead to radiative heating. Water and organic compounds such as proteins and starches, which are the main components of food, absorb FIR energy at wavelengths greater

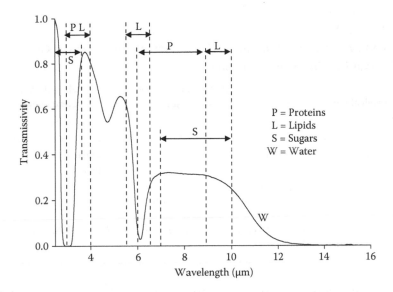

FIGURE 1.5 Principal absorption bands of the main food components compared with water.

than 2.5 μm. This finding is in good agreement with the previous work (Sandu, 1986), showing that most foods have high transmissivities (i.e., low absorptivities) at wavelengths shorter than 2.5 μm.

Due to a lack of information, data on absorption of IR radiation by the principal food constituents can be regarded as approximate values. Amino acids (Koegel et al., 1957), polypeptides, proteins, and nucleic acids (Blout, 1957) reveal two strong absorption bands localized at 3–4 and 6–9 μm. Lipids show strong absorption phenomena over the entire IR radiation spectrum, with three stronger absorption bands situated at 3–4, 6, and 9–10 μm (Schwarz et al., 1957; Freeman, 1957). Sugar gives two strong absorption bands centered at 3 and 7–10 μm (Manning, 1956). Recently, the functional groups corresponding to wavelengths (or frequencies) related with food components were identified using FTIR spectroscopy (Sivakesava and Irudayaraj, 2000).

The key absorption ranges of food components are as visualized in Figure 1.5 (Sandu, 1986). The figure shows the principal absorption bands of the major food components compared to the absorption spectrum of water, indicating that the absorption spectra of food components overlap with one another in the spectral regions considered. The water effect on absorption of incident radiation is predominant over all the wavelengths, suggesting that selective heating based on distinct absorptivities for a target food material can be more effective when predominant energy absorption of water is eliminated. The IR absorption bands characteristic of chemical groups relevant to the heating of food are summarized in Table 1.2 (Rosenthal, 1992).

Interactions of light with food material and the crucial optical principles such as regular reflection, body reflection, and light scattering have been discussed (Birth, 1978). Regular reflection takes place at the surface of a material. For body reflection, the light enters the material, becomes diffuse due to light scattering, and undergoes some absorption. The remaining light leaves the material close to where it enters.

TABLE 1.2
The Infrared Absorption Bands Characteristic of Chemical Groups Relevant to the Heating of Food

Chemical Group	Absorption Wavelength (μm)	Relevant Food Component
Hydroxyl group (O–H)	2.7–3.3	Water, sugars
Aliphatic carbon-hydrogen bond	3.25–3.7	Lipids, sugars, proteins
Carbonyl group (C=O) (ester)	5.71–5.76	Lipids
Carbonyl group (C=O) (amide)	5.92	Proteins
Nitrogen-hydrogen group (–NH–)	2.83–3.33	Proteins
Carbon-carbon double bond (C=O)	4.44–4.76	Unsaturated lipids

Source: Rosenthal, I. 1992. *Electromagnetic Radiations in Food Science.* Berlin: Springer-Verlag. With permission.

Regular reflection produces only the gloss or shine of polished surfaces, whereas body reflection produces the colors and patterns that constitute most of the information obtained visually. For materials with a rough surface, both regular and body reflection will be diffuse.

The IR optical characteristics of different media are also theoretically discussed, demonstrating the necessity of the scattered radiation during measurements (Krust et al., 1962). It was experimentally observed that as the thickness of the layer increases, a simultaneous decrease in transmittance and increase in reflection occurs. However, no theoretical explanation of this phenomenon was presented.

Il'yasov and Krasnikov (1991) validated the physical properties of irradiated foodstuffs. The phenomena of propagation and attenuation of IR radiation in foodstuffs and the principles governing the transfer of radiant energy in multilayered systems, together with the energy utilized for selectively absorbing and scattering substances, were studied. A method to produce modified starches and dextrin by treatment of starch using shortwave IR radiation was successfully developed. IR treatment increased the efficiency of dextrinization by eight to ten times compared to the conventional conduction method in industry.

Estimation of nutrition retention under IR radiation treatment of food was verified by demonstrating that intense IR radiation produced extensive dehydration, causing moisture content as low as 3%–7% (Keya and Sherman, 1997). While some deterioration of protein quality was found, full inactivation of hemagglutinins and trypsin inhibitors in beans and related foods allowed the application of IR radiation as preheating technique for reduction of the total cooking time. Research done by Krause et al. (1983) was initiated to determine the effect of both IR heating and forced air convective heating on the product yield and the nutrient content of food materials such as rib eye, hamburger patties, potato, and fresh tomatoes. Nutrient analysis for selected items such as thiamin, riboflavin, vitamin C, β-carotene, seven fatty acids, 18 amino acids, ammonia, phosphorous, iron, and sodium demonstrated that respective riboflavin and vitamin A content in hamburger and tomatoes produced about

1.17% and 11.18% of enhancements after IR heating. The total amino acid content increased by 1.3% and 1.4% for hamburger patties and cod fillets, respectively, after IR heating, compared to forced convective heating.

Effects of high-intensity IR heating on the hydration rate, the cooking time of slot peas, and the functional properties of protein and starch components were explored using the scanning electron microscopy (SEM) technique (Cenkowski, 1998). The SEM micrographs showed that a radiated sample was composed of clearly separated starch granules and unattached particles of other cell components, enlarged by water uptake. The cell structure of hulless and pearled barley after IR heat treatment analyzed using the SEM technique showed the changes in microstructure of radiated hulless and pearled barley (Fasina et al., 1999). The hulless barley samples expanded more when subjected to IR heating than with the conventional method. This was attributed to the bran of hulless barley providing additional resistance for removal of evaporated moisture from within the kernel during IR heating. This result might be caused by a buildup of hydrothermal pressure, eventually leading to more expansion of the hulless barley kernel. The starch-protein matrix characteristics of barley investigated using the SEM micrograph with more magnified scale verified that unprocessed kernels of barley samples had a starch portion consisting of small and large granules that were oval to round in shape with diameters ranging between 2 and 25 μm. However, radiated kernels swell up to 50 μm and eventually rupture, meaning starch gelatinization. Studies of cell structure change by IR radiation could have good potential for exploration of IR heating mechanism, nutritional change, and water retention of food system treated by IR radiation in the future.

1.6   SPECTRAL SELECTIVITY FOR IR HEATING

Very few attempts have been made to study selective heating in the food industry as well as in other similar areas of research. Some work on similar lines has been found in the literature applied to electronics

Most IR heaters consist of lamps emitting one specific peak wavelength corresponding to fixed surface temperature. The type of IR emitter and control of the accurate wavelength should be considered for optimal processing. Bischof (1990), from Heraeus Silica and Metals Ltd. (Bromborough, Wirral, U.K.), presented one example of their product, a new IR curing system especially invented for manufacturing technology of surface-mounted devices (SMDs). This system, the INFRA-DRY-CM, could be integrated into existing or new process lines without major alterations, with enhancement of line speeds up to 40%, where adhesives must be cured following automated placement of SMDs onto printed circuit boards (PCBs). Accurate wavelength control achieved by adjusting the source temperature validated that the lower-side components could be kept below 80°C while the board surface was cured up to 120°C. It was noted that tight control of IR wavelengths allowed IR radiation to emit the spectral spectrum where a target material had relatively high absorptivities.

Similarly, a new reflow soldering technique was developed using selective IR heating (Sakuyama et al., 1995). Note that both an aluminum oxide heater and a halogen heater were used for IR sources. The aluminum oxide heater radiates intensive IR rays between 5 and 8 μm that are readily absorbed by a glass-epoxy substrate, whereas the halogen heater radiates intensive IR rays between 1 and 2 μm that are readily absorbed by the resin used in quad flat packages (QFPs). Temperature differences between PCB substrate and large QFPs reduced using both the heat sources during the reflow demonstrated possible differential heating of the components, thereby leading to uniform heating of the reflow target. Results demonstrated 20°C to 30°C of soldering temperature lower than the conventional reflow soldering with enhancement of PCB substrate warping with increase in through-hole resistance and, most importantly, thermal safety of SMDs.

There have been no reports on selective heating in the food processing field. However, a few studies have been conducted to investigate the spectral control for optimal food process. Wavelengths above 4.2 μm are most desirable for optimal IR process of a food system due to predominant energy absorption of water in the wavelengths below 4.2 μm (Alden, 1992). Lentz et al. (1995) discussed the importance of IR-emitting wavelength for thermal process of dough. Excessive heating of the dough surface and poor heating of the interior were observed when the IR-emitting wavelength was not consistent with the wavelengths best absorbed for dough. Excessive surface heating, in the absence of corresponding heat removal to the interior, gave rise to crust formation, inhibiting heat transfer.

A study by Bolshakov et al. (1976) suggested that a maximum transmission of IR radiation should cover the spectral wavelength of 1.2 μm obtained by analysis of the transmittance spectrograms of lean pork for deep heating of pork. A two-stage frying process consisted of the first stage, to aim surface heat transfer by radiant flux with λ_{max} of 3.5 to 3.8 μm (FIR), and the second stage, to achieve greater penetration of heat transfer by radiant flux with a λ_{max} of 1.04 μm (NIR). Higher moisture content and sensory quality of the products were obtained using combined FIR and NIR heaters, compared to the conventional method. A similar study explored by Dangerskog (1979) used two alternative types of IR radiators for frying equipment, which were quartz tube heaters (Phillips 1 kW, type 13195X) whose filament temperature was 2340°C at 220 V rating, corresponding to λ_{max} of 1.24 μm as NIR region, and tubular metallic electric heaters (Backer 500 W, type 9N5.5) at a temperature of 680°C at 220 V, corresponding to λ_{max} of 3.0 μm as FIR region. It was observed from the study that both penetration capacity and reflection increased as the wavelength of the radiation decreased, indicating that although the shortwave radiation (NIR) had a higher penetrating capability than the longwave radiation (FIR), the heating effects were almost the same due to body reflection (Table 1.3).

For effective differential IR heating, the selection of the optimal spectral region of IR source covering absorptivity of a target food system is very crucial. However, there seems to be a lack of a consistent method to truly explore the intrinsic selective heating process in the area of food engineering. Dangerskog and Österström (1979) first used a bandpass filter (Optical Coating Laboratory, Inc., type L-01436-7) in his frying experiment of pork for transmitting only the wavelength above 1.507 μm, as a good forerunner for the design of a future system to specify the spectral regions considered.

Jun (2002) developed a prototype selective FIR heating system, demonstrating the importance of optical properties besides thermal properties when electromagnetic radiation is used for processing. The system had the capability to selectively

TABLE 1.3
Penetration Depth of NIR Energy into
Food Products

Product	Spectral Peak (μm)	Depth of Penetration (mm)
Dough, wheat	1.0	4–6
Bread, wheat	1.0	11–12
Bread, biscuit, dried	1.0	4
	0.88	12
Grain, wheat	1.0	2
Carrots	1.0	1.5
Tomato paste, 70%–85% Water	1.0	1
Raw potatoes	1.0	6
Dry potatoes	0.88	15–18
Raw apples	1.16	4.1
	1.65	5.9
	2.36	7.4

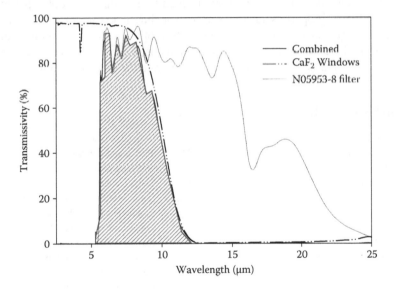

FIGURE 1.6 Transmission curves of the optical filters with CaF_2 windows (shaded area is for the final narrow bandpass filter after combination).

heat higher absorbing components to a greater extent using optical bandpass filters that can emit radiation in the spectral ranges as needed. Figure 1.6 shows that the original high-pass filter (N0953-8) with wide transmission region can trim out spectral regions greater than 10 μm with the aid of the additional CaF_2 cutoff window. The narrow bandpass filter was needed for differential heating of soy protein over glucose. It was found that combined spectral distribution could afford to enhance selectivity for heating due to the relatively narrow bandwidth (~5 µm). Hypothetical IR energy delivery at specific wavelength (or frequency) ranges has been successfully proved.

1.7 CONCLUSION

Fundamentals of IR radiation for food processing have been addressed. Estimation of the amount of IR energy needed for food products requires knowledge of the emissive power from IR lamps with different surface temperature, the travel pathways of photo particles, view factors, radiative properties of food products such as reflectivity and transmissivity at the interface, absorption and emission within the medium, and heat transfer incorporating the thermal properties of food products. As expected, the usefulness of radiation theory is extremely limited in practice for foods; food materials are rarely smooth and diffuse, and more complicated still are heterogeneous food materials. However, the theory still provides food engineers with an excellent tool to augment rough experimental data, which can be further tuned through better interpolation and extrapolation. Planck's law, Wien's displacement law, Stefan–Boltzmann's law, exponential decay, view factors, and interaction of IR radiation with food compositions are used when designing and operating IR heaters for food applications. Further studies are needed for a complete set of experimental data for radiative properties such as surface roughness, extinction coefficient, and penetration depth versus frequencies. This will resolve the problems with prediction errors caused by a large number of assumptions used for ideal, theoretical models and provide the realistic hybrid model to incorporate the empirical values.

NOMENCLATURE

A	Area	n	Refractive index
B	Absorbance	M	Total nodal number
C_1, C_2	Coefficients	m	Nodal number
C_p	Specific heat	q	Radiative heat flux
c_0	Speed of light	q_{abs}	Initial heat flux absorbed by sample
D	Thickness	q_{cond}	Conductive loss of heat flux
d_{glass}	Thickness of glass plate	q_{conv}	Convective loss of heat flux
E_b	Total emissive power	q_r	Radiative heat flux (sample)
$F_{i\text{-}j}$	View factor from i to j	$q_{rad,out}$	Radiative loss of heat flux
H	Transmissive heat flux	R	Gas constant
h	Planck's constant	S	Extinction coefficient
h_{conv}	Convective heat transfer coefficient	T	Temperature
k	Boltzmann's constant	t	Time
k	Thermal conductivity of sample	u	Mass of medium per unit area
k_d	Rate constant	x	Experimental factor
k_{glass}	Thermal conductivity of glass plate	y	Log reduction
N	Population of viable spores	Z	Distance

α	Absorptivity	**Subscripts**	
β	Regression coefficient	i	Sampling time
λ	Wavelength	i, j	Boundary
ρ	Density	∞	Ambient
σ	Stefan–Boltzmann constant		
σ*	Extinction coefficient		
τ	Transmissivity		

REFERENCES

Alden, L. B. 1992. Method for cooking food in an infra-red conveyor oven. U.S. Patent 5: 223–290.

Birth, G. S. 1978. The light scattering properties of foods. *Journal of Food Science* 43: 916–925.

Blout, E. R. 1957. Aqueous solution infrared spectroscopy of biochemical polymers. *Annals of the New York Academy of Sciences* 69: 84–93.

Bolshakov, A. S., V. G. Boreskov, G. N. Kasulin, F. A. Rogov, U. P. Skryabin, and N.-N. Zhukov. 1976. 22nd European Meeting of Meat Research Workers. Paper 15. (Cited by Dagerskog, 1979.)

Cenkowski, S., and F. W. Sosulski. 1998. Cooking characteristics of split peas treated with infrared heat. *Transactions of the ASAE* 41(3): 715–720.

Decareau, R. V. 1985. *Microwaves in the Food Processing Industry.* Orlando, FL: Academic Press.

Fasina, O. O., R. T. Tyler, M. D. Pickard, and G. H. Zheng. 1999. Infrared heating of hulless and pearled barley. *Journal of Food Processing and Preservation* 23: 135–151.

Freeman, N. K. 1957. Infrared spectroscopy of serum lipids. *Annals of the New York Academy of Sciences* 69: 131–144.

Halford, R. S. 1957. The influence of molecular environment on infrared spectra. *Annals of the New York Academy of Sciences* 69: 63–69.

Il'yasov, S. G., and V. V. Krasnikov. 1991. *Physical Principles of Infrared Irradiation of Foodstuffs.* New York: Hemisphere Publishing Co.

Jun, S. 2002. Selective far infrared heating of food systems. Ph.D. dissertation, the Pennsylvania State University.

Keya, E. L., and U. Sherman. 1997. Effects of a brief, intense infrared radiation treatment on the nutritional quality of maize, rice, sorghum, and beans. *Food and Nutrition Bulletin* 18(4): 382–387.

Koegel, R. J., R. A. McCallum, J. P. Greenstein, M. Winitz, and S. M. Birnbaum. 1957. The solid-state infrared absorption of the optically active and racemic straight-chain α-amino acids. *Annals of the New York Academy of Sciences* 69: 94–115.

Krause, G., N. Unklesbay, and M. E. Davis. 1983. Nutrient retention of portioned menu items after infrared and convective heat processing. *Journal of Food Science* 48: 869–873.

Krust, P. W., L. D. McGlauchlin, and R. B. Mcquistan. 1962. *Elements of Infra-Red Technology.* New York: John Wiley & Sons.

Lentz, R. R., P. S. Pesheck, G. R. Anderson, J. DeMars, and T. R. Peck. 1995. Method of processing food utilizing infra-red radiation. U.S. Patent 5382441.

Manning, J. J. 1956. Infrared spectra of some important narcotics. *Applied Spectroscopy* 10: 85–98.

Modest, M. F. 1993. *Radiative Heat Transfer.* New York: McGraw-Hill International Editions.

Nowak, D., and P. P. Lewicki 2004. Infrared drying of apple slices. *Innovative Food Science and Emerging Technologies* 5: 353–360.

Planck, M. 1901. Distribution of energy in the spectrum. *Annalen der Physik* 4(3): 553–563. (Cited by Modest, 1993; Sakai and Hanzawa, 1994.)

Rosenthal, I. 1992. *Electromagnetic Radiations in Food Science.* Berlin: Springer-Verlag.

Sakai, N., and T. Hanzawa. 1994. Applications and advances in far-infrared heating in Japan. *Trends in Food Science and Technology* 5: 357–362.

Sakuyama, S., H. Uchida, I. Watanabe, K. Natori, and T. Sato. 1995. Reflow soldering using selective infrared radiation. In *Proceedings of the IEEE/CPMT International Electronic Manufacturing Technology (IEMT) Symposium,* 393–396.

Sandu, C. 1986. Infrared radiative drying in food engineering: A process analysis. *Biotechnology Progress* 2(3): 109–119.

Schwarz, H. P., L. Dreisbach, R. Childs, and S. V. Mastrangelo. 1957. Infrared studies of tissue lipids. *Annals of the New York Academy of Sciences* 69: 116–130.

Sivakesava, S., and J. Irudayaraj. 2000. Analysis of potato chips using FTIR photoacoustic spectroscopy. *Journal of the Science of Food and Agriculture* 80: 1805–1810.

2 Infrared Radiative Properties of Food Materials

Griffiths Gregory Atungulu and Zhongli Pan

CONTENTS

2.1 Introduction 19
2.2 Characteristics of Thermal Radiation... 20
    2.2.1 Fundamental Properties Related to Radiative Heat Transfer of
    Foods... 20
        2.2.1.1 Emissivity 21
        2.2.1.2 Reflectivity, Absorptivity, and Transmissivity................... 22
        2.2.1.3 Absorptivity 25
        2.2.1.4 Attenuation.. 26
2.3 Interaction of Radiation with Foods 27
    2.3.1 Effect of Size and Form of Food Particles, Their Microstructure,
    and Optical Properties 27
    2.3.2 Effect of Water Content, Dimension, and Physicochemical
    Nature of the Product.. 29
    2.3.3 Food Processing Effects on Radiative Properties of Foods 32
2.4 Radiative Property Control for Optimal Processing 32
    2.4.1 Selective IR Absorption for Foods 32
    2.4.2 Modeling Selective IR Absorption for Food Heating........................ 33
    2.4.3 Advances of Selective IR Absorption of Foods................................. 36
2.5 Conclusion 37
References... 37

2.1 INTRODUCTION

Transfer of energy by radiation is significantly different from other energy transfer phenomena because it is not proportional to a temperature gradient and does not need a natural medium to propagate. Typically, any molecule possesses translation, vibrational, rotational, and electronic energy under quantum states. Passing from one energy level to another implies energy absorption or emission by a molecule. On the other hand, a molecule emits energy as radiation when passing to a lower energy level. The absorption or emission of energy is in the form of photons that have a double nature: particle–wave. Any body at a temperature higher than absolute zero can emit radiant energy, and the amount of energy emitted depends on the temperature of the body. A temperature increase implies that the radiation spectrum moves to shorter wavelengths or is more energetic.

To be precise, infrared (IR) radiation is electromagnetic radiation whose wavelength is longer than that of visible light, but shorter than that of terahertz radiation and microwaves. The IR portion of the electromagnetic spectrum spans roughly three orders of magnitude (750 nm to 100 μm) and has been very useful in food processing in a number of ways:

1. Food processes involving heating
2. Spectroscopic measurements of chemical composition (analytical application) of foods
3. Noncontact temperature measurement of foods

This chapter is primarily intended to address radiation absorption during applications of IR radiation in heating of foods and will discuss the relationship between IR wavelength and food compositions as well as the effect of physical and chemical properties on IR radiation heating. Subsidiary information on IR application in food processes involving heating can also be obtained in books such as in Datta and Almeida (2005) and Krishnamurthy et al. (2009). For detailed applications involving composition measurements, the reader is referred to literature reviews such as those by Williams and Norris (2001) or Buning-Pfaue (2003) and for noncontact temperature measurement to books such as Michalski et al. (2001).

2.2   CHARACTERISTICS OF THERMAL RADIATION

2.2.1   FUNDAMENTAL PROPERTIES RELATED TO RADIATIVE HEAT TRANSFER OF FOODS

Thermal radiation involves absorption, reflection, and transmission of incident radiation, as well as emission of radiation. The first three characteristics depend on the spectral composition of the radiation source and the degree of polarization and spatial properties of incident radiation; that is, on the irradiation conditions and the state and the properties of the irradiated substance itself. The emission of radiation by a substance is determined solely by the state and the optical properties of the substance. Under arbitrary conditions of irradiation, the sum of absorbed energy F_A, reflected energy F_R, and transmitted energy F_T per unit time according to the law of conservation of energy should be equal to the incident radiant power F:

$$F_A + F_R + F_T = F, \text{ or } A + R + T = 1 \tag{2.1}$$

where $A = F_A/F$ is absorptance (absorption coefficient), $R = F_R/F$ is reflectance (reflection coefficient), and $T = F_T/F$ is transmittance (transmission coefficient). The quantities A, R, and T depend on the state of polarization of the incident radiation.

Fundamentally, radiant energy is transported by photons. The photon is the quantum of the electromagnetic field and is also the force carrier for the electromagnetic force. Radiant energy is a function of frequency v, according to the expression:

$$E = hv \qquad (2.2)$$

in which the proportionality constant h is Planck's constant, whose value is $h = 6.6262 \times 10^{-34}$ J·s. Wave theory considers radiation to be an electromagnetic wave, relating frequency to wavelength according to the following equation:

$$v = c/\lambda \qquad (2.3)$$

where λ is the wavelength of radiation, and c is the value of the light speed under vacuum conditions (2.9979×10^8 m/s).

In the study of radiative heat transfer of foods, emissivity, reflectivity, absorptivity, and transmissivity of the product are crucial properties of interest. The reader is referred to undergraduate textbooks on heat transfer such as that by Incropera and Dewitt (1996) or specialized books on radiative heat transfer (e.g., Modest, 1993; Howell, 1982; Siegel and Howell, 1981; Sparrow and Cess, 1978) for further details on Stephan–Boltzmann's law, Planck's law of radiation, and basic definitions of properties such as emissivity, reflectivity, absorptivity, and transmissivity.

2.2.1.1 Emissivity

The emissivity of a surface is defined as the ratio of the radiation emitted by a real surface to the ratio of the radiation emitted by a perfectly radiating ideal surface (called a *blackbody*). Similar to other food radiative properties, emissivity depends on the wavelength of radiation incident on the food, which in turn depends on the emission characteristics of the source of radiation (emitter). The emissivity at a specified wavelength is called the *spectral emissivity*. For plant materials such as leaves, for example, the emissivity is typically above 0.97. Regarding the radiation source, Wien's displacement law indicates that the peak of the emission decreases with increasing temperature:

$$\lambda_{max} \times T = 2897.8 \ \mu m \cdot K \qquad (2.4)$$

where λ_{max} is the peak wavelength, and T is the absolute temperature of the emitter surface (see Figure 2.1). Accordingly, shorter-wavelength emitters (e.g., tungsten filament) operate at temperatures above 2000°C, medium-wavelength emitters (e.g., quartz tube) operate at around 700°C–1150°C, and long-wavelength emitters (e.g., ceramic) operate at below 800°C. The quantity (amount) and quality (spectral distribution) of the emitted energy depends on the temperature and the emissivity.

To be precise, the emissivity of any surface varies with temperature, wavelength, and direction of emitted radiation. Temperature-, wavelength-, and direction-dependent emissivity data are generally not available for food materials. Inclusion

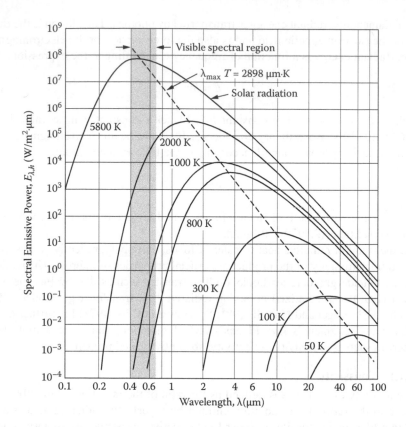

FIGURE 2.1 Spectral blackbody emissive power. (From Incropera, F.P., and Dewitt, D.P. 1996. *Introduction to Heat Transfer.* New York: John Wiley & Sons. Reprinted with permission of John Wiley & Sons.)

of such variation in radiative heat transfer analysis can make it quite complex and almost intractable. Thus, radiation heat transfer calculations commonly use two approximations called *gray* and *diffuse*. Diffuse approximation refers to properties independent of direction and gray approximation to properties independent of wavelength. By using bandpass optical filters, specific spectral regions can be obtained for selective heating of foods (Jun and Irudayaraj, 2003a,b).

2.2.1.2 Reflectivity, Absorptivity, and Transmissivity

When foodstuff is subjected to IR radiation, complex and irreversible biochemical processes take place continuously. Calculation of heat transfer in irradiation should take into account the fact that not all the absorbed heat goes toward heating. Some of it is used up in endothermic reactions, phase transitions, etc. Only that fraction of the absorbed energy which is lost by the excited electron in nonradiative transfer between the energy levels goes toward increasing the thermal energy of thermal motion of microparticles; that is, toward heating up the substance. For a detailed description on foodstuff as radiation-absorbing and radiation-scattering objects, energy transfer in foodstuffs under different conditions of IR irradiation, and the corresponding mechanistic and mathematical analyses of the phenomena, the reader is referred to the excellent book by Il'yasov and Krasnikov (1978).

The reflectivity, absorptivity, and transmissivity of food materials are dependent on wavelength and direction, as is emissivity. For most foods, the absorption and reflection of radiation take place throughout a layer and not only on the surface. Two limiting cases of reflection from surface area are called *specular* and *diffuse*. In specular reflection, the angle of incidence is equal to the angle of reflection. In diffuse reflection, on the other hand, the intensity of the reflected radiation is the same at all angles of irradiation and reflection. Although real surfaces are neither totally specular nor diffuse, but somewhere in between, they are assumed to be one of the two limiting cases for simplicity. Most food surfaces are likely to be rough, leading to diffuse behavior; that is, the intensity of reflected radiation will be about the same at all angles. From the point of view of wave optics, the reflection of electromagnetic waves from a polydispersed medium means a partial or complete return of wave motion from a diffracting surface of the interface of two media to the half-space of the first medium. Specular reflection can be observed only when the radius of curvature (r) of different inhomogeneities parallel to the interface change by $\Delta r \ll \lambda$. The surface is considered optically rough when $\Delta r > \lambda$; in this case, we have diffuse reflection (Kizel, 1973; Korolev, 1966). In foodstuffs exposed to IR radiation, the incident electromagnetic wave causes the particles to oscillate and consequently to emit secondary waves that are coherent with the incident wave. The interference of incident and secondary waves results in the appearance of reflected and refracted waves. An example of the variation of specular reflectivity with incident angle can be seen in Figure 2.2, and an example of reflectivity and transmissivity averaged over all wavelengths and directions can be seen in Figure 2.3. Although these properties can be defined for specific wavelengths and direction, only a small amount of data is available for wavelength dependence, and very little or no data are available for directional dependence over food surfaces. However, even if the data were available,

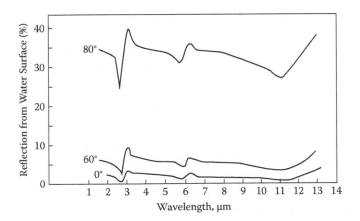

FIGURE 2.2 Spectral reflectance of water as a function of incident angle and wavelength. (From Wolfe, W.L. 1995. *Handbook of Military Infrared Technology*. Washington, DC: Office of Naval Research, Department of the Navy.)

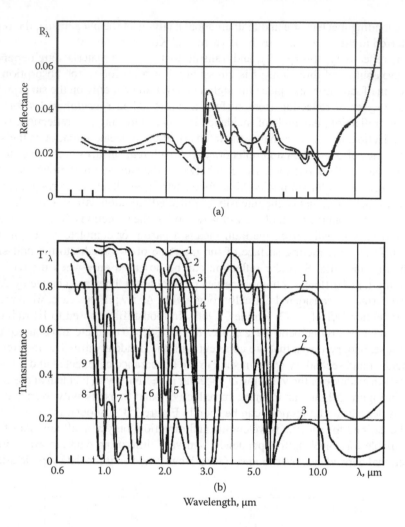

FIGURE 2.3 Spectral reflectance (a) and transmittance (b) of distilled water for various thicknesses (mm): (1) 0.005, (2) 0.01, (3) 0.03, (4) 0.1, (5) 0.3, (6) 1.0, (7) 10.0, (8) 30.0, (9) 100.0. The dashed line in (a) is for NaCl solution and for water from the Black Sea. (From Il'yasov, S.G., and Krasnikov, V.V. 1978. In *Physical Principles of Infrared Irradiation of Foodstuffs*. Washington, DC: Hemisphere Publishing Corporation. With permission.)

their inclusion in radiative heat transfer analysis would make it worse and probably lead to results of no practical significance in food processing. To some extent, data on wavelength dependence of food properties are scant. Wavelength dependence of food properties is called *spectral properties*. Thus, spectral reflectivity, absorptivity, and transmissivity are defined as:

$$\rho_\lambda = \frac{G_{\lambda,ref}}{G_\lambda} \qquad (2.5)$$

$$\alpha_\lambda = \frac{G_{\lambda,abs}}{G_\lambda} \tag{2.6}$$

$$\tau_\lambda = \frac{G_{\lambda,trans}}{G_\lambda} \tag{2.7}$$

where G_λ is the radiation energy incident at wavelength λ, and $G_{\lambda,ref}$, $G_{\lambda,abs}$, and $G_{\lambda,trans}$ are radiation energy reflected, absorbed, and transmitted, respectively. The related averages are thus:

$$\rho = \frac{\int_0^\infty \rho_\lambda G_\lambda d_\lambda}{\int_0^\infty G_\lambda d_\lambda} \tag{2.8}$$

$$\alpha = \frac{\int_0^\infty \alpha_\lambda G_\lambda d_\lambda}{\int_0^\infty G_\lambda d_\lambda} \tag{2.9}$$

$$\tau = \frac{\int_0^\infty \tau_\lambda G_\lambda d_\lambda}{\int_0^\infty G_\lambda d_\lambda} \tag{2.10}$$

For an example of spectral transmissivity through a slice of potato, see Figures 2.4 to 2.6.

2.2.1.3 Absorptivity

The quantum energy of IR photons is in the range 0.001 to 1.7 eV. IR is absorbed more strongly than microwaves but less strongly than visible light. The result of IR absorption is heating of the tissue since it increases molecular vibrational activity. Absorptivity, too, depends on the spectral distribution of incident radiation. Fundamentally, when the material and the source of incident radiation are at the same temperature, emissivity is equal to absorptivity (Incropera and Dewitt, 1996):

$$\varepsilon_\lambda(T) = \alpha_\lambda(T) \tag{2.11}$$

The preceding equation (Equation 2.11) disregards the directional dependence of emissivity or absorptivity. For all practical purposes, the average values (Equation 2.12) are generally considered equal and the approximation satisfactory when the

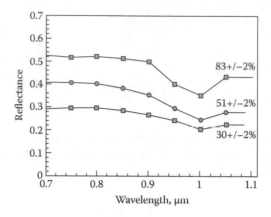

FIGURE 2.4 Spectral reflectance of potato tissue is shown as a function of moisture content. Measurements are at room temperature (25°C), in samples of 1 cm thickness and 2.5 cm diameter. (From Almeida, M.F. 2004. Combination microwave and infrared heating of foods. Ph.D. Dissertation, Cornell University, Ithaca, New York.)

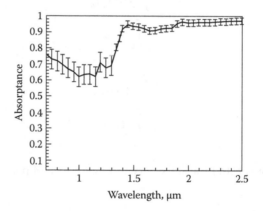

FIGURE 2.5 Spectral absorptance in potato tissue at 51% moisture content. (From Datta, A.K., and Almeida, M. 2005. *Engineering Properties of Foods*. Boca Raton, FL: CRC Press. pp. 209–235. With permission.)

temperatures of the two surfaces involved in exchanging radiation are equal, or the error is considered small when the temperature difference is less than a few hundred degrees kelvin (Cengel, 1998):

$$\varepsilon(T) = \alpha(T) \tag{2.12}$$

2.2.1.4 Attenuation

Attenuation is the gradual loss in intensity of electromagnetic energy as it passes through a medium. As the electromagnetic waves move through a food material part, the radiation energy is absorbed or scattered. Electromagnetic energy is attenuated

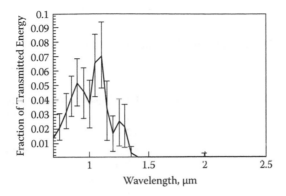

FIGURE 2.6 Fraction of transmitted energy in a sample of raw potato at 86% moisture content and 1 cm thickness. Sample diameter is 2.5 cm, and the temperature is approximately 26°C. (From Almeida, M.F. 2004. Combination microwave and infrared heating of foods. Ph.D. Dissertation, Cornell University, Ithaca, New York.)

due to the combined effect of absorption and scattering. This attenuation is also called *extinction* and is typically expressed as

$$q = q_o e^{-x/\delta} \tag{2.13}$$

where q_o is the incident energy flux, and q is the energy flux at distance x from the incident surface. The foregoing relationship (Equation 2.13) is also known as Beer–Lambert's law in many contexts. The penetration depth (δ) describes the attenuation and is a complex function of (Sandu, 1986)

1. The chemical composition of the food.
2. The physicochemical state of the irradiated medium; that is, solid, liquid, or powder; frozen or unfrozen; dispersion, emulsion, or solution; etc.
3. Physical properties such as density, porosity, and water content.

Table 2.1 shows typical penetration depths for some food materials from the work of Ginzberg (1969). The spectral variation of penetration depth in potato tissue is shown in Figure 2.7.

2.3　INTERACTION OF RADIATION WITH FOODS

2.3.1　Effect of Size and Form of Food Particles, Their Microstructure, and Optical Properties

The transfer of thermal radiation energy in foodstuffs takes place through the absorption and scattering of radiation. When an electromagnetic wave passes through a medium, it undergoes diffraction from the surface of a particle whose refractive index differs from that of the medium. This phenomenon is known as *scattering*

TABLE 2.1
Penetration Depth in Some Typical Food Materials

Material	Penetration Depth (mm)	λ_{max} of Incident Radiation (μm)
Apple	1.8	1.16
	2.6	1.65
	3.2	2.35
Bread, rye	3	~0.88
Bread, wheat	4.8–5.2	~1
Bread, dried	1.7	~1
	5.2	~0.88
Carrots	0.65	Not available
Dough, macaroni	1–1.1	Not available
Dough, wheat (445 moisture)	1.7	~1
Potato, dry	6.5–7.8	~0.88
Potato, raw	2.6	~1
Tomato, paste (70% to 85% moisture)	0.4	~1
Wheat, grains	0.9	~1

Note: Recomputed from Ginzberg (1969), where penetration depth is defined as the distance over which the flux drops to $1/e$ of its incident value.

Source: Datta, A.K., and Almeida, M. 2005. *Engineering Properties of Foods.* Boca Raton, FL: CRC Press. With permission.

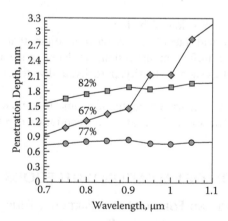

FIGURE 2.7   Spectral variation of penetration depth for potato samples at various moisture contents. Penetration depth is calculated by fitting an exponential curve through the transmittance versus thickness data. (From Datta, A.K., and Almeida, M. 2005. *Engineering Properties of Foods.* Boca Raton, FL: CRC Press, pp. 209–235. With permission.)

and the *attenuation* of radiation and depends on the size and form of particles, on their microstructure and optical properties, and on the wavelength of the radiation. Foodstuff consists mainly of moisture, protein, carbohydrates, and fats, whose physicochemical properties and structure are classified as capillary-porous colloidal systems. Studies on the transfer of thermal radiation energy into foods, various capillary-porous systems, pigments, and powdery, fibrous, cellular substances as well as other optically heterogeneous substances have revealed that all these substances can be regarded as energy-absorbing and energy-scattering polydispersed media (Avramenko et al., 1974; N.I. Angersbakh, 1986; A.K. Angersbakh, 1987; Ageenko et al., 1984; Ilyasov et al., 1977). The absorption and scattering of thermal radiation are determined mainly by the following four processes:

1. Resonance absorption of radiation by molecules of a dry substance and by the molecules of structural water and bound water.
2. Scattering radiation due to fluctuations in the density and concentration of a substance, and molecular scattering of radiation (e.g., scattering by the molecules of proteins, starch, polymer particles, and so on).
3. Scattering of radiation by suspended colloidal particles, granules of starch, plant cells, microfibrils, pigment particle, and so on.
4. Scattering of radiation by other optical inhomogeneities (capillaries and pores).

Comparison of the absorption spectra of a dielectric substance and those of foodstuff and other radiation-scattering substances is shown in Figure 2.8 (Selyukov, 1968). The fact that the cellular structure of foodstuff resembles that of wood, paper, and other substances on which radiation is scattered explains the similarity of the indicated optical properties. The appearance of the resulting continuous spectrum is, overall, strongly affected by the presence of water in the foodstuffs. Different types of water such as structural, adsorbed, capillary-condensed, and free water in foodstuffs have different physicochemical properties. During processing of foodstuffs with IR, the water contained in the capillary and pores can be in different aggregate states such as vapor, liquid, and ice. Foodstuffs such as vegetables and fruits consist mostly of water, and their optical properties are predominantly determined by water. Because of this, the absorption spectra of water are of great interest. Throughout the IR spectral region, water exhibits strong absorption and weak scattering of radiation.

2.3.2 EFFECT OF WATER CONTENT, DIMENSION, AND PHYSICOCHEMICAL NATURE OF THE PRODUCT

The radiation absorption of foodstuffs depends mainly on the water content, thickness, and physicochemical nature of the product. During IR heating, radiation energy is absorbed in foods by organic materials at discrete frequencies corresponding to intramolecular transitions between energy levels. The mechanisms for energy absorption are associated with the wavelength range of the incident radiation energy thus (Rosenthal, 1992; Sakai and Hanazawa, 1994):

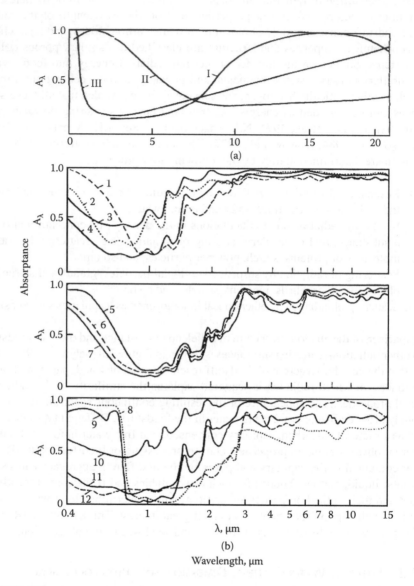

FIGURE 2.8 Characteristic dependence of absorptance. (a) For solids: I—dielectric; II—semiconductor; III—metal. (b) For foodstuffs and other radiation scattering substances: 1—dried apple, l (undried) = 10 mm; 2—apple, l = 10 mm, Wd = 86.6%; 3—potato, l = 10 mm, Wd = 74.5%; 4—dried potato, l (undried) = 10 mm; 5—redwood, l = 10 mm; 6—beech, l = 10 mm; 7—pine wood, l (air-dried state) = 10 mm; 8—dried tea leaves; 9—fresh green tea leaves; 10—potato starch, Wd = 76.5%; 11—potato starch, Wd = 11.8%; 12—enamel VL-55. Symbol l represents thickness of the layer of the sample. (From Lin, C.C., and Chan, S.H. 1975. *Journal of Heat Transfer*, C97(1), pp. 26–32. With permission.)

1. Changes in the electronic state correspond to wavelengths in the range of 0.2–0.7 μm (ultraviolet [UV] radiation and visible rays).
2. Changes in vibration state correspond to wavelengths in the range of 2.5–100 μm (middle-IR radiation [MIR] and far-IR radiation [FIR]).
3. Changes in the rotational state correspond to wavelengths above 100 μm.

Foodstuffs absorb medium-IR to far-IR energy most efficiently through stretching vibration, which leads to the radiative heating process. Even though complete information is not available, the approximate values for the strong absorption bands of major food constituents are proteins at 3–4 and 6–9 μm; lipids at 3–4, 6, and 9–10 μm; and sugars at 3 and 7–10 μm. The four principal absorption bands of liquid water are 3, 4.7, 6, and 15.3 μm (Sandu, 1986). Superimposing the IR absorption bands of the principal food constituents onto that of liquid water (Figure 2.9) shows that the absorption spectra of food components overlap with one another (Sandu, 1986). Since the strong absorption bands of major food constituents are close to the water absorption peaks, it remains a challenge for practical applications to efficiently use differential or selective heating for targeting water without heating up other components in a food material. It has been concluded that most foodstuffs showed high transmissivity at wavelength < 2.5 μm (Sandu, 1986). Studies also showed that the transmissivity of foodstuffs increases abruptly when their water content is lowered in the near-IR (NIR) range, while fresh and dry apple showed similar spectral absorptivity at wavelengths above 3 μm (Krust et al., 1962; Ginzberg, 1969).

Ginzberg (1969) also showed that as the thickness of foodstuffs increases, a simultaneous decrease in transmittance and increase in absorptivity occur. For different applications, the optimal thickness of a foodstuff and the selected radiation wavelength could be different, based on varying transmissivity and absorptivity. Thin products could be preferred for processes using IR energy since the high transmissivity would result in higher heating rates. On the other hand, NIR has advantages over MIR and FIR with its superior transmissivity. However, the temperature of NIR

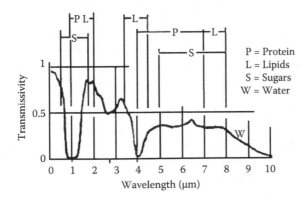

FIGURE 2.9 Principal absorption bands of the main food components compared with water. (From Sandu, C. 1986. *Biotechnology Progress* 2(3):109–119. With permission.)

radiation could be too high for processing food and agricultural products to maintain high quality of the products. In addition, the decrease in absorptivity and increase in transmissivity of NIR during drying could also be a problem to thin materials. As materials get dried, the shrinkage of thin materials can result in low absorption of NIR energy since most of the radiation energy can be reflected and transmitted through the thin layer. As the food material exposed to electromagnetic radiation undergoes changes in temperature, moisture content, structures, and other properties, the main characteristics of radiation also change (Ivanov, 1969; Ilyasov, 1977).

2.3.3 FOOD PROCESSING EFFECTS ON RADIATIVE PROPERTIES OF FOODS

In general, changes in radiative properties are influenced by the various transformations in structure and food chemistry during processing. As the food material undergoes processing, it is expected that such procedures will affect the radiative properties. For instance, heating potato with microwaves or boiling in water initiates the starch gelatinization process, which is likely to influence radiative properties such as reflectance. Similarly, such phenomena can be noted, for instance, in crumb and crust portions during bread making, which are due to structure and surface property differences resulting from the two states of gluten and starch matrix.

2.4 RADIATIVE PROPERTY CONTROL FOR OPTIMAL PROCESSING

2.4.1 SELECTIVE IR ABSORPTION FOR FOODS

An ideal IR heating system optimally raises the temperature of a target with the least energy consumption. Such a system may comprise a device that can directly convert its electrical power input to a radiant electromagnetic energy output, with the chosen single-band or narrowband wavelengths that are aimed at a target, such that the energy comprising the irradiation is partially or fully absorbed by the target and converted to heat. The higher the efficiency of conversion of electrical input into radiant electromagnetic output, the higher the overall system efficiency. Furthermore, the more efficiently the radiant electromagnetic waves are aimed to expose only the desired areas on the target, the more efficiently the system will accomplish its work. The radiation-emitting device chosen for use should have an instant "on" and instant "off" characteristic such that when the target is not being irradiated, neither the input nor the output energy is wasted. For an optimal system, care must be taken so that the set of system output wavelengths matches the absorptive characteristic of the target.

Limited research has focused on radiative property control for selective heating of food materials. The reader is referred to the excellent research by Jun and Irudayaraj (2003a,b) that described the design and evaluation of IR heating systems for selective heating. Subsidiary information may also be found in Chapter 1 of this book in the section dealing with spectral selectivity for IR heating. The type of IR emitter and control of accurate wavelength are key considerations for optimization of the process. In practice, the IR source emits radiation covering a very wide range and, hence, it is a challenge to cut off the entire spectral distribution to obtain a specific bandwidth. In the context of food processing, wavelengths above 4.2 μm are most desirable for an optimal IR process in a food system due to predominant energy absorption of water in the wavelength below 4.2 μm (Alden, 1992).

Some studies that have emphasized the significance of considering spectral selectivity for IR heating include research by Lentz et al. (1995), which noted the importance of IR-emitting wavelength for thermal processing of dough to avoid crust formation. In their study, excessive heating of dough surface and poor heating of the interior were observed when the IR spectral emission was not consistent with the wavelengths best absorbed for dough. The study of Shuman and Staley (1950) clearly demonstrated the importance of spectral control of the IR source to manipulate the delivery of heat amounts to specific food materials. They indicated that orange juice has a minimum absorption in the range between 3 and 4 μm, whereas dried orange solids have a maximum absorption in the same region and, therefore, the IR source was controlled to emit a spectral range between 5 and 7 μm to obtain the desirable absorption of orange juice. Other studies of similar interest include those by Bolshakov et al. (1976) and Dagerskog (1979).

Overall, there seems to be a lack of consistent methods to explore the intrinsic selective heating process in the area of food engineering. Dagerskog and Osterstrom (1979) first used a bandpass filter (Optical Coating Laboratory, Inc., type L-01436-7) in their frying experiments of pork to transmit only wavelengths above 1.507 μm, which turned out to be a good example of the design of selective IR heating systems to emit spectral regions of interest.

2.4.2 Modeling Selective IR Absorption for Food Heating

Mathematical modeling is a very useful tool to quickly and inexpensively ascertain the effect of different system and process parameters on the outcome of a process. The understanding of changes in optical properties when a food material is subject to thermal radiation is crucial. Recently, Jun and Irudayaraj (2003a,b) developed an excellent and novel selective FIR heating system (Figure 2.10), demonstrating the importance of optical properties besides thermal properties when electromagnetic radiation is used for processing. Their contribution in this area is reviewed extensively in this section. The authors (Jun and Irudayaraj, 2003a,b; Krishnamurthy et al., 2009) developed a system that had the capability to selectively heat higher absorbing components to a greater extent using optical bandpass filters that can emit radiation in the spectral ranges as needed. They further performed simulation of the heat and mass transfer phenomena in the food domain in one dimension (Figure 2.11), using the following governing differential equation described:

$$\rho C_p \frac{\partial T}{\partial t} = k \frac{\partial^2 T}{\partial z^2} - \frac{\partial q_r}{\partial z} \qquad (2.14)$$

where q_r is the heat flux (W/m²), T is the temperature (°C), ρ is the density (kg/m³), C_p is the specific heat of food sample (J/kg.°C), k is the thermal conductivity (W/m.°C),

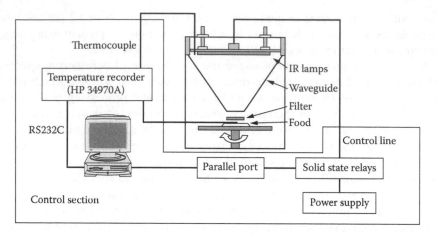

FIGURE 2.10 Schematic diagram of the FIR heating system with control part. (From Jun, S., and Irudayaraj, J. 2003a. *Drying Technology* 21(1):51–67. With permission.)

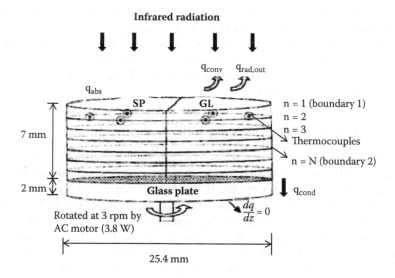

FIGURE 2.11 Schematic of the discretized food domain with food holder. (From Jun, S., and Irudayaraj, J. 2003a. *Drying Technology* 21(1):51–67. With permission.)

t is the time, and z is the distance (m). Based on the assumption that the initial temperature (T_o) is uniform, the initial boundary condition is given by

$$T(z,t) = T_o \text{ for } 0 \le z \le D; \, t = 0 \tag{2.15}$$

where T_o is the initial sample temperature (°C), and D is the sample thickness (m). By considering convection (q_{conv}) and radiation losses ($q_{rad,out}$) at boundary 1 and conduction loss (q_{cond}) at boundary 2, the boundary conditions can be given by

$$k\frac{\partial T}{\partial z}=h_{conv}(T_1-T_\infty)-q_r(1)+\varepsilon(\lambda)\Big[E_b(T_1)-E_b(T_\infty)\Big]\quad \text{at the top } (z=0),\ t>0$$

and

$$k\frac{\partial T}{\partial z}=q_r(N)-\frac{k_{glass}}{d_{glass}}(T_N-T_\infty)\quad \text{at the bottom } (z=D),\ t>0 \qquad (2.16)$$

The expression $E_b(T)$ is the total emissive power at a given source temperature obtained using Planck's integral:

$$E_b(T)=\int_0^\infty E_{b\lambda}(T,\lambda)d\lambda=\int_0^\infty \frac{2\pi hc_o^2}{n^2\lambda^5\Big[e^{hc_o/n\lambda kT}-1\Big]}d\lambda \qquad (2.17)$$

The IR radiation is not only absorbed by the surface but also penetrates into the food sample on the system, causing a local radiative heat flux of

$$q_r(n)=q_{abs}\exp\Big[-S\cdot\Delta z\cdot(n-1)\Big] \qquad (2.18)$$

where q_{abs} is the initial radiant heat flux absorbed by the food sample on the surface, S is the extinction coefficient (m^{-1}), and Δz is the grid size (m) (Dagerskog, 1979). Simplification of an FIR heating system is shown in Figure 2.12, where boundary 1 denotes an assembly of FIR source lamps, boundary 2 denotes a cone-shaped waveguide to keep IR radiation from dispersing out into the air, boundary 3 denotes an opening outlet, and boundary 4 denotes a sample surface (Jun and Irudayaraj, 2003a;

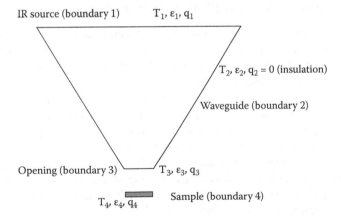

FIGURE 2.12 Simplified gray and diffuse enclosure of the heating chamber. (From Jun, S., and Irudayaraj, J. 2003a. *Drying Technology* 21(1):51–67. With permission.)

Krishnamurthy et al., 2009). Under the conditions that the IR lamps and the wave-guide are gray and diffuse reflectors, the total heat flux absorbed by the food sample can be calculated using the energy balance equation for each surface (Modest, 1993) and the equations for each boundary can be solved simultaneously. The amount of heat flux absorbed by food surface, q_{abs}, is dependent upon the spectral absorptivity (α) of food, the spectral distribution of filtered IR radiation, and the view factor (F) as obtained from the opening (boundary 3) to the food surface. This relationship can be expressed as

$$q_{abs} = F_{3-4} \cdot \alpha(\lambda) \cdot \tau(\lambda) \cdot q_3 \qquad (2.19)$$

where τ is the filter transmissivity, which is a function of the wavelength (λ). The incoming IR heat flux in the boundary 3, q_3 is spectral dependent, and, hence, food absorptivity (α) and filter transmissivity (τ) can be combined into an integral form of q_3 with respect to the wavelength. F_{3-4} can be calculated using the equation set formulated for disk to parallel coaxial disk with the same radius (Modest, 1993). An explicit (forward in time) finite difference method is applied to solve Equation 2.14 through Equation 2.16 because this technique is relatively simple and very accurate for highly transient problems (Nowak and Levicki, 2004). If the food domain is sub-divided by the grid points for each layer (Figure 2.11), the infinitesimal differentials are replaced by differences of finite size (time and space), and the degree of accuracy of the representation is determined by the step size of these differences. Finite dif-ference formulas obtained from Taylor series expansions such as forward difference and centered difference are used to approximate time and space derivatives in the partial differential equation (Togrul, 2005). The reader is referred to the work by Jun and Irudayaraj (2003a) for the detailed solution methodology and comparison between the simulated temperatures of various foods and their measured data at the food surface during IR heating with and without filters.

2.4.3 ADVANCES OF SELECTIVE IR ABSORPTION OF FOODS

Some advances have been made in research to directly output substantial quantities of IR radiation at selected wavelengths for the purpose of replacing broadband-type devices with narrowband irradiation sources. Recent advances in semiconductor pro-cessing technology have resulted in the availability of direct electron-to-photon solid-state emitters that operate in the general mid-IR range above 1 μm (1000 nm). These solid-state devices operate analogous to common light-emitting diodes (LEDs), only they do not emit visible light but emit thermal IR energy at the longer mid-IR wave-lengths. In one form, these are an entirely new class of devices that utilize quantum dot technology that have broken through the barriers which have prevented use-able, cost-effective solid-state devices from being produced that could function as direct electron-to-photon converters whose output is pseudo-monochromatic and in the mid-IR wavelength band. To distinguish this new class of devices from the con-ventional shorter-wavelength devices (LEDs), these devices are more appropriately described as radiance or radiation-emitting diodes (REDs). These devices have the property of emitting radiant electromagnetic energy in a tightly limited wavelength range. Furthermore, through proper semiconductor processing operations, REDs can be tuned to emit at specific wavelengths that are most advantageous to a particular radiant treatment application. REDs may take a variety of forms, including diode forms or laser diode forms or, in some cases, laser forms. In addition, innovations in RED technology related to the formation of a doped planar region in contact with an oppositely doped region formed as a randomly distributed array of small areas of material or quantum dots for generating photons in the targeted IR range, and potentially beyond, has evolved. This fabrication technique, or others, such as the development of novel semiconductor compounds, adequately applied would yield suitable pseudomonochromatic, solid-state mid-IR emitters. RED devices and the technology with which to make them are subjects of a patent application (Sinharoy and Wilt, 2004).

2.5 CONCLUSION

It is evident from this chapter that an understanding of the radiation adsorption properties of food materials is crucial for development of IR selective heating, which would improve IR heating energy efficiency as well as the quality of processed products. Presently, there is not much literature on selective heating using IR in foods. However, it is recognized that IR can be controlled or filtered to allow radiation within a specific spectral range to pass through using suitable optical bandpass filters. Such a controlled radiation can stimulate the maximum response of the target object when the emission band of IR and the peak absorbance band of the target objects are identical. The ability to generate only wavelength-specific radiant energy output might hold the promise of greatly improving the efficiency of various process heating applications.

REFERENCES

Ageenko, I.S., Ilyasov, S.G., Krasnikov, V.V., and Tyurev, E.P. 1984. Methods of measurement of thermal radiational characteristics of foodstuffs subjected to infrared red radiation by a hemispherical flux. *Ing.–Fiz. Zhur.*, 46(6).

Alden, L.B. 1992. Method for cooking food in an infra-red conveyor oven. U.S. Patent 5, 223–290.

Almeida, M.F. 2004. Combination microwave and infrared heating of foods. Ph.D. Dissertation, Cornell University, Ithaca, New York.

Angersbakh, A.K. 1987. Intensification of the process of thermal radiational convective drying of apples and quince. Summary of dissertation (Cand. Tech. Sci.). Moscow, 24 pp.

Angersbakh, N.I. 1986. Thermal radiational convective drying of grapes using solar energy. Summary of dissertation (Cand. Tech. Sci.). Moscow.

Avramenko, V.V., Eselson, M.P., and Zaika, A.A. 1974. *Infrared Spectra of Foodstuffs*. Moscow: Pishchevaya Prom.

Bolshakov, A.S., Borreskov, V.G., Kasulin, G.N., Rogov, F.A., Skryabin, U.P., and Zhukov, N.N. 1976. 22nd European meeting of meat research workers. Paper 15. (Cited by Dagerskog, 1979.)

Buning-Pfaue, H. 2003. Analysis of water in foods by near infrared spectroscopy. *Food Chemistry* 82(1), 107–115.

Cengel, Y.A. 1998. *Heat Transfer: A Practical Approach*. New York: McGraw-Hill.

Dagerskog, M. 1979. Infra-red radiation for food processing II. Calculation of heat penetration during infra-red frying of meat products. *Lebensmittel Wissenschaft und Technologie*, 12(5), 252–257.

Dagerskog, M., and Osterstrom, L. 1979. Infra-red radiation for food processing I. A study of fundamental properties of infrared radiation. *Lebensmittel Wissenschaft und Technologie*, 12(4), 237–42.

Datta, A.K., and Almeida, M. 2005. Properties relevant to infrared heating of foods. In Rao, M.A., Rizvi, S.S.H., Datta, A.K. (Eds.), *Engineering Properties of Foods*. Boca Raton, FL: CRC Press, pp. 209–235.

Ginzberg, A.S. 1969. *Application of Infrared Radiation in Food Processing*. London: Leonard Hill Books.

Howell, J.R. 1982. *A Catalog of Radiation Configuration Factors*. New York: McGraw-Hill.

Ilyasov, S.G. 1977. Theoretical principles of infrared radiation of foodstuffs. Summary of doctoral dissertation. Moscow.

Ilyasov, S.G., and Krasnikov, V.V. 1978. Foodstuffs as radiation absorbing and radiation scattering objects. In *Physical Principles of Infrared Irradiation of Foodstuffs*. Washington, DC: Hemisphere Publishing Corporation, 1–24.

Ilyasov, S.G., Krasnikov, V.V., and Tyurev, E.P. 1977. Methods of investigating the spreading of the effective cross section in the case of directional irradiation of absorbing and scattering substances. *Ing.–Fiz. Zhur.*, 32(2), pp. 264–269.

Incropera, F.P., and Dewitt, D.P. 1996. *Introduction to Heat Transfer*. New York: John Wiley & Sons.

Ivanov, V.V. 1969. *Optics of Scattering Media*. Minsk: Nauka i Tekhnika.

Jun, S., and Irudayaraj, J. 2003a. Selective far infrared heating system—Design and analysis (Part 1). *Journal of Drying Technology*, 21(1), 51–67.

Jun, S., and Irudayaraj, J. 2003b. Selective far infrared heating system—Spectral manipulation (Part 2). *Journal of Drying Technology*, 21(1), 69–82.

Kizel, V.A. 1973. *Reflection of Light*. Moscow: Nauka.

Korolev, A.F. 1966. *Theoretical Optics*. Moscow: Vyssh. Shkola.

Krishnamurthy, K., Khurana, H.K., Jun, S., Irudayaraj J., and Demirci, A. 2009. *Food Processing Operations Modeling: Design and Analysis*. In Jun, S. and Irudayaraj, J.M. (Eds.), Food Science and Technology Series, 2nd ed. Boca Raton, FL: CRC Press, pp. 113–142.

Krust, P.W., McGlauchlin, L.D., and Mcquistan, R.B. 1962. *Elements of Infrared Technology*. New York: John Wiley & Sons.

Lentz, R.R., Pesheck, P.S., Anderson, G.R., DeMars, J., and Peck, T.R. 1995. Method of processing food utilizing infra-red radiation. U.S. Patent 5382441.

Lin, C.C., and Chan, S.H. 1975. An analytical solution for a planar nongray medium in radiative equilibrium. *Journal of Heat Transfer*, C97(1), 26–32.

Michalski, L., Eckersdorf, K., Kucharski, J., and McGhee, J. 2001. *Temperature Measurements*. Chichester: John Wiley & Sons.

Modest, M.F. 1993. *Radiative Heat Transfer*. New York: McGraw-Hill.

Nowack, D., and Levicki, P.P. 2004. Infrared drying of apple slices. *Innovative Food Sciences Emerging Technology*, 5, 353–360.

Rosenthal, I. 1992. *Electromagnetic Radiation in Food Science*. New York: Springer-Verlag.

Sakai, N., and Hanzawa, T. 1994. Applications and advances in far-infrared heating in Japan. *Trends in Food Science and Technology*, 5, 357–362.

Sandu, C. 1986. Infrared radiative drying in food engineering. *Biotechnology Progress*, 2(3), 109–119.

Selyukov, N.G. 1968. Optical properties of foodstuffs subjected to thermal radiational treatment. Summary of dissertation (Cand. Tech. Sci.). Moscow. 34 pp.

Shuman, A.C., and Staley, C.H. 1950. Drying by infra-red radiation. *Food Technology*, 4, 481–484.

Siegel, R., and Howell, J.R. 1981. *Thermal Radiation Heat Transfer*. New York: McGraw-Hill.

Sinharoy, S., and Wilt, D. 2004. Quantum Dot Semiconductor Device. Application Ser. No. 60/628,330. http://www.faqs.org/patents/app/20090102083.

Sparrow, E.M., and Cess, R.D. 1978. *Radiation Heat Transfer*. Washington, DC: Hemisphere Publishing Corp.

Togrul, H. 2005. Simple modeling of infrared drying of fresh apple slices. *Journal of Food Engineering*, 71(3), 311–323.

Williams, P., and Norris, K.H. 2001. *Near Infrared Technology in Agricultural and Food Industries*. St. Paul, MN: American Association of Cereal Chemists.

Wolfe, W.L. 1995. *Handbook of Military Infrared Technology*. Washington, DC: Office of Naval Research, Department of the Navy.

3 Heat and Mass Transfer Modeling of Infrared Radiation for Heating

Fumihiko Tanaka and Toshitaka Uchino

CONTENTS

3.1 Introduction 41
3.2 The IR Heating Model.. 42
    3.2.1 Radiation Model 42
    3.2.2 Airflow Model 45
    3.2.3 Conductive Heat Transfer Model... 45
    3.2.4 Numerical Simulation... 47
3.3 Model Prediction 48
    3.3.1 Wood Heating.. 48
    3.3.2 Thermal Treatment of Fresh Fruit for Surface Decontamination 49
        3.3.2.1 Figs.. 49
        3.3.2.2 Strawberry... 50
        3.3.2.3 Peach.. 51
3.4 Conclusion 53
References... 54

3.1 INTRODUCTION

Infrared (IR) radiation heating operation is relevant to many industrial heating and drying processes because it can achieve rapid and contactless heating. Specific advantages of IR heating compared with conduction and convection heating are as follows: reduced heating time, uniform heating, reduced quality losses, absence of solute migration in material, versatility, simplicity, compact equipment, and energy saving (Krishnamurthy et al., 2008). In recent years, IR heating has also been applied to disinfection of food and agricultural produce (Hamanaka et al., 2006).

Geometries of materials are generally complex and different from one another. Factors affecting IR heating widely vary, depending on the circumstances surrounding the materials used. In addition, the combination with conductive and/or convective heat transfer complicates the problem. Therefore, it is very difficult to find optimum conditions for IR heating only by experiment. In order to design and control the IR heating system appropriately, a mathematical modeling technique becomes one of the most useful tools. As for food operations, many IR heating models have been provided (Datta and Ni, 2002; Shilton et al., 2002), and these predictions were conducted by solving the heat and mass transfer equation as a deterministic model with appropriate boundary and initial conditions. Generally, the value of the incident radiative heat flux at the surface of the food is specified simply. In complex geometrical configurations, this approach is not enough to formulate the boundary condition, because of the difficulty of determining the view factors of the contributing surfaces (Incropera and De Witt, 1990). While the random nature of photons is considered to play an important role in radiation heat transfer, it was neglected or simplified in the process of model development in previous work. In this case, the Monte Carlo (MC) modeling technique is useful for determining the view factors in the radiation heating operation. It is also useful in computing radiation heat transfer in complex configurations, and the process has been widely described in the literature (Tanaka et al., 2007; Guilbert, 1989; Howell, 1988, 1998; Maltby and Burns, 1991). The Monte Carlo ray-tracing technique is useful to simulate the behavior of photons, which carry energy between physical surfaces. In Monte Carlo simulations of radiation heating problems, the energy emitted from a physical surface is simulated by the propagation of a large number of photons. The photon is followed as it proceeds from one interaction to another, which are described as random events. This continues until the photon is absorbed or leaves the computational domain. A large number of trajectories are required to ensure that the variation due to the random events is small. The results are used to determine the fraction of energy that has been absorbed on each surface in the geometry. We focus on this method for modeling radiative exchange between physical surfaces. The model considers both internal heat transfer (conduction) and external heat transfer (convection and surface radiation). We also explain mass transfer during IR heating.

In this chapter, typical heating and mass transfer modeling methods used for IR heating or IR heating combined with other heating methods are introduced. Selected modeling samples of food and agricultural products are given.

3.2 THE IR HEATING MODEL

Heat transfer is the thermal energy in transit due to a temperature difference. In thermal processing, there are three general ways that heat can be transferred from one material to another. Heat transfer in food processing is generally a complicated phenomenon involving conduction, convection, and radiation. A transient simulation model for strongly coupled solution of three-dimensional conduction, convection, and radiation is required to operate the IR heating processes in food industries. By combining the Monte Carlo IR model with convention–diffusion airflow and heat transfer models, the complex heat transfer phenomenon can be predicted successfully.

3.2.1 RADIATION MODEL

In order to model FIR radiation, the Monte Carlo ray-tracing technique (Tanaka et al., 2007) was useful to simulate the behavior of photons, which carry energy between physical surfaces. This technique can be described as follows.

The surfaces in the geometry of the IR heating cell (oven) and food were sub-divided into sets of nonoverlapping primitive surfaces, the total number being N. The source location of photons on the primitive surface and the initial direction of travel were determined stochastically with a pseudo-randomizer $(0 \leq \xi_i \leq 1)$. The random sampling of the source location on a primitive surface with local curvilinear coordinates $(r_1 \leq r \leq r_2, s_1 \leq s \leq s_2, t_1 \leq t \leq t_2)$ consists of three random choices of the pseudorandom variable $0 \leq \xi_i \leq 1$ with $i = 1–3$.

$$r = r_1 + \zeta_1(r_2 - r_1)$$
$$s = s_1 + \xi_2(s_2 - s_1) \tag{3.1}$$
$$t = t_1 + \xi_3(t_2 - t_1)$$

The birth of a photon requires not only a location but also the initial direction of travel. We generated direction cosines from

$$r' = \sqrt{\xi_4} \cos \varphi$$
$$s' = \sqrt{\xi_4} \sin \varphi \tag{3.2}$$

where

$$t' = \sqrt{1 - \xi_4}$$
$$\varphi = \pi(2\xi_5 - 1) \tag{3.3}$$

where ξ_4 and ξ_5 are two choices of the pseudorandom variable $0 \leq \xi_i \leq 1$ with $i = 4$ and 5. The initial traveling direction can be expressed as the following unit vector:

$$\mathbf{e} = r'\mathbf{e}_x + s'\mathbf{e}_y + t'\mathbf{e}_z \tag{3.4}$$

The absorbed ratio of photons is simply proportional to the emissivity of the sample, assuming a gray body. A photon traveling across the geometry is absorbed or reflected at a physical surface in the following cases:

$$\xi_6 \leq \varepsilon \rightarrow \text{absorption} \tag{3.5}$$

$$\xi_6 > \varepsilon \rightarrow \text{reflection}$$

where ξ_6 is the pseudorandom variable $0 \leq \xi_6 \leq 1$. The absorbed ratio of photons is simply proportional to the emissivity of the sample (ε). In general, the emissivity of foods can be defined as the ratio of the radiation emitted by the surface to the radiation emitted by a blackbody.

IR power absorption by foods has been treated by two formulations: one with zero penetration and the other with finite penetration depth. In regard to a finite penetration model, Ginzburg (1969) and Datta and Ni (2002) treated IR power deposition as an exponential decay from the surface into the material. IR power flux is modeled as an exponential decay similar to microwave power, which leads to the volumetric heat source term due to IR heating as

$$Q_{rad}(x) = \frac{dq_{rad}}{dx} = \frac{q_{rad,0}}{\delta_{rad}(x)} \exp\left(-\int_0^x \frac{dx}{\delta_{rad}}\right) \tag{3.6}$$

where $q_{rad,0}$ is the surface IR flux and δ_{rad} is the IR penetration depth, defined precisely as the microwave penetration depth; that is, the distance over which the magnitude of the flux drops to $1/e$ of its value (Metaxas and Meredith, 1983; Tanaka et al., 2001). If the IR penetration depth is large, IR energy is absorbed, quite similar to microwave energy. However, Parroufe et al. (1992) considered all energy to be absorbed at the surface and noted that IR had no special influence beyond increased heat flux in the heat and mass transfer study of IR drying, as compared to convective heating. For heat transfer only, Sakai et al. (1993) found no significant differences between zero and finite penetration models. In any case, it is comparatively easy to build both models because it depends on whether the internal heat transfer model as described in the following text includes the volumetric heat source term due to IR heating as shown in Equation 3.6.

In the case of reflection on a diffuse surface of a gray body, a new direction of travel is determined according to Equations 3.2 and 3.3. As for a real surface, the directional distribution for photon travel can be given. In general, a gray surface is assumed to simulate the behavior of the photon. Tracking photons across the geometry of IR heating cell is the most computationally intensive task. Finally, the number of photons absorbed by a surface is obtained.

The resulting irradiation on surface i is given by

$$G_i = \frac{1}{S_i} \sum_{j=1}^{N} \frac{n_j^i}{n_j} S_j \varepsilon_j \sigma T_j^4 \tag{3.7}$$

with n_j^i being the number of photons emitted by surface j and incident on surface i. Each primitive surface S_j is treated as a separate uniform photon source with emissivity ε_j and temperature T_j. The Boltzmann constant is indicated by σ. The number of photons n_j emitted by surface j is

$$n_j = N_p = \frac{S_j \varepsilon_j \sigma T_j^4}{\sum_{x=1}^{N} S_x \varepsilon_x \sigma T_x^4} \tag{3.8}$$

Thus, the following equation is derived:

$$G_i = \frac{1}{S_i} \frac{\sum\limits_{j=1}^{N} n_j^i}{N_p} \sum\limits_{x=1}^{N} S_x \varepsilon_x \sigma T_x^4 = \frac{1}{S_i} \frac{n^i}{N_p} \sum\limits_{x=1}^{N} S_x \varepsilon_x \sigma T_x^4 \tag{3.9}$$

It therefore suffices to add all incident photons on surface i to the photon current n_i and multiply the relative photon current by the available emissive power in the system to obtain the irradiation. The net radiation flux $q_{rad,i}$ leaving the surface i is equal to

$$q_{rad,i} = \varepsilon_i \sigma T_i^4 - \varepsilon_i G_i \tag{3.10}$$

In order to obtain good estimates of this quantity of interest, many tracks of photons need to be generated.

3.2.2 AIRFLOW MODEL

The airflow model in an IR heating oven is based on governing continuity, momentum, and energy equations with random photon events, which play a key role in radiation heat transfer.

$$\frac{\partial u_i}{\partial x_i} = 0 \tag{3.11}$$

$$\rho_a \frac{\partial u_j}{\partial \theta} + \rho_a \frac{\partial}{\partial x_i}(u_i u_j) = \frac{\partial}{\partial x_i} \mu \frac{\partial u_j}{\partial x_i} + \frac{\partial}{\partial x_i} \mu \frac{\partial u_i}{\partial x_j} - \nabla p + \rho_a g \beta(T - T_{ref}) \tag{3.12}$$

$$\rho_a \frac{\partial H}{\partial \theta} + \rho_a \frac{\partial}{\partial x_i}(u_i H) = k_a \nabla^2 H \tag{3.13}$$

where u, x, T, ρ_a, θ, μ, p, g, β, H, and k_a are airflow rate, direction, temperature, density, time, dynamic viscosity, pressure, gravitational force, thermal expansion coefficient, enthalpy, and thermal conductivity of air, respectively.

These governing equations with appropriate boundary and initial conditions can be solved numerically.

3.2.3 CONDUCTIVE HEAT TRANSFER MODEL

By combining the Monte Carlo IR model with convection–diffusion airflow and heat transfer models, the IR heating process can be predicted successfully.

Heat conduction is a transient process in a three-dimensional space. When conductive heat transfer occurs inside the conductive solid and conductive and radiation transfer take place at the surface, conservation of energy is required at the surface of a solid. The thermal diffusion equation is used to provide the temperature distribution in the conductive solid.

$$\rho_s c_{ps} \frac{\partial T}{\partial \theta} = k_s \nabla^2 T \tag{3.14}$$

where ρ_s, c_{ps}, and k_s are density, heat capacity, and thermal conductivity of the heated product as specified earlier. As for the finite penetration model, the energy generation term, Q_{rad}, is added to the right side in Equation 3.14. When IR heating is assisted by microwaving, the similar energy generation term, Q_{mic}, is added to the preceding equation (Tanaka et al., 2001). As for the thawing process with IR heating, latent heat of melting is released according to thawing rate (Tanaka et al., 2000).

The boundary condition is

$$k_s \frac{\partial T}{\partial \mathbf{n}} = h(T_a - T) - q_{rad} - q_{evap} \tag{3.15}$$

where $\mathbf{n}$, h, and q_{evap} are the outward normal to the heated product surface, heat transfer coefficient, and energy loss due to moisture evaporation, respectively.

Mass transfer is mass in transit as a result of a species concentration difference in the medium. In general, the mass diffusion equation is useful for simulating transitional heat flow in food material (Crank, 1985). In the drying process, many mass transfer models have been provided (Afzal and Abe, 1998; Ranjan et al., 2002; Togrul, 2005). These models have been derived theoretically or empirically. Constant and falling rate drying models are popular in the food-drying operation (Henderson and Perry, 1980). Datta and Ni (2002) treated food as a porous media to describe multiphase transport of liquid water, vapor, and air. Such a multiphase porous media model can accommodate a wide range of moist solid foods such as potato, meat, and cookie dough. Tanaka et al. (2008) have predicted the drying process of high-moisture material in pneumatic dryer by using the following model (The Society of Chemical Engineers of Japan [SCEJ], 1984):

$$\dot{w} = k_m \left[0.622 \frac{a_w(T_p, M) p_{sat}(T_p)}{P - a_w(T_p, M) p_{sat}(T_p)} - X \right] \tag{3.16}$$

where $\dot{w}$, a_w, T_p, M, p_{sat}, P, and X are the drying rate, water activity, sample temperature, moisture content, saturated pressure, total pressure, and absolute humidity, respectively. The saturated pressure, $p_{sat}(T_p)$, was calculated according to the procedures given in the IFC Formulation for Industrial Use (The Japan Society of Mechanical Engineers [JSME], 1999). Water activity can be expressed as a function of sample temperature and moisture content (Chen and Clayton, 1971; Murata et al.,

1988, 1995). The mass transfer coefficient can be related to the heat transfer coefficient as follows (Bennet and Myers, 1983):

$$k_m = \frac{h}{C_a(1+X)\left(\dfrac{\alpha}{D_v}\right)^{2/3}} \tag{3.17}$$

where C_a, D_v, and α are the specific heat of air, diffusivity of water vapor in air, and thermal diffusivity. The latent heat for vaporization of water in foods, L, depends on temperature, T_p, and moisture content, M, and increases with decreasing temperature and moisture content.

$$L(T_p, M) = (V_v - V_l)T_p\left(\frac{dp_p(T_p, M)}{dT_p}\right)_V \tag{3.18}$$

where V_v and V_l are the specific volumes for vapor and liquid, respectively.

$$V_v - V_l = \frac{461.5T_p}{p_p(T_p, M)} - 0.001 \tag{3.19}$$

$$p_p(T_p, M) = a_w(T_p, M)p_{sat}(T_p) \tag{3.20}$$

$$\left(\frac{dp_p(T_p, M)}{dT_p}\right)_V = \left(\frac{da_w(T_p, M)}{dT_p}\right)_V p_{sat}(T_p) + a_w(T_p, M)\left(\frac{dp_{sat}(T_p)}{dT}\right)_V \tag{3.21}$$

The energy loss due to moisture evaporation can be obtained by

$$q_{evap} = \dot{w}\, L(T_p, M) \tag{3.22}$$

Since the moisture removal rate is associated with heat transfer flux, it is usually desired to solve the heat and mass transfer model as a set of simultaneous equations.

3.2.4 NUMERICAL SIMULATION

Modeling of IR heat transfer inside food has been an intensive research area because of the complexity of optical characteristics, radiative energy extinction, and combined conductive and convective heat transfer phenomena (Krishnamurthy et al., 2008). In order to explain drying behavior during IR heating, a suitable mass transfer model is required for every product.

Generally, numerical methods applied to solve the set of governing equations with the appropriate initial and boundary conditions are finite elements, finite difference, and finite volume or control volume method. Time discretization can be accomplished by using first- or second-order implicit methods.

A lot of computational fluid dynamics (CFD) commercial software, which includes the MC ray-tracing technique for estimating IR heat transfer, is sometimes used in the food processing field. Some software can solve a time-dependent, scalar, anisotropic conduction–convection–radiation problem with mass transfer phenomenon. While large computation time is required to solve a complex IR heating model, we are encouraged by the significant progress made by computer technology. Selected modeling samples of products will be given in the next section.

3.3 MODEL PREDICTION

3.3.1 Wood Heating

Tanaka et al. (2006) have investigated the IR heating process of wood in laminar flow. The radiation geometry of the IR heating cell for experiment and prediction is shown in Figure 3.1. The IR oven cavity includes a plate-type IR heater (heating surface area = 0.2 m × 0.3 m; PLR-620, Noritake Co., Nagoya, Japan). A rectangular-shaped wood material (0.03 m × 0.03 m × 0.2 m; ρ_s = 700 kg m^{-3}; C_{ps} = 1.3 kJ kg^{-1} K^{-1}; k_s = 0.2 Wm^{-1} K^{-1}; ε = 0.9) was set at the center of the IR heating oven. A forced airflow of 0.12 ms^{-1} at 20°C was imposed, and the IR heater temperature was set at 200°C. Initial temperatures of air and conductive solid were set to 20°C. A rectangular-shaped material was heated from the top surface by the an IR heater. The data were collected by an IR thermographic camera (Therma CAM3000SC, FLIR Systems, Inc., Wilsonville, Oregon). Figure 3.2 shows the actual and predicted surface temperature distributions for a rectangular-shaped material at the central section (z = 0.1 m) at steady state. As shown in Figure 3.2, the model predicted well the actual temperature distribution around a rectangular-shaped material with a root mean square error of ±3.8°C. A large variation of temperature around the rectangular shape was found, and the maximum difference of temperature was 34°C for

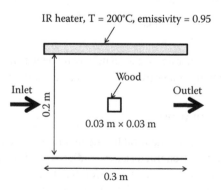

FIGURE 3.1 Radiation geometry of the IR heating cell for experiment and simulation.

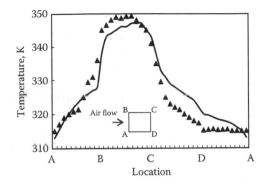

FIGURE 3.2 Comparison of observed data and predicted surface temperature distribution around a rectangular-shaped wood shown in Figure 3.1. (▲: observed, –: predicted).

simulation because of one-sided heating. The lowest heating occurred at the bottom, which was poorly exposed to the radiation field. One of the distinctive trends of IR heating appears well in this simulation.

3.3.2   THERMAL TREATMENT OF FRESH FRUIT FOR SURFACE DECONTAMINATION

Generally, fresh fruit and vegetables can be contaminated with different amounts and kinds of microorganisms. Postharvest heat treatments have recently received attention as a means to prevent fungal spoilage and to inactivate pathogens living on the surface of products. The IR heating method is useful for surface decontamination of products because it is regarded as a highly efficient method for surface heating without chemicals (Krishnamurthy et al., 2008). Compared to water submersion heating, the IR heating rates were small, and the IR surface temperature uniformity was inferior. However, there is an advantage to using no water with large heat capacity and avoiding discharge wastewater. Furthermore, the cross-contamination might be caused by microbial contaminated water as heat medium. It is expected to become an alternative technique for surface decontamination of fresh fruit and vegetables (Tanaka et al., 2007).

3.3.2.1   Figs

The common fig (*Ficus carica* L.) is a traditional fruit in western Asia, and it historically originated presumably in southern Arabia (Ikegami et al., 2009). Figs are one of the richest plant sources of calcium and fiber. They also contain copper, manganese, magnesium, potassium, and vitamins. Since figs are perishable fruit, it is difficult to extend their shelf-life. In order to extend their shelf-life and ensure food safety, microbial decontamination treatment is very important in the early postharvest stage. We have developed a six-fig array model to confirm the possibility of simultaneous heat treatment in a pack house (Figure 3.3). This configuration model includes a far-IR (FIR) heater (glass-covered type; glass length: 160 mm; diameter: 15 mm; surface temperature: 700°C; emissivity: 0.75) with a reflector (emissivity: 0.35), a

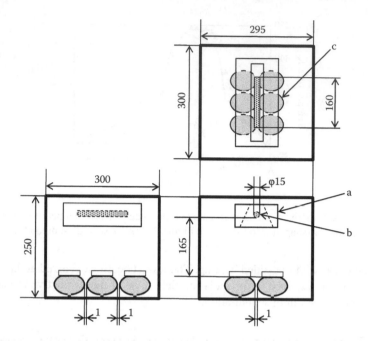

FIGURE 3.3 Schematic representation of the experimental apparatus (unit: mm). (a) Reflector (150°C), (b) IR heater (700°C), (c) fig (initial temperature: 20°C; emissivity: 0.98).

sample holder for arranging six figs with their ostiole side up, and a metal frame for fixing the parts in the prescribed position. Six lateral open air boundaries of the calculation domain were assigned free slip conditions, and all physical surfaces were specified nonslip walls. At the fig surface, a conservative interface flux condition was applied. The initial temperatures of air and fig were assumed to be 20°C for simulation. IR heating simulations without mass transfer were carried out using CFX 11.0 (ANSYS Inc.). This software used the Navier–Stokes equations for fluid flow in combination with radiation heat transfer using the Monte Carlo ray-tracing technique described in Section 3.21. The developed model included both internal heat transfer (conduction) and external heat transfer (coupled radiation and natural convection). All equations were implemented in the instructed finite volume code ANSYS CFX 11.0. The predicted time progress of temperature contours on the surface during 120 s of IR heating is presented in Figure 3.4. By comparing the surface temperature distribution of figs located in the middle and both ends after 120 s heating, the ostiole temperatures for each fig reached 46.8°C and 46.1°C, respectively. Uniform heating could be achieved in this six-array model. It was suggested that simultaneous heat treatment could be achieved in this configuration. The developed model is useful for evaluating IR heating systems.

3.3.2.2 Strawberry

Microbial risk assessment can be carried out by coupling the developed IR model with a microbial kinetics model. For example, the first-order destruction model is used

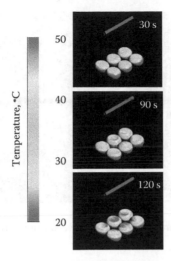

FIGURE 3.4   Predicted temperature distributions as a function of heating time.

extensively for predicting microbial population during thermal treatment (Marquenie et al., 2003).

$$\frac{dN_b}{d\theta} = -k(T)N_h$$

$$k(T) = k_{ref}\exp\left[\frac{F_u}{R}\left(\frac{1}{T_{ref}} - \frac{1}{T}\right)\right]$$

(3.23)

where N_b, $k(T)$, k_{ref}, E_a, R, T, and T_{ref} are microbial population, specific inactivation rate varying as a function of the temperature, inactivation rate at the reference temperature, Arrhenius activation energy, universal gas constant, temperature, and reference temperature, respectively.

The model was solved for the geometry of the IR heating cell, including the real shape of the strawberry fruit (see Figure 3.5). IR heater temperatures were set at 250°C, and the strawberry fruit was heated from both sides. Figure 3.6 shows the survival curve of *Monilinia fructigena* during 300 s of the IR heating with two heaters. The maximum temperature reaches the reference temperature (43°C, for *M. fructigena*) at about 180 s, and the microorganism starts to become inactivated. *M. fructigena* is almost inactivated within 300 s at the maximum temperature point. It is expected that the developed technique will become a powerful tool to evaluate food safety and quality.

3.3.2.3   Peach

Thermal treatments and exposure to ultraviolet (UV) light have already been used for disinfection of strawberries and other fruits; however, few reports on the effect of combined treatments are found in the literature. Marquenie et al. (2003) have reported on the combination of a UV light pulse and thermal treatment for

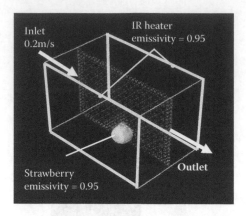

FIGURE 3.5 Model geometry and air velocity vector in the IR heating cell (size: 0.3 m × 0.2 m × 0.2 m).

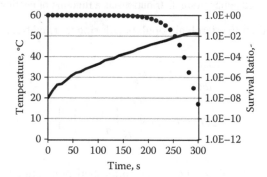

FIGURE 3.6 Predicted inactivation curve of *Monilia fructigena* at the maximum temperature point on a strawberry fruit during IR heating.

Botrytis cinerea and *Monilia fructigena*. They showed that a higher temperature and longer duration of the heat treatment seem to induce more sensitivity to the subsequent pulsed light treatment, resulting in a synergistic effect of both treatments. Hamanaka et al. (2007) have investigated the effect of IR and UV combined treatment on mold ascospore inactivation. In this section, we introduce the IR–UV combined treatment of peach. Figure 3.7 illustrates the predicted temperature contours on the surface of peach by using the IR heating cell as shown in Figure 3.3. If we can apply the MC method for UV radiation modeling, the intensity distribution of UV on the peach surface can be estimated. By using the IR (30 s heating)–UV (30 s lighting) treatment method (Figure 3.8), *Moinlinia fructicola* and *Phomopsis* sp., which are cultivated on the surface of peach at ambient temperature overnight, were inactivated to preserve the product from decay. Figure 3.9 shows comparisons of the effect of treatment with IR–UV irradiation versus untreated products (control). It is revealed that IR–UV treatment had a significant effect on microbial inactivation for both microorganisms.

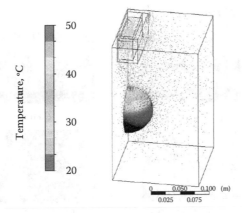

FIGURE 3.7　Model geometry for peach IR heating (one-quarter divided model).

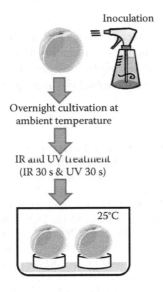

FIGURE 3.8　Procedure for the investigation of IR–UV treatment for surface decontamination of peach.

Further studies on the development of an IR–UV irradiation model, including the synergistic effect of both treatments, will be required to optimize the microbial inactivation process.

3.4　CONCLUSION

In this chapter, we introduced typical heating and mass transfer modeling methods used for IR heating or IR heating combined with other heating methods. It is expected that the IR heating model will be useful for food process optimization;

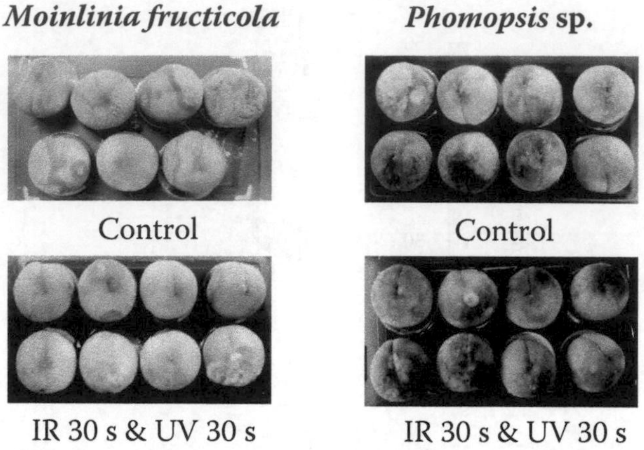

Moinlinia fructicola   *Phomopsis* sp.

Control   Control

IR 30 s & UV 30 s   IR 30 s & UV 30 s

FIGURE 3.9 Comparison of the effect of treatment with IR–UV irradiation versus untreated products (control).

however, the application and understanding of IR heating in food processing is still in its infancy. Some novel techniques introduced in this chapter offer many advantages over the utilization of IR heating in food processing, especially in the area of drying, minimal processing, and microbial risk management.

REFERENCES

Afzal, T. M., and Abe, T. (1998). Diffusion in potato during far infrared radiation drying. *Journal of Food Engineering*, 37(4), 353–365.

Bennet, C. O., and Myers, J. E. (1983). *Momentum, Heat and Mass Transfer* (3rd ed.). Tokyo: McGraw-Hill Book Company.

Chen, C. S., and Clayton, J. T. (1971). The effect of temperature on sorption isotherms of biological materials. *Transactions of the ASAE*, 14, 927–929.

Crank, J. (1985). *The Mathematics of Diffusion* (2nd ed.). New York: Oxford University Press.

Datta, A. K., and Ni, H. (2002). Infrared and hot-air-assisted microwave heating of foods for control of surface moisture. *Journal of Food Engineering*, 51, 355–364.

Ginzburg, A. S. (1969). *Application of Infrared Radiation in Food Processing*. Chemical and Process Engineering Series, London: Leonard Hill Books.

Hamanaka, D., Tanaka, F., and Uchino, T. (2007). Effect of the combination of infrared heating with ultra violet irradiation on the inactivation of mold ascospores. In *Proceedings of International Workshop on Agricultural and Bio-Systems Engineering (IWABE)*, Ho Chi Minh City, 44–50.

Hamanaka, D., Uchino, T., Furuse, N., Han, W., and Tanaka, S. (2006). Effect of the wavelength of infrared heaters on the inactivation of bacterial spores at various water activities. *International Journal of Food Microbiology*, 108(2), 281–285.

Henderson, S. M., and Perry, R. L. (1980). *Agricultural Process Engineering* (3rd ed.). Westport, CT: The AVI Publishing Company, Inc.

Ikegami, H., Nogata, H., Hirashima, K., Awamura, M., and Nakahara, T. (2009). Analysis of genetic diversity among European and Asian fig varieties (*Ficus carica* L.) using ISSR, RAPD, and SSR markers. *Genetic Resources Crop Evolution*, 56(2), 201–209.

Incropera, F. P., and De Witt, D. P. (1990). *Fundamentals of Heat and Mass Transfer* (3rd ed.), John Wiley & Sons, New York.

The Japan Society of Mechanical Engineers. (1999). *JSAM Steam Tables*. Tokyo: Maruzen.

Krishnamurthy, K, Khurana, H. K., Jun, S., Irudayaraj, J., and Demirci, A. (2008). Infrared heating in food processing: An overview. *Comprehensive Reviews in Food Science and Food Safety*, 7, 2–13.

Maltby, J. D., and Burns, P. J. (1991). Performance, accuracy, and convergence in a 3-dimensional Monte-Carlo radiative heat-transfer simulation. *Numerical Heat Transfer Part B—Fluid Amentals*, 19(2), 191–209.

Marquenie, D., Geeraerd, A. H., Lammertyn, J., Soontjens, C., Van Impe, J. F., Michiels, C. W., and Nicolaï, B. M. (2003). Combinations of pulsed white light and UV-C or mild heat treatment to inactivate conidia of *Botrytis cinerea* and *Monilia fructigena*. *International Journal of Food Microbiology*, 85(1–2), 185–196.

Metaxas, A. C., and Meredith, R. J. (1983). *Industrial Microwave Heating*. London: Peter Pererinus.

Murata, S., Tagawa, A., and Ishibashi, S. (1988). An equation for calculating the latent heat of vaporization of water in cereal grain. *Journal of Japanese Society of Agricultural Machinery*, 50(3), 85–93.

Murata, S., Tanaka, F. Amaratunga, K. S. P., Shibuya, K., and Uchino, T. (1995). Measurement of the adsorption equilibrium moisture content of polished rice. *Journal of Japanese Society of Agricultural Machinery*, 57(6), 45–52.

Parroufe, J.-M., Dostie, M., Mujumdar, A. S., and Poulin, A. (1992). Convective transport in infrared drying. In A. S. Mujumdar (Ed.), *Drying '92* (pp. 695–703). The Netherlands: Elsevier.

Ranjan, R., Irudayaraj, J., and Jun, S. (2002). Simulation of three-dimensional infrared drying using a set of three-coupled equations by the control volume method. *Transactions of the ASAE*, 45, 1661–1668.

Sakai, N., Fujii, A., and Hanzawa, T. (1993). Heat transfer analysis in a food heated by far-infrared radiation. *Nippon Shokuhin Kogyo Gakkaishi*, 40(7), 469–477.

Shilton, N., Mallikarjunan, P., and Sheridan, P. (2002). Modeling of heat transer and evaporative mass losses during the cooking of beef patties using far-infrared radiation. *Journal of Food Engineering*, 55, 217–222.

The Society of Chemical Engineers, Japan. (1984). *Kagaku Kougaku*. Tokyo: Maki Shuppan.

Tanaka, F., Maeda, Y., and Morita, K. (2000). Mathematical modeling of three-dimensional heat transfer during food freezing process. *Journal of Japanese Society of Agricultural Machinery*, 62(4), 120–126.

Tanaka, F., Mallikarjunan, P., and Hung, Y. -C. (2001). Mathematical modeling of microwave heating of chicken breast meat. *Journal of Japanese Society of Agricultural Machinery*, 63(1), 48–54.

Tanaka, F., Morita, K., Iwasaki, K., Verboven, P., Scheerlinck, N., and Nicolaï, B. M. (2006). Monte Carlo simulation of far infrared radiation heat transfer. *Journal of Food Process Engineering*, 29(4), 349–361.

Tanaka, F., Verboven, P., Scheerlinck, N., Morita, K., Iwasaki, K., and Nicolaï, B. M. (2007). Investigation of far infrared radiation heating as an alternative technique for surface decontamination of strawberry. *Journal of Food Engineering*, 79(2), 445–452.

Tanaka, F., Uchino, T., Hamanaka, D., and Atungulu, G. G. (2008). Mathematical modeling of pneumatic drying of rice powder. *Journal of Food Engineering*, 88(4), 492–498.

Togrul, H. (2005). Simple modeling of infrared drying of fresh apple slices. *Journal of Food Engineering*, 71, 311–323.

4 Emitters and Infrared Heating System Design

Ipsita Das and S.K. Das

CONTENTS

4.1   Introduction 58
4.2   Classification of IR Emitters 58
     4.2.1   Mechanism of Infrared Absorption by Foods 60
     4.2.2   Electric IR Emitter 60
           4.2.2.1   Reflector-Type IR Incandescent Lamps 61
           4.2.2.2   Quartz Tube IR Emitter 64
           4.2.2.3   Ceramic IR Emitters 65
           4.2.2.4   Tubular Metal-Sheathed Elements.................................66
           4.2.2.5   Radiant Panels...66
     4.2.3   Gas-Fired IR Emitters 66
           4.2.3.1   Direct Flame IR Radiator 68
           4.2.3.2   Ceramic Burner...68
           4.2.3.3   Metal Fiber Burner 68
           4.2.3.4   High-Intensity Porous Burner..68
           4.2.3.5   Catalytic Gas-Fired IR Emitter 68
     4.2.4   Carbon Twin IR Emitter...69
4.3   Selection Criteria of IR Emitters 70
     4.3.1   Temperature of the Emitter...70
     4.3.2   Working Life Span..71
     4.3.3   Response Time or Speed of Heating 71
     4.3.4   Coefficient of Radiation Efficiency 71
     4.3.5   Stability of Distributed Radiation...72
4.4   Specification of IR Emitters 72
4.5   Comparison of Different Forms of IR Heat 72
4.6   Design Variables of IR Heating Systems...73
     4.6.1   Power Required for Heating 73
     4.6.2   Efficiency of IR Heating System 80
     4.6.3   Design Arrangement for Components of an IR Heating System
           (Electrical IR Emitter)..82
           4.6.3.1   Distance between the Emitter and the Object 82
           4.6.3.2   Distance between the Emitters 85
Nomenclature...86
References...87

4.1   INTRODUCTION

Infrared (IR), or radiant heat, is one of the most widely used heating techniques in various industries. Radiant heat transfer is often said to be more efficient and cost-effective than convective heat transfer, largely because radiant heat transfer reduces cycle times, focuses energy on the target, and produces virtually no volatile organic compounds, carbon monoxide (CO), and nitrogen oxides (NO_x). Moreover, this technique does not require an expensive and insulated enclosure to trap and recirculate heat.

IR radiant energy, emitted by the sun, can be generated even from a small fire. In fact, any object or material at temperature above absolute zero emits IR radiation. IR radiant energy covers a wide spectrum of wavelengths, ranging from 0.7 μm to a very long wavelength of 400 μm. The quality and intensity of radiation are related to the temperature of the radiator by the Stefan–Boltzmann law, which states that the energy emitted by a radiator roughly depends on the fourth power of its absolute temperature.[1] The amount of IR radiant energy emitted from a given source can be calculated easily from Figure 4.1. For a given temperature, the wavelength at which maximum radiation occurs (peak wavelength) can be determined using Wien's displacement law.[2] Figure 4.2 shows the relationship between wavelength and temperature of filament. Color can be used as a good indicator of heat output. Emitters can radiate a range of colors from normal color at ambient condition to maximum glowing white. However, the best radiation heat transfer efficiency occurs when the heater glows red.[3]

4.2   CLASSIFICATION OF IR EMITTERS

According to the wavelength of radiation, emitters are broadly classified into three categories, namely, shortwave, medium-wave, and longwave emitters, though the boundary between them is not very precise (Table 4.1). Shortwave emitters emit

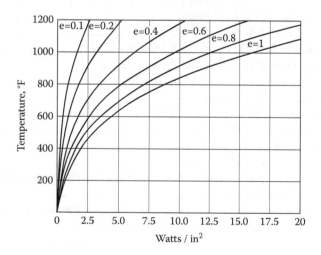

FIGURE 4.1   Surface temperature versus radiation emission at various emissivity values. (From Mor Electric Heating Association, U.S. With permission.)

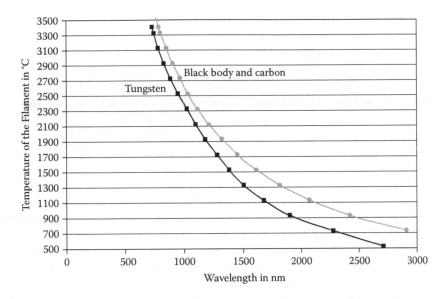

FIGURE 4.2 Relation between wavelength and temperature. (From Heraeus Noble Light, Germany. With permission.)

TABLE 4.1
Comparison of Different Forms of Infrared Wave Emitter

	Emitter Type with Spectrum Category		
	Shortwave	Medium-Wave	Longwave
Infrared parameters	Halogen and incandescent lamp	Quartz emitter	Resistance material
Material	Quartz tube with tungsten coil inside sealed	Quartz tube with Fe-Cr-Al alloy as heating element	Steel tube with Fe-Cr-Al alloy as heating element enclosed
Radiation efficiency	High: ~92%	Medium: ~60%	Low: ~40%
Peak wave length	1.2 μm	2.2 μm	4.0 μm
Corresponding radiation temperature	2500 K	1300 K	800 K
Major heating mode	Radiation	Combined radiation and convection	Convection
Response time to 90% output	1 s	30 s	300 s
Convergence of radiation	Possible with good focusing	Fairly possible	Fairly possible

Source: Data adapted from Anonymous, 2009. http://www.ercis.lu/default.asp?contentID=680.

IR rays of wavelength ranging from 0.7 to 1.4 μm, which is closest to the visible spectrum, giving a bright white appearance with the corresponding radiator temperature being around 1300–2600 K. This type of emitter is often considered the most powerful because it can achieve the highest power densities, up to or more than 300 kWm^{-2}, and are very responsive as they can attain the maximum temperature within a few seconds. They are used extensively for industrial processes such as preheating, metal castings, powder coating, and adhesive bonding. Medium-wave IR emitters emit a radiation spectrum of wavelength ranging from 1.4 to 3.0 μm at radiator temperatures of 850–1200 K and power densities of up to 90 kWm^{-2}; they appear as a bright orange glow with a response time of about a minute. Medium-wave emitters are extensively used for drying and curing of food product. Longwave IR emitters have a radiation temperature of 500–800 K, which corresponds to a spectrum of wavelength more than 3.0 μm. It attains a power density up to 40 kWm^{-2}. The emitter appears dull orange and in many cases is not even perceivable to the human eye. This type of emitter takes more than 5 min to reach its peak emissive power.[4] Longwave IR emitters create a stream of hot air that is useful for processes requiring a combination of convection and IR heating.

Basically, two types IR emitters are conventionally used in industry, namely, electric IR emitters and gas-fired IR emitters. Electric IR emitters can emit any wavelength beginning from short- to longwave, depending on the voltage supplied to the emitters. In contrast, gas-fired IR emitters are restricted to medium- or long-wavelength-range heaters.

4.2.1 MECHANISM OF INFRARED ABSORPTION BY FOODS

When radiant energy impinges upon a food surface, it induces a change in the electronic, vibrational, and rotational states of atoms and molecules. The mechanism of energy absorption is determined by the wavelength range of the incident radiation flux. Changes in the electronic state correspond to a wavelength range of 0.2–0.71 μm (ultraviolet and visible rays), while changes in the vibrational and rotational states correspond to wavelength ranges of 2.5–100 μm (far-IR, FIR) and above 100 μm (microwaves), respectively. Substances absorb FIR energy most efficiently through the mechanism of molecular vibration, which leads to heating of the product.[5] The IR emitters emitting radiation between the medium- to far-infrared region of 2.5 to 7 μm were found suitable for drying of agricultural and food product. The main components of foodstuff, that is, water and organic compounds such as proteins and starches, absorb FIR energy at wavelengths greater than 2.5 μm.[6]

4.2.2 ELECTRIC IR EMITTER

Electrical IR emitters consist of a metal filament placed inside a sealed enclosure that is either filled with inert gas or evacuated. The radiant energy in electric IR heaters is generated by passing electric current through a high-resistance wire (nichrome wire, iron–chromium wire, or tungsten filament), where the element and the material surrounding the element get heated to an incandescent temperature. Electric IR emitters could be operated up to a temperature of 2600 K, corresponding to near-IR radiation of 0.7 to 1.4 μm. The spectrum of wavelength, or conversely the emissive powers of these electric IR emitters, is moderated with power input to the system. Several types of electric IR radiant emitters with different specifications are available for specific purpose and use. These include the reflector-type incandescent lamp, quartz tube, and resistance elements such as metallic tube, ceramic tube, and nonmetallic rod.

Based on their wavelength and temperature, incandescent lamps are classified as shortwave emitters, while quartz tubes and resistance elements are classified as medium-wave and longwave emitters, respectively. Lamp codes are designated by the American National Standards Institute (ANSI) based on mechanical, electrical, and performance characteristics. Ready availability, ease of control, instrumentation, and the possibility of wide selection of heating elements make electrical IR emitters a very attractive proposition. They transfer heat directly to the product without heating the surrounding air. This gives benefits such as faster heat transfer and greater control for spot heating. Unlike natural gas-, diesel-, and propane-fueled emitters, the electric IR emitter generates zero emissions.

4.2.2.1 Reflector-Type IR Incandescent Lamps

IR reflective emitters use a reflective shield (reflectors) to direct radiant heat onto a desired surface very precisely by converging or concentrating the energy into a choice of wide, medium, or narrow patterns. The reflector efficiency in transferring energy to the surface is influenced by the reflector's material, shape, and contour.[7] Anodized aluminum is used as a common reflector material that can achieve approximately 92% reflectivity for normal spectrum of IR radiation. Gold-anodized aluminum is the best reflector material, considering the combined factors of cost, workability, and weight, and is reported to give approximately 95% reflectivity and reasonable durability.[8]

IR incandescent lamp (Figure 4.3) life, expressed in hours, is calculated at design voltage under ideal laboratory conditions. Deviations from design voltage will alter values of current consumption, radiation output, and temperature (Figure 4.4) besides decreasing or increasing the lamp life.[9] The reflector-type IR lamp primarily consists of a glass enclosure and a tungsten wire filament. The inside parabola surface of the bulb is glazed with a thin silver coating that acts as a reflector. The parabola shape of the reflector facilitates distributing the radiation energy isotropically. The filament acts as an electrical resistor that dissipates power, proportional to the voltage applied and duration of current flowing through it (Figure 4.5). A wide variety of reflector-type incandescent lamps such as vacuum, gas-filled, and halogen lamps are available for different technical and commercial applications. Clusters of lamps can be mounted to generate high intensity. These kinds of emitters are mostly used in the food industry. IR incandescent lamps have several benefits (Table 4.2).

4.2.2.1.1 Incandescent Vacuum Lamp

Typical vacuum lamps may have filament temperatures ranging from 1800 to 2600 K, emit a large portion of energy as heat in the wavelength corresponding to the near-IR

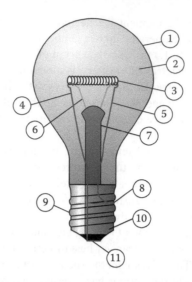

FIGURE 4.3 Schematic of infrared incandescent lamp. 1. Glass enclosure; 2. low-pressure inert gas inside (argon, neon, nitrogen); 3. tungsten filament; 4, 5, and 8. contact wires; 6. support wires; 7. stem (glass mount); 9. cap for electrical contact and sealing; 10. insulation; 11. electrical contact.

region, and convert almost 88% of electrical energy to radiation energy (Figure 4.6). The high temperature of the tungsten filament leads to extremely fast heating that further depends on the filament type, shape, size, and amount of current drawn. This type of IR heat lamps has low thermal inertia,[10] which means that the time lag in delivering heat and returning back to the initial state is low, thus enabling the system to be controlled precisely either manually (on–off control) or electronically (automatic close loop control). Additionally, these heaters are clean from the environmental standpoint.

4.2.2.1.2   Incandescent Gas-Filled Lamps

Gas-filled lamps produce light from an incandescent filament operated in an inert gas atmosphere such as nitrogen, argon, krypton, and xenon. These gases suppress evaporation of the metal from the filament, consequently increasing the life of the lamp even when operated at a filament temperature as high as 3000 K. The gas further helps in cooling the filament and reduces migration of evaporated tungsten to the wall of the lamp. The cost of such lamps increases significantly when filled with inert gases, particularly xenon, since it is scarce. The higher operating temperature of gas-filled lamps produces high luminous energy per unit power supplied, a characteristic that justifies their use in critical applications.

4.2.2.1.3   Tungsten Halogen Lamps

They are more efficient than standard incandescent vacuum and gas-filled lamps in a quartz envelope. The halogen IR lamp is similar to an inert gas-filled lamp, except

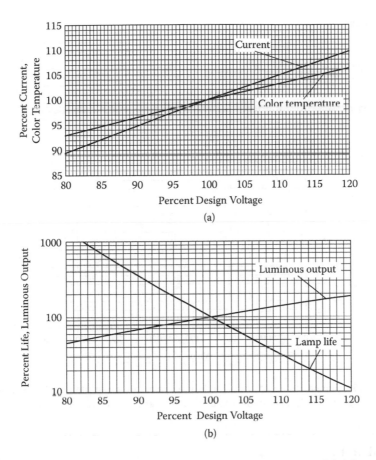

FIGURE 4.4 Percentage variations in current, color temperature, and brightness when operating voltage differs from design voltage. (From International Light Technologies, U.S. With permission.)

that it contains a small quantity of an active halogen gas such as bromine. At a sufficiently high wall temperature of 500 K, the halogen gas in the lamp reacts with the tungsten deposit in quartz envelope and forms unstable tungsten halide. When the filament temperature attains approximately 2500 K, the halide compound in turn dissociates to tungsten metal and halogen again. The metal vapor deposits onto the colder part of the filament. These lamps are designed to maintain a required wall temperature exceeding 500 K when operated at rated voltage.

The temperature rises between 1300 and 2600 K when the heat intensity varies between 150 and 300 kWm^{-2}. Halogen lamps are very quick to attain peak intensities within 1 to 3 s, with typical radiant efficiency of 70% to 80%, and are resistant to thermal and physical shock. These types of lamps are used for curing, baking, dehydration of foods, and processing of material for the semiconductor and aerospace industries. The radiation output of a halogen lamp is more stable than that of a nonhalogen gas lamp due to the cleaning action, as stated earlier.

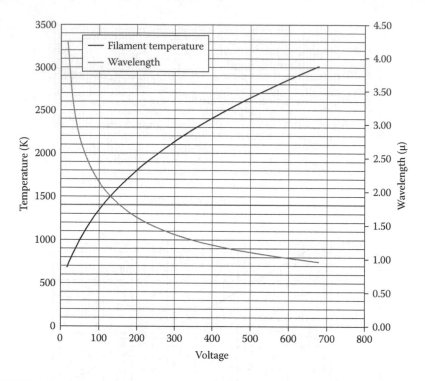

FIGURE 4.5 Temperature and wavelength of a typical tungsten filament lamp at different applied voltage. (From Energy Radiation Curing Industrial Solutions, U.K. With permission.)

TABLE 4.2
Benefits of Infrared Incandescent Lamps

Benefits	Features
Heating-up speed	High emission; more than 90% emission is possible within 1 s
Clean	No pollution; environment friendly, with no emission by-products
Resistant to shock	The lamps are heat thermal shock resistant
Energy conversion	High; more than 85% of consumed energy transferred into IR heat
Working life	As high as 5000 h
Compactness	Construction is compact. It can be as narrow as a lamp

Source: Data adapted from Anonymous, 2009. http://www.slideshare.net/marktenheuw/ 070716-warmterras-nl.

4.2.2.2 Quartz Tube IR Emitter

The quartz tube IR emitter contains a coiled nickel–chrome wire lying unsupported within a fused quartz tube that is more transparent to IR and allows higher heat intensities.[11] Typically, the source attains a temperature of around 900 to 1200 K with power density of 60 kWm^{-2}. Replacing nickel–chrome with a tungsten filament

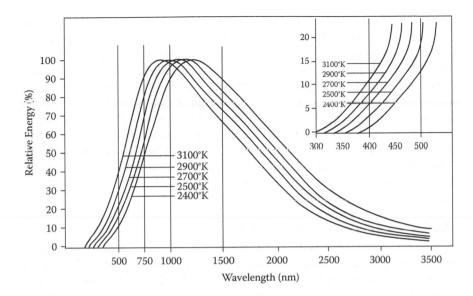

FIGURE 4.6 Spectral radiation output for tungsten filament lamps. (From International Light Technologies, U.S. With permission.)

inside the hermetically sealed vacuum tube permits a temperature around 1500–2000 K. The quartz tube heater appears red to orange during operation in 20 to 60 s, with a radiant efficiency of 70% to 80%, and emits peak energy radiation at wavelengths between 2.5 and 4.0 μm at rated voltage.

Since quartz tubes are not sealed or filled with inert gas, oxidation of metal wire limits the operating temperature. Consequently, it restricts the mounting distance among the IR heating tubes in an array placed inside the curing oven for intense heating of the target. The lamp life depends primarily on closeness between the element operating temperature and its oxidation temperature. These emitters are resistant to thermal shock due to the extremely low coefficient of thermal expansion. Quartz glass IR emitters are preferred to conventional heating sources, ceramic, gas, or metal IR emitters, as quick transfer of large amounts of energy is possible, precisely matching the product to be heated.

4.2.2.3 Ceramic IR Emitters

Ceramic IR emitters contain finely coiled heating element wires (iron, chrome, and aluminum) embedded directly inside ceramic materials that make the system resistant to thermal cracking and damage and protect it from oxidation and corrosion. When energized, the entire surface of the ceramic body glows red hot and acts as an emitter of radiant heat energy. It emits a medium-wave to longwave spectrum of radiation, with peak values between 3.3 and 5.7 μm, producing temperatures between 600 and 900 K with power densities ranging from 10 to 60 kWm^{-2} and an energy efficiency of about 95%. Ceramic heaters can be positioned in a row or side by side for generating a high-intensity emitting surface. The modular approach provides a uniform heat profile over the target area. Individual ceramic heater units can be programmed and are best for zone and control heating (even gentle heating) by regulating the power density. The superior control of programmed temperature outweighs the cost factor. It is extensively used in the plastic industry.

The colors of the emitters differentiate the temperature range of a particular heater. Some heaters are designed to change color when heated, thus facilitating easy identification of damaged or nonfunctioning elements in the heater array. Among the three available basic emitter faces, for example, concave, flat, and convex, the concave surface emits a concentrated radiant pattern that is highly effective for zone heating. Flat surface emitters are suitable for uniform heating at close proximity between the emitter and the target. The convex-shaped emitters radiate a wide area pattern suitable for comfort heating and other applications that require isotropic emission of radiation.

4.2.2.4   Tubular Metal-Sheathed Elements

Metal sheath elements are basically resistance heating wire embedded in an electrically insulating ceramic material, enclosed in a steel or alloy tube. Metal sheath elements are rugged; resistant to thermal shock, vibration, and impact; and more durable than lamps and tubes. They can be mounted in any position. This category of heaters generates IR radiation in the wavelength range of 2.5–3 µm with a maximum sheath temperature of 1000 K and radiant efficiency of 60% to 75%. They typically get heated up within 60 to 180 s, giving higher radiant efficiency when shielded from direct airflow.

4.2.2.5   Radiant Panels

Radiant panels contain a grid of resistance wire or ribbons sandwiched between a thin plate of electrical insulation on the emitting side and thermal insulation on the back side, with typical dimensions of 25 to 76 cm width and 30 to 240 cm length. Low-temperature panels often use thin ceramic papers, boards, or steel, while for high-temperature panels, alloy, quartz, or ceramic plates are used. Typically, at relatively high temperatures, they generate energy in the wavelength range of 2.5–5 µm, with source temperature around 800 to 1100 K, heat intensity of 20 kWm^{-2}, and heat-up period within 60 to 300 s. They are resistant to thermal and physical shock and possess a radiant efficiency of 55% to 70%.

Since the entire surface emits radiation, high IR intensities can be achieved at lower source temperatures compared to the IR lamp or tube and cover a wider area with uniform radiation without any need for a reflector. These IR emitters are durable, typically of 5000 to 10,000 h, unless elements are overheated or physically damaged.

4.2.3   Gas-Fired IR Emitters

Natural gas or liquefied gas (propane) can be used for heating the perforated ceramic surface (luminous heaters) or steel tube (tube heater) that emits radiation. In gas-fired IR emitters, combustion of the air and fuel stream takes place on the burner surface, which raises the surface temperature to about 1000–1200 K and emits radiation in the wavelength[12] of 1.6 to 10 µm, precisely matching the IR absorption

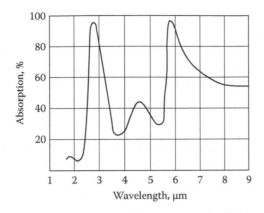

FIGURE 4.7 Spectral absorption characteristics for water. (From Mor Electric Heating Association, U.S. With permission.)

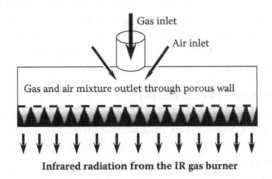

FIGURE 4.8 Schematic of gas-fired IR emitter construction.

characteristics of water (Figure 4.7). Thus, products having high moisture can absorb more radiation emitted by this class of IR emitters. Its operating efficiency ranges from 30%–50%. Although gas-fired IR systems are emitters of longwave radiation, in recent years, shortwave emitters generating temperatures of around 1600–1700 K have been reported.

The gas-fired IR emitter (Figure 4.8) is often paired with a gas convection oven for faster heating than what the convection oven alone can achieve. The advantages of these emitters are that they are independent of electricity, cheaper in terms of operating cost, capable of saving energy, and are more reliable and durable than electric IR emitters. The initial cost of the gas IR emitter is, however, higher. The benefits of gas-fired IR emitters include uniform heating of the target with high thermal efficiency and faster control. However, gas-fired IR emitters have serious limitations; for example, the burning area must be well ventilated to reduce the hazards of evolved carbon monoxide. This type of IR emitter is not beneficial for high level of heating demand, but moderate heat output is feasible with an extended heating period. There are different types of gas-fired IR emitters: direct flame IR radiator, ceramic burner, metal fiber burner, high-intensity porous burner, and catalytic burner.

4.2.3.1   Direct Flame IR Radiator

The radiating surface is directly heated with gas flames in this burner. Typically, in this type of radiator, the surface temperature rises to 500–700 K, which corresponds to a peak wavelength around 4 μm. This direct flame IR radiator is mostly used in modern conveyor-type bread-baking ovens with a provision of recirculation of exhaust gas.

4.2.3.2   Ceramic Burner

In a ceramic burner, combustion occurs inside a perforated ceramic panel that turns reddish orange and radiates medium-wave IR spectrum of radiations around a peak value of 2.4 μm. Its source temperature rises to about 1200 K and develops maximum power density[13] of about 120 kWm^{-2}. These IR emitters are used in industries such as textile, paint, powder coating, process heating, etc.

4.2.3.3   Metal Fiber Burner

In this burner, combustion takes place inside a thin wire mesh made of special steel that turns reddish orange (surface temperature around 1300 K) and radiates a medium-wave spectrum of IR radiation with a peak value of about 2.2 μm, giving a maximum power density of about 200 kWm^{-2}. The metal fiber burner finds similar applications as the ceramic burner, including paper drying and coil coating.

4.2.3.4   High-Intensity Porous Burner

High-intensity porous burners consist of porous or perforated refractory plates with a fine cellular structure mounted on cast iron or formed steel plenum chambers. In this burner, combustion takes place within the perforated panel that has two levels of perforations: lower-region medium-size pores through which the gas–air mix enters and gets preheated and the upper-region large-size pores where the fuel burns uniformly. Heat produced due to this combustion process is transferred to the exit face of the ceramic surface, which incandesces to produce shortwave IR radiations of 1.7 μm with a source temperature of 1400 to 1700 K and maximum power density of 1000 kWm^{-2} within 15 to 30 s of heating. Recently, a porous matrix of woven ceramic fiber has been developed[14] to heat up and cool down within a few seconds. Porous matrix burners are resistant to thermal and physical shock and have the highest radiant efficiency among all the gas-fired IR emitters; they are used in the same industries as other gas-fired emitters.

4.2.3.5   Catalytic Gas-Fired IR Emitter

The relatively new type of apparatus is a diffusion-type heater, which operates on a catalytic exothermic chemical oxidation reduction principle, converting methane or propane gas to moisture and carbon dioxide in the presence of platinum and oxygen and thus releasing IR energy. The catalytic IR heater emits radiation heat over a wide range of wavelengths without any visible flame. The burners are modular in nature, where a dimension of individual burner varies from 175 × 330 mm or as big as 120 × 300 mm.

Catalytic burners are similar to porous matrix burners in construction, appearance, operation, and tolerance to thermal and physical shock, but the refractory material

TABLE 4.3

Characteristic Features of Different Gas-Fired IR Emitters

	Types of Burner			
Parameters	Catalytic Burner	Ceramic Burner	Metal Fiber Burner	Porous Burner
Wavelength category	Longwave	Medium-wave	Medium-wave	Shortwave
Peak wavelength	3.5 µm	2.4 µm	2.2 µm	1.7 µm
Maximum burner temperature attained	850 K	1100 K	1200 K	1400 K
Maximum thermal load	30 kWm^{-2}	120 kWm^{-2}	200 kWm^{-2}	1000 kWm^{-2}

Source: Data adapted from Anonymous, 2009. http://www.sulekhab2b.com/altotec/default.htm.

is usually glass or ceramic wool.[14] Typically, the heating sources are raised to 600 to 800 K with power density varying from 6 to 28 kWm^{-2}, emitting longwave radiation (5.6 to 10 µm) with the heating-up period varying between 180 and 300 s and radiation efficiency ranging from 30% to 75%. These heaters use a capsulated chamber to oxidize the gas, thereby limiting the amount of green exhaust gas (water vapor and carbon dioxide) generated in the burning process. Catalytic burners are augmented with an alternative heating source, such as electric heating elements, to preheat the catalyst before the combustion process is initiated. This burner is not suitable for any desired or programmed heating with fluctuating heat levels and duration. The characteristics of different gas-fired IR emitters are presented in Table 4.3.

4.2.4 CARBON TWIN IR EMITTER

This new-generation IR emitter is a unique combination of the carbon IR emitter and shortwave IR emitter in twin quartz glass tubes. This unique combination allows their application in a product that requires both shortwave and medium-wave IR radiation and more depth of penetration; for example, simultaneous cooking and browning of foodstuffs. Twin-tube carbon IR emitters can be easily retrofitted into existing IR systems and improve the system with decreased operating costs. All carbon IR emitters deliver effective medium-wave radiation at high power densities and speed up the drying of water-based products extremely efficiently. Specifically, IR emitters with carbon technology provide a power density of up to 150 kWm^{-2} and have a response time in seconds, which allows excellent control over both level and duration of radiation as desired. Comprehensive tests have shown that carbon emitters dry the water-based coatings significantly more efficiently than shortwave IR emitters, consuming only up to 30% of the energy.

In the medium-wave range, existing carbon emitters have been replaced by twin-tube carbon emitters. This type of emitter improves the system efficiency and decreases the operating cost. Carbon twin emitters have a maximum power density

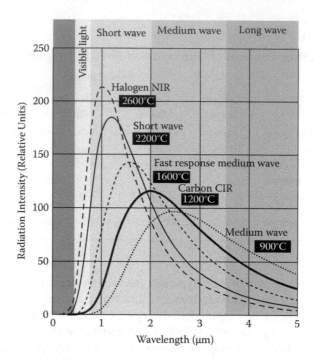

FIGURE 4.9 Spectral distribution of different IR emitters. (From Heraeus Noblelight GmbH, Germany. With permission.)

that is twice than that of conventional medium-wave carbon emitters. The spectrum of the carbon IR emitter is compared with other IR emitters in Figure 4.9. The spectra indicate that halogen emitters radiate the most power in the shortwave region (wavelengths less than 2 μm). Typically, at 3 μm, carbon twin emitter output is higher than that of the halogen emitter by approximately 200%.

4.3 SELECTION CRITERIA OF IR EMITTERS

The selection of IR heater is important for the success of the heating process. The efficiency of IR heaters essentially depends on matching between the spectra of emission wavelength from the emitter to the absorption spectrum of the material to be heated. This results in optimum efficiency, speeding up the process, and reduction of energy cost up to 50%. The following paragraphs discuss certain technical characteristics of IR emitters for their appropriate selection.

4.3.1 TEMPERATURE OF THE EMITTER

The penetration depth of IR radiation varies with the emission spectra of wavelength and temperature of the emitter. Generally, shorter-wavelength radiation has greater penetrating power. The radiation intensity of shortwave emitting heaters tends to be higher with high filament temperature. If the target material absorbs more toward the longer-wave radiation, gas catalytic heaters (achieve temperatures up to 650 K) should be considered. Ceramic panels, quartz tubes, and metal sheath heaters can achieve temperatures in the range of 800–1100 K. Shortest-wavelength radiation can be obtained from a quartz lamp operating at 2600 K.

4.3.2 Working Life Span

Emitters with minimum depreciation or longest possible working life are desirable. The rated life of most IR lamps is 5000 h. High-temperature lamps have shorter rated life. The life expectancy depends on filament temperature, end seal temperature, quartz temperature, voltage control, etc.

4.3.3 Response Time or Speed of Heating

Response time is the time required for heating up and cooling down of IR emitters in response to changes in applied voltage. If an application needs to achieve the process temperature within a very short time, then shortwave emitters such as the quartz lamp would be the appropriate device; an alternate option could be the quartz tube. If an application allows more time, then ceramic panels will be useful. Medium-wave and longwave emitters have higher thermal inertia.

4.3.4 Coefficient of Radiation Efficiency

Radiation efficiency is defined as the percentage of radiant output from the heat source to the energy input. It is highest for short-wavelength sources. The most important problem of IR heating is the match between the spectral distribution of the emitting radiation and the optical properties of the irradiated object. With proper matching, maximum energy transfer will be possible. Factors that affect the efficiency of the IR heater are as follows:

a. The thermal and radiation characteristics of the irradiated object determine the required output efficiency of the emitter and the depth of radiation. In IR processing, the effect of the optical property of food products has been demonstrated recently.[15] The heating system should be capable of delivering the desired spectral band of IR radiation using an optical bandpass filter that facilitates high absorbance of energy by the components present in the irradiated body.
b. The geometrical parameters describe the relative positions of the IR emitter and the irradiated object. It significantly influences the density and uniformity of radiation.
c. The physical properties of the medium in the working chamber significantly influence the rate of heat transmission between the radiator and the irradiated body.

4.3.5 STABILITY OF DISTRIBUTED RADIATION

The distribution of the spectral emissivity of the IR emitter should be uniform at each wavelength, for homogeneous heat distribution on the food product leads to uniform heating. This depends mainly on the temperature of the source.

4.4 SPECIFICATION OF IR EMITTERS

The following parameters need to be considered when specifying IR heaters:

a. *Operating voltage*: Each IR lamp has a rated voltage. Filament temperature, peak wavelength, and energy emission are controlled by adjusting the applied voltage. Even lamp power, measured in watts, depends on the applied voltage.
b. *Filament temperature*: The filament temperature decides the evaporation rate and redeposition of tungsten metal when the lamp is under operation. Increasing input voltage increases the filament temperature, which in turn increases the heat output but shortens the lamp life.
c. *Maximum power density*: Maximum power density is the amount of power per square meter that the emitter is capable of delivering per second. It is a good measure of how quickly the heater can transfer heat to a surface receiving radiation. High power-density heaters should not be used with extremely viscous materials or volatile materials due to the risk of fire hazards.

4.5 COMPARISON OF DIFFERENT FORMS OF IR HEAT

Some of the more familiar forms of IR emitter seen today are quartz lamps, metal-sheathed tubular heaters, quartz tubes, gas-fired catalytic, flat-faced panels, and ceramic emitters. Each source has its own distinctive properties (Table 4.4). Metal-sheathed elements are best for convection heating, such as ovens. They are rugged, cost-effective, and efficient. Quartz tube IR emitters are best for radiant applications,

TABLE 4.4
Comparisons among Different Types of Infrared Emitters

	Type of IR Emitters					
Characteristic Parameters	Quartz Lamp	Quartz Tube	Metal Sheath	Catalytic	Radiant Panel	Ceramic
Radiation emission efficiency	86%	61%	56%	80%	88%	96%
Ease of handling in relation to physical strength	Very low	Low	High	High	Medium	Medium
Qualitative heating and cooling speed	Very fast	Fast	Slow	Very slow	Slow	Slow

Source: Data adapted from Anonymous, 2009. http://www.infraredheaters.com/basic.htm.

particularly for heat-sensitive materials that need instant switch on and switch off of input radiation. Quartz lamps are also capable of instant on and off controls and possess extremely high power density. These are effective for high-speed production processes. Ceramic IR heaters are most efficient in converting electrical energy to IR heat energy. When comparing all the different types of IR heaters on parameters such as efficiency, physical strength, working life, zoning ability, and other factors, ceramic elements and quartz tubes are the most preferred IR heaters.

4.6 DESIGN VARIABLES OF IR HEATING SYSTEMS

The efficiency of an IR heating system vis-à-vis design depends on the mode of heat transfer and the duration for which the body receives energy from the emitting source. The amount of heat transferred from the emitter to the object depends directly on the power density of the emitter and temperatures of both and indirectly on the distance between them. Physically, heat flux can be increased with additional reflectors. The design of the total system is of paramount importance for energy efficiency.[16] However, large heat losses may occur by noneffective implementation of the heaters and process chamber design, making that heating method very power intensive. The basic approach to design of IR emitter is based on the following aspects:

Calculation of power density of the emitter and total power required for the job.
Radiation efficiency of the IR heating system.
Construction features of the systems such as spacing between the emitters, number of emitters in the array, and distance between the emitter and the receiving surface.

4.6.1 POWER REQUIRED FOR HEATING

The overall power requirement for the process is the sum of the energy involved in heating the product and evaporating the moisture along with heat loss per unit time. The following are the factors that must be carefully considered in calculating power input:

a. Boundary conditions related to the material to be heated, such as initial moisture content, final moisture content, and maximum allowable temperature of the finished product.
b. Mass flow rate of the material to be processed; upper limit for variable flow condition.
c. Mass balance on moisture in the case of the drying process.
d. Thermophysical properties of the material such as specific heat, density, heat of vaporization, conductivity, and diffusivity.
e. Radiation properties in the IR spectrum of radiation such as absorptivity, reflectivity, emissivity, and transmissivity for calculation of efficiency of the system.
f. Temperature gradient in heat transfer.

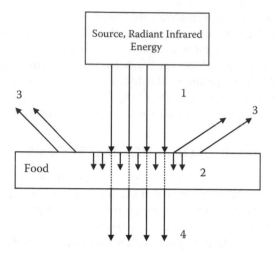

1 = Incoming infrared radiation energy
2 = Absorbed component of infrared energy
3 = Reflected components of radiant energy
4 = Energy loss due to natural convective

FIGURE 4.10 Schematics of energy balance for a piece of food receiving infrared radiation.

The thermodynamic equilibrium method has been suggested[17] to calculate the heat requirement in a heating chamber for both sensible heating and drying of food product using IR (Figure 4.10). However, in practice, the amount of IR energy required for the process is higher, considering combined efficiency in heat transfer. The efficiency of the system is dependent on losses in conversion of electrical energy to thermal energy and also while transferring thermal energy from the emitter to the product. The latter is dependent on design of heating enclosure and materials used in constructions.

Thermodynamically, the radiation heat transfer between source and target (Q_{1-2}) is the sum of the heat absorbed by the product to increase the sensible heat (Q_{HM}) in the product, heat of vaporization of the water (Q_{EVP}) present in the product, and total heat loss (Q_{LOSS}) from the system. Q_{1-2} is also equal to the sum of the energy flux radiating from the source surface (Q_{surf}) and the amount of heat that penetrates the product (Q_z)

$$Q_{1-2} = Q_{HM} + Q_{EVP} + Q_{LOSS} \qquad (4.1)$$

$$Q_{1-2} = Q_{surf} + Q_z \qquad (4.2)$$

The quantity Q_{surf} is related to the absorptivity of the irradiated product (α_2) in the spectrum of IR radiation.

$$Q_{surf} = \alpha_2 \, Q_{1-2} \tag{4.3}$$

According to Bouguer's law,[18] Q_z decays exponentially with the penetration depth (z) and absorption coefficient (α_c)

$$Q_z = Q_{1-2} \times e^{-\alpha_c z} \tag{4.4}$$

IR can penetrate significantly into the food product (Table 4.5), and such penetration may have a dramatic effect on surface moisture and surface temperature (Figure 4.11).[19] As the IR penetration depth decreases, the IR energy gets absorbed closer to the surface, thereby increasing surface temperature. The largest increase in temperature has been observed for zero penetration depth, which corresponds to low surface moisture, while an increase in surface moisture is attributed to increase in penetration depth.

The absorption coefficient in Equation 4.4 is determined from the weighted average of absorption coefficient of water (α_w) and that of dry solid[15] (α_s) using Equation 4.5.

$$\alpha_c = \alpha_s \left(1 - P_V\right) + \left(VMC \times \alpha_w\right) \tag{4.5}$$

where porosity (P_v) and volumetric moisture content (VMC) of the product are determined experimentally. The value of α_w is obtained from the literature. However, α_s in terms of total absorptive and reflectivity of solids can be determined[20] using Equation 4.6:

TABLE 4.5
Infrared Penetration Depth for Different Foods

Product	Spectral Peak (μm)	Penetration Depth (mm)
Dough, wheat	1.0	4–6
Bread, wheat	1.0	11–12
Biscuit	1.0	4
	0.88	12
Grain, wheat	1.0	2
Carrot	1.0	1.5
Tomato paste, 70%–85% MC	1.0	1
Potatoes, raw	1.0	6
Potatoes, dry	0.88	15–18
Apples, raw	1.16	4.1
	1.65	5.9
	2.36	7.4

Source: Data adapted from Ginzburg, A.S. 1969. *Application of Infrared Radiation in Food Processing*, London: Leonard Hill Books.

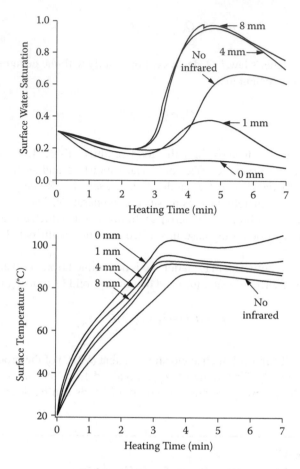

FIGURE 4.11 Effect of infrared penetration on surface moisture and surface temperature. (From Nowak, D., and Lewicki, P.P. 2004. *Innovative Food Science and Emerging Technologies* 5: 353–360. With permission.)

$$\alpha_s = \frac{1}{L} \ln \left(1 - \frac{\alpha_2}{1 - R_s} \right) \tag{4.6}$$

where L is the product thickness, and R_s is reflectivity of the product in the IR spectrum of radiation.

Under thermodynamic equilibrium, the fraction of heat flux (Q_{1-2}) emitted by body 1 (source) in unit time having temperature T_1 to body 2 (object) having temperature T_2 is

$$Q_{1-2} = F_{1-2} \times \varepsilon_{1-2} \times A_1 \times \sigma \times \left(T_1^4 - T_2^4 \right) \tag{4.7}$$

The emissivity factor (ε_{1-2}) of both bodies in the case of the parallel plane is equal to the product[21] of emissivity of source (ε_1) and emissivity of the object (ε_2)

$$\varepsilon_{1-2} = \varepsilon_1 \times \varepsilon_2 \tag{4.8}$$

The geometric view factor term F_{1-2} is the fraction between 0 and 1 that quantifies the amount of radiant energy emitted from the source that is incident upon the target. The following relation exists[1] valid for both black and nonblack surfaces:

$$A_1 F_{1-2} = A_2 F_{2-1} \tag{4.9}$$

a. In the case of infinite parallel black planes:

    For perfect blackbody, $\varepsilon_1 = \varepsilon_2 = 1$

    $F_{1-2} = F_{2-1} = 1.0$, the geometric factor can be omitted   (4.10)

b. In case of infinite parallel gray planes:

    Both source and object are gray bodies with emissivities and absorptivities assumed to be $\varepsilon_1 = \alpha_1$ and $\varepsilon_2 = \alpha_2$, we can proceed as follows:

In unit time, the radiation flux emitted by body 1 is $\varepsilon_1 A_1 \sigma T_1^4$. Of this, the fraction α_2 ($\varepsilon_2 = \alpha_2$) is absorbed by body 2, and the amount absorbed is $(\varepsilon_2)(\varepsilon_1 A_1 \sigma T_1^4)$. The amount reflected back to surface A_1 is $(1 - \varepsilon_2)(\varepsilon_1 A_1 \sigma T_1^4)$ and, of this amount, the fraction reflected back to A_2 is $(1 - \varepsilon_1)$; or the amount is $(1 - \varepsilon_1)(1 - \varepsilon_2)(\varepsilon_1 A_1 \sigma T_1^4)$. Body 2 absorbs the fraction ε_2

$$Q_{2-1} = \varepsilon_2 \left(1 - \varepsilon_1\right)\left(1 - \varepsilon_2\right)\left(\varepsilon_1 A_1 \sigma T_1^4\right) \tag{4.11}$$

This continues, and the resultant series[22] becomes

$$Q_{1-2} = A_1 \sigma T_1^4 \left[\varepsilon_1 \varepsilon_2 + \varepsilon_1 \varepsilon_2 \left(1 - \varepsilon_1\right)\left(1 - \varepsilon_2\right) + \varepsilon_1 \varepsilon_2 \left(1 - \varepsilon_1\right)^2 \left(1 - \varepsilon_2\right)^2 + \ldots\right] \tag{4.12}$$

By simplifying Equation 4.12, we get

$$Q_{1-2} = A_1 \sigma T_1^4 \frac{\varepsilon_1 \varepsilon_2}{1 - \left(1 - \varepsilon_1\right)\left(1 - \varepsilon_2\right)} = A_1 \sigma T_1^4 \frac{1}{\dfrac{1}{\varepsilon_1} + \dfrac{1}{\varepsilon_2} - 1} \tag{4.13}$$

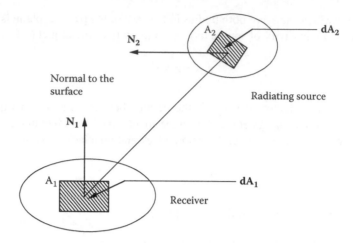

FIGURE 4.12   Two bodies receiving radiation from each other.

where

$$\varepsilon_{1-2} = \frac{1}{\dfrac{1}{\varepsilon_1} + \dfrac{1}{\varepsilon_2} - 1} \tag{4.14}$$

c. The general equation for the view factor between blackbodies (Figure 4.12) is calculated by means of a formula derived from Lambert's law[23] as follows:

$$F_{1-2} = \frac{1}{A_1} \int_{A_1} \int_{A_2} \frac{\cos\theta_1 \ \cos\theta_2 \ dA_1 \ dA_2}{\pi \ r^2} \tag{4.15}$$

d. The radiant energy transferred inside a multilayer body, between surface 1 and surface 2 (cf. Figure 4.13) can be written as:

$$Q_{1-2} = \left[\sigma \ A_1 \left(T_1^4 - T_2^4\right)\right]\left(\frac{1}{\dfrac{1}{\varepsilon_1} + \dfrac{1}{\varepsilon_2} - 1}\right)\left(\frac{1}{y+1}\right) \tag{4.16}$$

where y is the number of radiation planes (shields) between the two surfaces.

Case I

For a perfect blackbody, $\varepsilon_1 = \varepsilon_2 = \varepsilon = 1$; then, the equation is transformed into

$$Q_{1-2} = \left[\sigma \ A_1 \left(T_1^4 - T_2^4\right)\right]\left(\frac{1}{y+1}\right) \tag{4.16a}$$

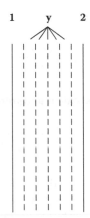

FIGURE 4.13 Diagram of a cross section of a solid porous body, 1 and 2—boundary surfaces, y—intermediate layers.

Case II
When two bodies have equal emissivity in the respective spectrum of radiation, then $\varepsilon_1 = \varepsilon_2 = \varepsilon$, and Equation 4.14 changes to

$$Q_{1-2} = \left[\sigma A_1\left(T_1^4 - T_2^4\right)\right]\left(\frac{\varepsilon}{2-\varepsilon}\right)\left(\frac{1}{y+1}\right) \qquad (4.16b)$$

Case III
$y = 0$ indicates that there are no planes between two surfaces (source and object), $\varepsilon = 1$, and Equation 4.16 becomes

$$Q_{1-2} = \left[\sigma A_1\left(T_1^4 - T_2^4\right)\right]\left(\frac{1}{2}\right) \qquad (4.16c)$$

The amount of energy required for sensible heat gain by the product can be calculated with Equation 4.17:

$$Q_{HM} = \frac{G_1 C_p'\left(T_M - T_0\right)}{3600} \qquad (4.17)$$

where G_1 is the mass flow rate of the product being irradiated and C_p' is the heat capacity of the product, which can be obtained from Equation 4.18.

$$C_p' = m_c C_{pw} + \left(1 - m_c\right)C_{pd} \qquad (4.18)$$

where C_{pw} and C_{pd} are the heat capacities of water and dry solid, respectively. m_c is the fraction of moisture present (wet basis) per unit mass of the initial material.

The amount of energy consumed to evaporate moisture from product at T_m

$$Q_{EVP} = \frac{G_1 \, m_c \, \lambda}{3600} \qquad (4.19)$$

The amount of energy loss could be calculated from Equation 4.20:

$$Q_{Loss} = h_c A_c \left(T_{wall} - T_{air}\right) + \sigma A_c \, \varepsilon \left[\left(T_{wall}\right)^4 - \left(T_{air}\right)^4\right] + Q_{L1} \qquad (4.20)$$

where h_c is the convective heat transfer coefficient, and T_{wall} and T_{air} are the absolute temperatures of the radiating wall and air, respectively. A_c is the total outer surface area of the chamber wall. Q_{L1} is the loss due to conversion for electrical energy to thermal energy.
Q_{Loss} also can be determined using Equation 4.21:

$$Q_{Loss} = \left(h_c + h_r\right) A_c \left(T_{wall} - T_{air}\right) \qquad (4.21)$$

where h_r is the radiative heat transfer coefficient. It can be estimated from Equation 4.22:

$$h_r = \sigma\left(T_1^2 - T_2^2\right)\left(T_1 + T_2\right) \qquad (4.22)$$

The practical installed capacity or the electric power of the radiators to be installed in the heating chamber is calculated using the following equation.
The power required[24] for heating in the drying chamber is

$$P = \frac{Q_{1-2}}{1000} \times \frac{K_2}{K_1} \qquad (4.23)$$

where K_1 and K_2 are the coefficient of voltage fluctuations and the power reserve coefficient, respectively. The values of K_1 and K_2 are 1 and 1.5, respectively. An additional 20%–25% has to be added to the estimated power requirement in order to compensate for IR source efficiency and loss.

4.6.2 EFFICIENCY OF IR HEATING SYSTEM

The ratio of sensible heat divided by the total radiant energy striking the product is a measure of the utilization of the radiant energy and is designated as the performance ratio (η) of the radiant heating system without considering the energy required for evaporation. Mathematically, it is expressed as[25]

$$\eta = \frac{C_p' \rho L A_2}{Q_{1-2} t}\left(T_M - T_0\right)\left(1 - e^{\frac{h_c t}{C_p' \rho L A_2}}\right) \qquad (4.24)$$

The differential heat balance can be written as

$$\alpha_2\, Q_{1-2} = C'_p \rho L A_2 \frac{dT}{dt} + h_c A_2 \left(T - T_{air} \right) \tag{4.25}$$

With no change in thickness or area of the product, as well as density, the rate of temperature rise for the process will be

$$\frac{dT}{dt} = \frac{\alpha_2\, Q_{1-2}}{\rho\, L A_2\, C'_p} - \frac{h_c \left(T - T_{air} \right)}{\rho L A_2 C'_p} \tag{4.26}$$

The irradiated object attains a maximum temperature when there is a balance between the radiant heat absorbed and heat lost by convection to the atmosphere. This means the instantaneous rate of temperature rise (dT/dt) becomes zero. Equation 4.26 becomes

$$\frac{\alpha_2\, Q_{1-2}}{\rho L A_2 C'_p} = \frac{h_c \left(T_M - T_{air} \right)}{\rho L A_2 C'_p} \tag{4.27}$$

or

$$T_M = \frac{\alpha_2\, Q_{1-2}}{h_c} + T_{air} \tag{4.28}$$

Combining Equations 4.26 and 4.28 yields

$$\frac{dT}{dt} = \frac{h_c \left(T_M - T \right)}{\rho L A_2 C'_p} \tag{4.29}$$

Integrating Equation 4.29 results in

$$\int_{T_0}^{T_M} \frac{dT}{T_M - T} = \frac{h_c}{\rho L A_2 C'_p} \int_0^t dt \tag{4.30}$$

$$T = T_M - \left(T_M - T_0 \right) e^{-\frac{h_c t}{\rho\, L C_p A}} \tag{4.31}$$

Substituting Equation 4.31 in Equation 4.24 and rearranging yields the following expression for η:

$$\eta = \frac{C'_p \rho L A_2}{Q_{1-2} t} \left(T - T_0 \right) \tag{4.32}$$

The heat utilization efficiency of an IR food-drying system (considering both heating and evaporation) can be estimated[26] using Equation 4.33.

$$\eta = \frac{\rho \, L \, A_2 \lambda \left(M_i - M_o\right)}{3600 \, F \, t \,\left(100\text{-}M_o\right)} \tag{4.33}$$

where M_i and M_o are initial and final moisture contents (wet basis), F is the energy emitted per unit time by the heating source, and t is the drying time.

4.6.3   DESIGN ARRANGEMENT FOR COMPONENTS OF AN IR HEATING SYSTEM (ELECTRICAL IR EMITTER)

In food dehydration, temperature is one of the important parameters. It can be easily controlled in an IR heating system by adjusting the gap between the emitters and surface of the material and the distance between emitters in the array of heaters.

4.6.3.1   Distance between the Emitter and the Object

The radiant intensity at the receiving surface varies inversely with the square of its distance from the source of radiation. Assuming total radiative energy to be constant, as the distance from the emitter to object increases, the total amount of radiation gets distributed over a larger area, thereby reducing the energy flux,[27] which is expressed by Equation 4.34.

$$Q_{1\text{-}2} = {Q_r}\big/{r^2} \tag{4.34}$$

where Q_r is radiation received at distance r from the source.

The typical distributions of radiation energy for two lamps (1000 and 500 W), placed 10, 20, and 30 cm away from the surface of the irradiated object, are indicated in Figure 4.14. According to Lambert's cosine law of radiation,[23] the intensity of radiation falling onto a flat surface from a small radiant source is maximum when the receiving surface is placed normal to the source. However, if the receiving surface is moved away from the normal by an angle θ, the intensity of radiation received is proportional to the cosine of the angle θ—the angle between the normal to the receiving surface at that point and the direction of the radiation. The typical radiation energy distribution at different distances from the axis of two 100 and 150 W reflector-type IR lamps is shown in Figure 4.15.

In a convective drying system equipped with near-IR emitter with peak wavelength at 1.2 µm, it is reported[28] that water removal flux is dependent on the distance between emitters and the product and air velocity on the surface. The system was equipped with a near-IR emitter delivering peak wavelength at 1.2 µm. Figure 4.16 reveals that the flux of water evaporated becomes lower as the distance between the emitter and the surface of product increases. The air velocity also affects the rate of evaporation at a given distance between emitters and the surface of water

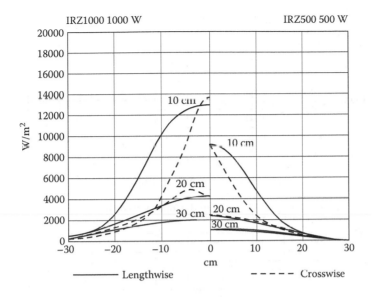

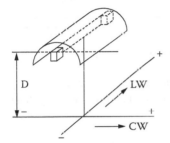

FIGURE 4.14 Radiation energy distribution at 10, 20, and 30 cm, cross-wise and lengthwise for: 1000 W (left) and 500 W (right) infrared radiation lamps. (From Philips, India. With permission.)

(Figure 4.17). The time to reach constant evaporation rate was influenced significantly by air velocity and distance between the emitters and the water surface. The effect of air velocity is substantial, while the influence of the distance between emitters and surface of water is evident but of lesser importance. The effect of air velocity (V) and distance between the emitters and the material (r) on the drying time t is described by the empirical Equation 4.35.

$$t = 0.45 + 0.0001(r/100)^{2.2} + 1.8V^{5.0} \qquad (4.35)$$

It may be noted that, for large-size IR ovens, there is a deviation from the usual inverse square law. Even the radiation heat transfer between two infinitely large parallel plates becomes independent of the distance between them.

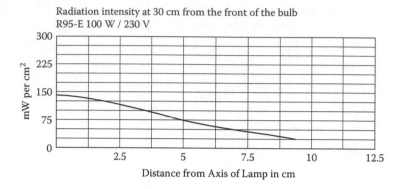

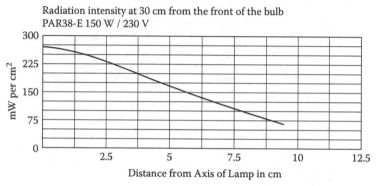

FIGURE 4.15 Typical radiation energy distribution at different distance from the axis of two 100- and 150-W reflector-type IR lamps. (From TechniLamp, South Africa. With permission.)

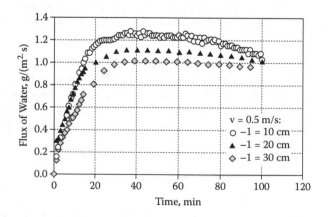

FIGURE 4.16 Flux of water evaporated at an air velocity of 5 ms^{-1} for different distances between the IR emitter and the heated surface. (From Datta, A.K., and Ni, H. 2002. *Journal of Food Engineering* 51: 355–364. With permission.)

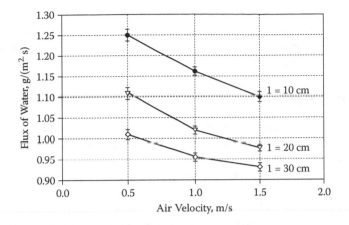

FIGURE 4.17   Relationship between air velocity and the flux of evaporated water at different distances between the infrared radiators and the surface of water. (From Datta, A.K., and Ni, H. 2002. *Journal of Food Engineering* 51: 355–364. With permission.)

4.6.3.2   Distance between the Emitters

The effect of spacing between the emitters on energy density received on the irradiated body is very significant. For a specific radiant emission pattern, the distance between the source and the object is dependent on the spacing between the emitters. It has been reported[29] that at a sufficiently small spacing between the lamps, the irradiation fields are superimposed on one another, thus providing a uniform temperature field. Therefore, the optimum distance between the emitters has immense practical importance. It has been reported by a previous researcher that the ratio E/r should be less than or equal to 1.2.

On the basis of the geometrical characteristics of the emitter, such as width, height, and focal length of the reflector, the correlation among different parameters using dimension graph analysis has been established.[30] The distance between the emitters and energy characterization have been used to express various positional arrangements of the IR heating system (Equation 4.36):

$$(r \times 100)^{0.55} = \left[\frac{(h_c / 1.163)}{\left(Q_r / 10^4 \right)} \right] \left[\frac{T_1 \, P^{0.30} \left(v / 1.163 \right)^{0.46} \left(Q_{1-2} / 10^4 \right)^{1.02}}{(E / 100)^{0.51} \left(h_c / 1.163 \right)^{0.8} \left(T_{air} - 273 \right)^{1.161}} \right] \quad (4.36)$$

where T_1 is the temperature of the source, v is the coefficient of heat conductivity of the air, and E is the distance between the emitters. The effect of distance between the emitters on density of irradiation[28] can be visualized by Equations 4.36 or 4.37:

$$Q_{1-2} = \frac{P_1 \, n \, \eta}{0.87 L^2} \quad (4.37)$$

where P_1 is the radiation flux of a single emitter in an array of n number of lamps. The relationship indicates that decrease in distance between the lamps will make the power of the emitter increase. This is attributed to complete overlapping of the radiation contour of the individual emitter.

NOMENCLATURE

A_1, A_2	Area of the emitter and object exchanging heat by radiation, m²
A_c	Total outer surface area of the chamber wall, m²
C_p'	Heat capacity of product, J kg⁻¹K⁻¹
C_{pw}, C_{pd}	Heat capacity of water and dry matter, respectively, J kg⁻¹K⁻¹
E	Distance between the emitters, m
F	Power emitted by the source, kW
F_{1-2}	View factor fraction of radiation from surface 1 intercepted on surface 2
F_{2-1}	View factor fraction of radiation from surface 2 intercepted on surface 1
G_l	Mass flow rate of the product leaving the irradiated chamber per hour, kgh⁻¹
h_c	Convective heat transfer coefficient, Wm⁻² K⁻¹
h_r	Radiation heat transfer coefficient, Wm⁻² K⁻¹
K_1	Coefficient of voltage fluctuations, dimensionless
K_2	Power reserve coefficient, dimensionless
L	Thickness of the product, m
M_i, M_o	Initial and final moisture content, % wet basis
m_c	Fraction of moisture present (wet basis) in the product
P_1	Radiation intensity from a single lamp, W
P_v	Porosity, ratio of volume of voids to total volume
Q_{1-2}	Heat transfer between the source and target, W
Q_{2-1}	Fraction of heat absorbed by irradiated object. W
Q_{EVP}	Heat absorbed for evaporating the moisture, J s⁻¹
Q_{HM}	Heat absorbed by the product for increase in temperature, J s⁻¹
Q_{LOSS}	Sum of the heat losses, J s⁻¹
Q_{L1}	Loss due to conversion for electrical energy to thermal energy, J s⁻¹
Q_r	Energy flux at distance r from the infrared source, W
Q_{surf}	Energy flux absorbed at the surface, W
Q_z	Energy that penetrates into the product, W
R_s	Reflectivity of the product
r	Distance between source and object, m
T_1, T_2	Absolute temperature of the source and target, K
T_{air}	Temperature of the air inside heating chamber, K
T_M	Maximum temperature attained by the object under irradiation, K
T_o	Initial temperature of the product subject to irradiation, K
T_{wall}	Average temperature measured at the outer surfaces of the walls, K
t	Drying time, s
V	Air velocity, ms⁻¹
VMC	Volumetric moisture content, m³ m⁻³

y	Number of radiation planes between the surfaces
z	Depth of radiation penetration into the product, m
α_c, α_s, α_w	Absorption coefficient of product, solids, and water, respectively, m^{-1}
α_2	Absorptivity of the irradiated product
$\varepsilon_1 \, \varepsilon_2$	Emissivity factor of source and receiving body, respectively
η	Heat utilization efficiency of the system
λ	Latent heat of vaporization at T_m, kJ kg^{-1}
v	Coefficient of heat conductivity of the air, W m^{-2} K^{-1}
σ	Stefan–Boltzmann constant, 5.56×10^{-8} (W m^{-2} K^{-4})

REFERENCES

1. McCabe, W.L., Smith, J.C., and Harriot, P. 1993. *Unit Operation of Chemical Engineering*, 5th ed. New York: McGraw-Hill.
2. Modest, M.F. 2003. *Radiative Heat Transfer*, 2nd ed. San Diego, CA: Academic Press.
3. Florian, J. 1996. *Practical Thermoforming: Principles and Applications*, 2nd ed. New York: Marcel Dekker.
4. Anonymous, 2009. http://www.drc.co.uk/DRC/Brochures/Infrared.pdf (accessed August 13, 2009)
5. Krishnamurthy, K., Khuarana, H.K., Jun, S., Irudayaraj, J., and Demiric, A. 2008. Infrared heat in food processing, an overview. *Comprehensive Reviews in Food Science and Food Safety* 7: 1–12.
6. Sakai, N., and Hanzawa, T. 1994. Application and advances in far infrared heating in Japan. *Trends in Food Science and Technology* 5(11): 357–362.
7. Pettersson, M., and Stenstrom, S. 2000. Modelling of an electric IR heater at transient and steady state conditions. Part I: Model and validation, *International Journal of Heat and Mass Transfer* 43: 1209–1222.
8. Anonymous, 2009. http://www.infraredheaters.com/basic.htm (accessed August 13, 2009).
9. Anonymous, 2009. http://www.radiantenergy.com/TechnicalData/T-3-ProductDataSheet.pdf (accessed August 13, 2009).
10. Anonymous, 2009. http://www.cpu.com.tw/kh/elec/pl/blb/ir250_series.pdf (accessed August 13, 2009).
11. Anonymous, 2009. http://www.aceheattech.com/medium-wave-heaters.html (accessed August 13, 2009).
12. Pettersson, M., and Strenstrom, S. 1998. Absorption of infrared radiation and radiation transfer mechanism in paper, *Journal of Pulp and Paper Science* 24(11): 349–355.
13. Bédard, N. 1998. Laboratory testing of radiant gas burners and electric infrared emitters. *Experimental Heat Transfer* 11(3): 255–279.
14. Anonymous, 2009. http://www.thermovation.com/pages/gasinfrared.htm (accessed August 13, 2009).
15. Jun, S., and Irudayaraj J. 2003. Selective far infrared heating system—Design and evaluation. I. *Drying Technology* 21(1): 51–67.
16. Ratti, C., and Mujumdar, A.S. 2007. Infrared drying. In *Handbook of Industrial Drying*, 3rd ed., vol. 1: 423–437. New York: Marcel Dekker.
17. Cheng, L.M. 1992. *Food Machinery for the Production of Cereal Foods, Snack Foods and Confectionary*. New York: Ellis Horwood Series in Food Science and Technology.
18. Siegel, R., and Howell, J.R. 2001. *Thermal Radiation Heat Transfer*, 4th ed. Philadelphia: Taylor & Francis.

19. Datta, A.K., and Ni, H. 2002. Infrared and hot-air-assisted microwave heating of foods for control of surface moisture, *Journal of Food Engineering* 51: 355–364.
20. Yagoobi, J., Sikirica, S.J., and Counts, K.M. 2001. Heating/drying of paper sheet with gas-fired infrared emitters-pilot machine trials. *Drying Technology* 19 (3–4): 639–648
21. Incropera, F.P., and DeWitt, D.P. 1990. *Fundamentals of Heat and Mass Transfer.* 3rd ed. New York: John Wiley & Sons.
22. Geankoplis, C.J. 1993. *Transport Processes and Unit Operations*, 3rd ed. NJ: Prentice-Hall.
23. Anonymous, 2009. http://www.noblelight.net/resources/pdf_downloads/technical_notes/ engineering_aspects of radiation theory.pdf (accessed August 13, 2009).
24. Hebbar, U.H., Vishwanathan, K.H., and Ramesh, M.N. 2004. Development of combined infrared and hot air dryer for vegetables. *Journal of Food Engineering* 65: 557–563.
25. Tiller, F.M., and Garber, H.J. 2002. Infrared radiant heating. *Industrial and Engineering Chemistry* 34(7): 773–781.
26. Singh, K.K. 1994. Development of a small capacity dryer for vegetables. *Journal of Food Engineering* 21: 19–30.
27. Anonymous, 2009. Instruction manual and experiment guide for the PASCO scientific, Model TD-8553/8554A/8555, Thermal radiation system. www.pasco.com (accessed August 13, 2009).
28. Nowak, D., and Lewicki, P.P. 2004. Infrared drying of apple slices. *Innovative Food Science and Emerging Technologies* 5: 353–360.
29. Ginzburg, A.S. 1969 *Application of Infrared Radiation in Food Processing*, 1st ed. London: Leonard Hill Books.
30. Fay, G. 1964. Theoretische fragen von infrarot. Proceedings of the 2nd Conference on Electro-Thermics, Budapest, Hungary.
31. Anonymous, 2009. http://www.ercis.lu/default.asp?contentID=680 (accessed August 13, 2009).
32. Anonymous, 2009. http://www.slideshare.net/marktenheuw/070716-warmterras-nl (accessed August 13, 2009).
33. Anonymous, 2009. http://www.sulekhab2b.com/altotec/default.htm (accessed August 13, 2009).

5 Infrared Drying

Caleb Nindo and Gikuru Mwithiga

CONTENTS

5.1 Introduction 89
5.2 IR Radiation and Its Propagation during Drying 89
5.3 Mechanism of IR Drying..91
5.4 Advantages of IR Radiation for Drying 91
5.5 Characteristics of IR Radiation 92
5.6 Application of IR Radiation in Industrial Drying 93
5.7 Some Specific Applications of IR Drying of Foods......................................95
5.8 Future Directions of IR Drying..97
References...97

5.1 INTRODUCTION

This chapter will present the mechanism, techniques, and advantages of infrared (IR) drying for various food and agricultural products. The propagation of IR radiation from a source (considered as any material at a temperature above absolute zero Kelvin) to other materials is discussed, and emphasis is placed on the subsequent transformation of the IR radiation into thermal energy for the drying of foods. The advantages and disadvantages of IR drying are briefly reviewed, followed by a discussion of selected applications in the processing of food and fiber. A novel drying system that utilizes IR radiation, thermal conduction, and convection from hot water for drying of pureed products is discussed.

5.2 IR RADIATION AND ITS PROPAGATION DURING DRYING

The IR radiation band of the electromagnetic spectrum encompasses wavelengths from 0.75 to 1000 μm and is usually subdivided into sections, namely, near-IR (0.75–3 μm), mid-IR (3.0–25 μm), and far-IR (25–1000 μm; Figure 5.1). This classification is based on vibrational fundamentals and the interest of physicists (Chantry, 1984). Chemists tend to define mid-IR from 3.0 to 40 μm because it is where characteristic absorption bands of organic molecules occur. In the electromagnetic spectrum, thermal radiation is mainly considered to be from 0.1 to 100 μm (Incropera and DeWitt, 2002). IR radiation is propagated essentially like visible light. In food materials, near- and mid-IR radiation is transferred to the molecules by vibration, while the interaction is principally rotational in the far-IR range.

Although there is no complete agreement on where each of the subsets end or start, IR wavelengths from 2.5 to 200 μm are the most commonly used for drying

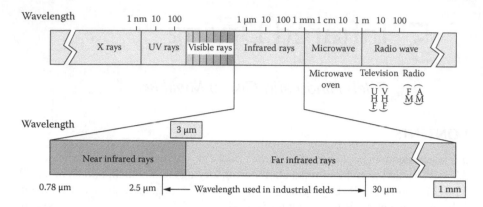

FIGURE 5.1   The electromagnetic spectra showing the location of IR radiation. (From Japan Far Infrared Association: http://www.enseki.or.jp/e_ippo.html. With permission.)

purposes. Water has strong absorption of IR energy around 3, 6, and 12, and 15 μm (Il'asov and Krasnikov, 1991; Datta and Almeida, 2005). Most ceramic heaters have peak spectral emissivity around 3.0 μm, and this explains why they are commonly used in many situations for drying.

The IR radiation band relevant to heating of foods, and especially drying, was given by Sandu (1986). Water has very strong absorption of IR radiation at around 2.7–3.3, 6.0, and greater than 12.5 μm. The O-H bonds in water absorb IR energy and start to rotate with the same frequency as the incident radiation. This transformation of IR radiation to rotational energy causes the evaporation of water. When IR radiation strikes a surface, part of it may be reflected (ρ), absorbed (α), or transmitted (τ). If the transmissivity is infinitesimal, then the material will reflect or absorb IR depending on the nature of the radiation and the surface characteristics of the material. This is termed *emissivity* (ε) and usually ranges from 0 to 1. Blackbody materials absorb all the radiation falling on them and therefore have an emissivity of 1.0 as opposed to completely reflective surfaces ($\varepsilon = 0$). Figure 5.2 illustrates how IR waves are propagated from source to and through a material that is semitransparent.

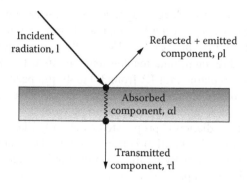

FIGURE 5.2   Propagation of IR radiation through absorbing and transmitting medium.

5.3 MECHANISM OF IR DRYING

In IR radiation drying, also called thermal radiation drying, heat is transferred to materials to be dried in the form of radiant energy. Artificial radiation drying involves the use of IR radiation generators such as special electric lamps and ceramic or metallic panels heated by electricity or gas. In natural radiation drying (solar drying), radiation from the sun is tapped either directly or indirectly for drying purposes. The major distinguishing feature of IR drying is that it does not require a medium for transmission of energy from source to the target. The materials being dried can themselves be regarded as absorbers of IR radiation. Therefore, to enhance the efficiency of drying, high absorption (and less scattering) of incident IR and coupling of absorbed energy with water in food are important considerations. The scattering of radiation may become significant when particulate materials (~0.1 mm and larger) are being dried. It is also important that the incident energy be uniformly applied on the receiving surface, a situation that requires that the configuration and characteristics of the IR emitter and any reflectors be optimized. These considerations have been applied in the drying of paint and paper for many years because coatings and paper webs are suited to IR drying. An IR source may be enclosed in a chamber with a highly reflective surface to take advantage of the multiple reflections within the enclosure and hence improve energy efficiency.

5.4 ADVANTAGES OF IR RADIATION FOR DRYING

IR radiation heat transfer has the following advantages that make it suitable for drying:

1. No gas resistance for heat flux as in the case of convection; energy is transferred directly to the product according to the laws of optics. Unless the intervening medium is saturated with water vapor, there is usually very little absorption within the space separating the bodies.
2. No direct contact with material is required as in conduction drying.
3. IR radiation, similar to visible radiation (light), can be focused to increase heating intensity, provide fast treatments, or target a particular area. Multiple reflections from the inside of cooking ovens is an example of how IR radiation can be manipulated to achieve the desired effect.
4. Very high heat transfer rates are achievable with compact heaters.
5. Fast response times (low thermal inertia) allow easy and rapid process control.
6. No pollution of the environment when compared to fossil fuels, which are also nonrenewable.

However, IR radiation is not always easily applicable. The treatment is homogeneous only in the case of flat and thin materials such as paper webs and coatings. For food materials that are several millimeters in thickness, differential absorption by protein, fat, carbohydrate, and water may influence the uniformity of drying. As already noted, surface characteristics, spectral composition, and direction of incident radiation also determine the absorption of IR. Since foods usually come in complex shapes and sizes, the application of IR in such situations may be limited because energy impinging on the material will be different from place to place. To achieve maximum uniformity of radiant flux density on the surface of foodstuffs being irradiated, there should be proper spacing between individual IR generators (if more than one unit is used) and the distance between the radiation sources and the foodstuff (Il'yasov and Krasnikov, 1991).

5.5   CHARACTERISTICS OF IR RADIATION

On the application of thermal radiation in the drying of paints and other thin sheet materials with film coating, Vainberg and Grabovsky (1986) indicated that it is possible to make the best use of all the advantages of radiant energy supply if the spectral radiation range of the IR radiator is matched with the spectral thermoradiation of the film and substrate. Such a radiation transfer process does not result in formation of crust on the surface (case hardening), which normally restricts the removal of moisture in certain materials undergoing heated air drying. In their experiments on paper drying, Lampinen et al. (1991) found that optimal efficiency was achieved when most of the radiation emitted by the radiator is within the same wavelength range where the material to be heated absorbs the most; and that absorption of radiation by water in paper was better at longer wavelengths (that is, in the far-IR region).

Figure 5.3 illustrates the transformation of incident IR radiation into various components during drying of small grains such as rough rice (Kudo et al., 1995). To minimize heat losses, the bottom and sides of the tray may be covered with aluminum film, as shown in Figure 5.4. The magnitudes of reflected and emitted radiation plus convected heat and heat of vaporization will vary depending on the surface properties and the water status of the rice.

Based on measurements of ambient air and rice surface temperatures, values of I_r, Q_r, Q_c, and Q_v were determined mathematically, taking the average emissivity of rough rice as 0.95. With an average drying time of 6.2 h, the energy exchange on rice surface on five different days with clear sky conditions was 3.5 ± 0.2, 3.1 ± 0.2, 2.2 ± 0.1, and 0.5 ± 0.1 kJ/m^2 for I_r, Q_r, Q_c, and Q_v, respectively. The sum of these calculated components was close to 9.4 ± 0.4 kJ/m^2 and was measured for those days using the far-IR sensor (Kudo et al., 1995). In closed thermal radiation chambers not open to the atmosphere as in Figure 5.4, the distribution of the radiation flux on the surface of the foodstuffs is determined by the optical–geometrical parameters

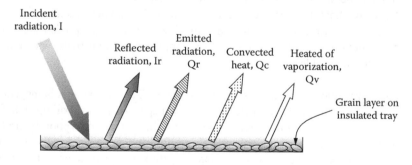

FIGURE 5.3   Energy balance on a thin layer of rough rice exposed to IR radiation.

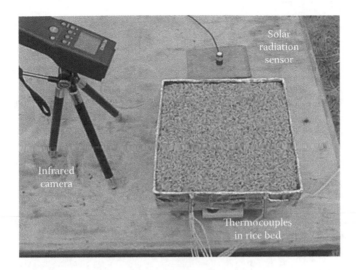

FIGURE 5.4 Example of rough rice drying using IR radiation from the sun.

of the IR source, the reflectors or reflecting surfaces, the foodstuff being irradiated, the closure of compartments, and by how these are located in relation to one another (Nindo et al., 1995; Il'yasov and Krasnikov, 1991).

5.6 APPLICATION OF IR RADIATION IN INDUSTRIAL DRYING

The IR wavelength spectrum ranges from 0.72 to 1000 μm. Although there is no complete agreement on where one subset ends and another starts, these subsets generally fall in the following regions of the electromagnetic spectrum. The shortwave is also referred to as near-IR, which ranges from 0.72 to 2 μm and covers the temperature range of 3900°C–1200°C. Middle-IR refers to the wavelength of 2–4 μm, which coincides with an emitting surface temperature of 1200°C–450°C. The last subset is referred to as the far-IR, and it represents the wavelength range of 4–1000 μm, which fits the temperature range of 450°C–0°C. It is far-IR that is most commonly used for drying purposes.

It can be observed from Figure 5.5 that the highest absorptivity for water occurs at a wavelength near to 3 μm and again at a wavelength near 6 μm. Clearly, these are the best wavelengths at which we should produce IR energy for drying purposes. It is for this reason that IR industrial elements intended for drying applications products are designed to emit energy at either lower or higher wavelengths excluding the 3.5–5.5 μm range. This is not to say that an emitter operating at this range of wavelength will not dry a substance, it just implies that the efficiency will be much lower due to the lower absorptivity.

There are numerous dryers that are designed to take advantage of this selective absorption on IR radiation by wet materials. Refractance Window™ drying (Nindo et al., 2003) is an example of a surface contact drying process that utilizes thermal energy from hot water at temperatures very near boiling to dry materials thinly spread on an endless plastic conveyor belt. The belt moves while in contact with

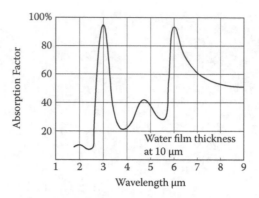

FIGURE 5.5 IR spectral absorption curve for water. (From Orfeuil, M., 1987. *Electric Process Heating.* Battelle Press, Columbus OH, p. 348. With permission.)

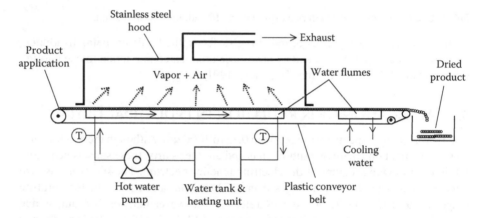

FIGURE 5.6 Schematic of Refractance Window™ dryer. (From Nindo et al., 2003. *J. Food Processing and Preservation*, 27:117–136. With permission.)

the hot water and results in very rapid drying (Figure 5.6). The dry product is then scraped off the conveyor using a doctor blade that spans the full width of the belt. Far-IR radiation from hot water is transmitted through the transparent plastic to the material with very large surface-to-thickness ratio, and this effect in addition to heat transfer by conduction and convection results in high water removal rates.

In industrial IR heating systems, the energy is mainly derived from chemical or electrical sources. Energy from chemical sources such as gas fuels is usually converted into IR radiation through a combustion process, with the efficiency of conversion ranging from 40% to 46%. Electrically heated IR emitters normally have a conversion efficiency of 78% to 85% (Ramaswamy and Marcohe, 2005). These IR-emitting radiators can operate a single filaments or as large modules with equivalent emitting power in excess of hundreds of kilowatts. However, the amount of energy that a module can deliver per square meter is dependent on the type. Gas-fired emitters

(a)   (b)

FIGURE 5.7 Single IR emitters: (a) showing a low-temperature ceramic emitter and (b) showing a quartz emitter. (From infraredheaters.com. With permission.)

will normally not exceed a power density of 22 kW/m^2, while electrical emitters can reach a density of up to 400 kW/m^2. In general, electric heaters are more expensive, but they deliver power intensities that are approximately three times higher when compared to gas-fired heaters, which gives them an advantage. The electrical heaters are also much easier to manipulate, and the power intensity of a given system can easily be controlled or adjusted according to requirement.

Normally, the IR modules are used in the thin layer drying of foods and have been applied in the finish drying of cookies. The product can be in motion when drying, and the continuous operation is facilitated by a conveyor belt or a vibrating plate. The advantage of IR modules is that they can be focused to deliver energy to a particular area or location, and the response time required to reach target temperature is very short when compared to purely convective drying systems. Considering gas-fired IR systems, Lia et al. (2004) found that efficiency depends on the size of the heater element and the separation distance between the element and the object being heated.

The ceramic emitter shown in Figure 5.7 can be used for spot or localized heating, especially when fitted with a hood. Each such emitter can be made with standard energy values of 60, 100, 150, 200, and 250 W. A number of such emitters can be mounted on a panel at regular intervals so as to make a module, with the spacing between each emitter being chosen to give the desired energy intensity. Each filament in quartz emitters can have a power rating of up to 1 kW.

Intermittent heating has been shown to reduce the amount of energy required to achieve a certain amount of drying. This is because the surface temperature of products rises rapidly under IR heating, and drying can continue even during the interim period when the power is off. Chou and Chua (2001) disclosed that by combining IR with other conventional drying systems some considerable savings in both length of drying time and amount of energy required can be achieved. They further stated that quality can considerably be improved by as much as a 20%–50% reduction in loss of ascorbic acid when drying fruits.

5.7 SOME SPECIFIC APPLICATIONS OF IR DRYING OF FOODS

A number of recent studies on IR drying of specific commodities have been documented. These studies range from laboratory experiments to large-scale industrial setups for drying of grains, vegetables, and fruits.

There are many investigations on IR drying of grains that have been reported. For example, Das et al. (2004) designed and tested a vibration-aided IR dryer using radiation intensities of 3100–4290 W/m^2 and rough rice grain bed depths of 12–16 mm. They found that for this range of thickness and for a given IR intensity, the grain depth had an insignificant effect on the drying rate for all the three rice varieties that were used in the study. Nindo et al. (1995) sought to restrict convective heating during far-IR radiation drying of rough rice by using cooled air only as a means to carry away moisture evaporated from the rice. They also investigated the effect of other IR bands, such as those exhibited by halogen lamps, popularly called *solar lamps* (Figure 5.8). Faster drying of rough rice was achieved when the intensity of heating was increased by varying the voltage input to the IR source (160 to 200 V) and by decreasing the distance between the lamp and the rice surface. Various models for describing IR drying of rough rice have been developed, including those suitable for combined IR-convective (Abe and Afzal, 1997) and those applicable to intermittent IR heating conditions (Cihan and Kahvec, 2007). Researchers have also looked at the effect of intermittent heating on head-rice yield, color, and gelatinization during rough rice drying.

Other crops that have been dried using IR include cashew kernels (Hebbar and Rastogi, 2001; Hebbar and Ramesh, 2005). The activation energy for the drying of cashew kernels at temperatures between 100°C and 120°C was found to be 28.7 kJ/mol. When they dried cashew kernels between 55 and 95°C, the optimum IR

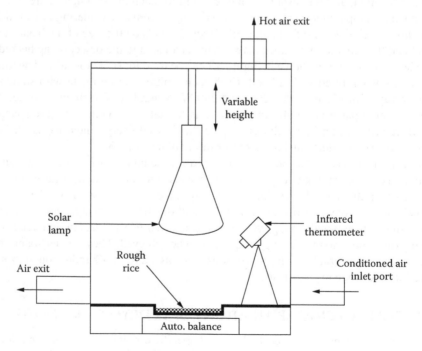

FIGURE 5.8 Setup for drying of rough rice using solar lamp.

drying condition for development of desirable color and kernel brittleness level was 55°C for 55 min.

The drying of onions using IR radiation has also received considerable attention in recent years. Mongpraneet et al. (2002) dried sliced welsh onions under IR conditions and observed an initial increase in the drying rate of the onions. This was followed by a constant rate drying period, which in turn gave way to a falling rate of drying. The intensity also affected both the drying rate and the quality as determined by product color. Other studies on IR drying of onions include those done by Sharma et al. (2005), Gabel et al. (2006), and Kumar et al. (2005). In most studies, it is evident that allowing the temperature to rise too high and/or heating for extended periods of time is detrimental to the quality of the onions.

Many other studies have been done on IR drying of fruits and vegetables. Among these are Hebbar et al. (2004) and Togrul (2006) for carrots, and Afzal and Abe (1998) for potatoes. Nasiroglu and Kocabiyik (2009) studied the drying of sliced red pepper in thin layers under varied IR radiation intensities and airflow through the dryer. The method of preparing fruits and vegetables for drying has mainly been by slicing and spreading on trays made with wire mesh or flat metal plates. Notable examples include IR drying of sliced apples (Sun, 2007; Nowak and Lewicki, 2004; Togrul, 2005), peaches (Wang and Sheng, 2006), and grapes (Caglar et al., 2009; Goudar and Shar, 2009).

5.8 FUTURE DIRECTIONS OF IR DRYING

Applications of IR energy for drying purposes have expanded from the earlier industrial applications such as drying of paints, coatings, paper, and wood; to space heating; to processing of food and fiber. Materials containing high moisture can be dried rapidly using IR radiation if they can be spread thinly on a flat surface or presented as thin layers of particles. Selecting IR generators with peak IR emission at wavelengths that match the absorption bands of water in foods is key to achieving high efficiency in IR drying. IR application for drying facilitates very easy heating control, with fast heat-up and cool-down times that are important for energy conservation process management. IR energy has the potential to be applied extensively in the food industry to process high-value and value-added food products, such as fruits, vegetables, spices, juices, and other biomaterials.

REFERENCES

Abe, T. and T. M. Afzal (1997). Thin-layer infrared radiation drying of rough rice. *J. Agric. Eng. Res.*, 67(4): 289–297.

Afzal, T. M. and T. Abe (1998). Diffusion in potato during far infrared radiation drying. *J. Food Eng.*, 37(4): 353–365.

Caglar, A., I. T. Togrul, and H. Togrul (2009). Moisture and thermal diffusivity of seedless grape under infrared drying. *Food Bioproducts Proc.*, doi:10.1016/j.fbp.2009.01.003.

Chantry, G. W. (1984). *Long-Wave Optics: Volume 1—Principles*. London: Academic Press.

Chou, S. K. and K. J. Chua (2001). New hybrid drying technologies for heat sensitive food-stuffs. *Trends Food Sci. Tech.*, 12: 359–369.

Cihan, A. and K. Kahvec (2007). Modeling of intermittent drying of thin layer rough rice. *J. Food Eng.*, 79(1): 293–298.

Das, I., S. K. Das, and S. Bal (2004). Drying performance of a batch type vibration aided infrared dryer. *J. Food Eng.*, 64(1): 129–133.

Datta, A. K. and M. Almeida (2005). Properties relevant to infrared heating of foods. In *Engineering Properties of Foods* (3rd ed.). M. A. Rao, Rizvi, S. S. H. and A. K. Datta (Eds.). Boca Raton, FL: CRC Press. pp. 209–237.

Gabel, M. M., Z. Pan, K. S. P. Amaratunga, L. J. Harris, and J. F. Thompson (2006). Catalytic infrared dehydration of onions. *J. Food Science, E: Food Eng. Physical Properties*, 71(9): 1750–3841.

Goudar, M. D. and V. H. Shar (2009). Microcontroller based optimized controller for fruit drying. *J. Instrum. Soc. India*, 39(1): 13–17.

Hebbar, U. H. and Ramesh M. N. (2005). Optimization of processing conditions for infrared drying of cashew kernels with testa . *J. Science Food and Agriculture*, 85(5), 865–871.

Hebbar, H. U. and N. K. Rastogi (2001). Mass transfer during infrared drying of cashew kernel. *J. Food Eng.*, 47(1): 1–5.

Hebbar, H. U., K. H. Vishwanathan, and M. N. Ramesh (2004). Development of combined infrared and hot air dryer for vegetables. *J. Food Eng.*, 65(4): 557–563.

Il'yasov, S. G. and V. V. Krasnikov (1991). *Physical Principles of Infrared Radiation of Foodstuffs*. New York: Hemisphere Publishing Corp.

Incropera, F. P. and D. P. DeWitt (2002). *Fundamentals of Heat and Mass Transfer*. New York: John Wiley & Sons.

Kudo, Y., C. I. Nindo, and E. Bekki (1995). Studies on sun drying of raw rough rice (Part 3): Drying characteristics and heat balance in plane surface bed. *J. Jap. Soc. Agr. Mach.*, 57(4): 17–25. (in Japanese)

Kumar, D. G. P., H. U. Hebbar, D. Sukumar, and M. N. Ramesh (2005). Infrared and hot-air drying of onions. *Journal of Food Processing and Preservation*, 29(2): 132–150.

Lampinen, M. J., K. T. Ojala, and E. Koski (1991). Modeling and measurement of infrared dryers for coated paper. *Drying Tech.*, 9: 973–1017.

Lia, B. X., Y. P. Lua, L. H. Liua, K. Kudo, and H. P. Tana (2004). Analysis of directional radiative behavior and heating efficiency for a gas-fired radiant burner. *Journal of Quantitative Spectroscopy Radiative Transfer*, 92: 51–59.

Mongpraneet, S., T. Abe, and T. Tsurusaki (2002). Accelerated drying of welsh onion by far infrared radiation under vacuum conditions. *J. Food Eng.*, 55(2): 147–156.

Nasiroglu, A. and B. Kocabiyik (2009). Thin-layer infrared radiation drying of red pepper slices. *J. Food Process Eng.*, 32(1): 1–16.

Nindo, C. I., H. Feng, G. Q. Shen, J. Tang, and D. H. Kang (2003). Energy utilization and microbial reduction in a new film drying system. *Journal of Food Processing Preservation*, 27: 117–136.

Nindo, C. I., Y. Kudo, and E. Bekki (1995) Test model for studying sun drying of rough rice using far-infrared radiation. *Drying Tech.*, 13(1–2): 225–238.

Nowak, D. and P. P. Lewicki (2004). Infrared drying of apple slices. *Innovative Food Sci. Emerging Technol.*, 5(3): 353–360.

Orfeuil, M. (1987). *Electric Process Heating*. Columbus, OH: Battelle Press.

Ramaswamy, H. S. and M. Marcohe (2005). *Food Processing Principles and Applications*. Boca Raton, FL: CRC Press.

Sandu, C. (1986). Infrared radiative drying in food engineering: A process analysis. *Biotechnol. Progr.* 2: 109–119.

Sharma, G. P., R. C. Verma, and P. B. Parthare (2005). Thin layer infrared radiation drying of onion slices. *J. Food Eng.*, 67(1): 361–366.

Sun, J. (2007). Characteristics of thin-layer infrared drying of apple pomace with and without hot air pre-drying. *Food Sci. Tech. Int.*, 13(2): 91–97.

Togrul, H. (2005). Simple modeling of infrared drying of fresh apple slices. *J. Food Eng.*, 71(3): 311–323.

Togrul, H. (2006). Suitable drying model for infrared drying of carrot. *J. Food Eng.*, 77(3): 610–619.

Vainberg, R. S. and V. V. Grabovsky (1986). Thermoradiation drying of a thin film of translucent substrate. *Drying Tech.*, 4: 101–109.

Wang, J. and K. Sheng (2006). Far infrared and microwave drying of peach. *Food Sci. Tech.*, 39(3): 247–255.

6 Combined Infrared and Hot Air Drying

Habib Kocabiyik

CONTENTS

6.1 Introduction 101
6.2 Drying Time, Drying Curve, and Drying Rate 103
6.3 Drying Energy 106
6.4 Product Quality... 108
6.5 Intermittent and Continuous Uses of IR and Convective Heating 109
6.6 Modeling Studies.. 110
6.7 Optimization of Combined IR and Hot Air Drying.................................... 110
6.8 Conclusion 113
References.. 114

A combined infrared (IR) and hot air drying system is an efficient and rapid drying method compared to the usage of IR and hot air drying separately. It has some advantages such as improved quality of dried product and good drying characteristics. Heat and mass transfer can take place more efficiently and, consequently, drying time dramatically reduces, energy efficiency increases, and specific energy consumption decreases. Because the temperature of the product being dried is kept relatively low during the drying process, thermal degradation of heat-sensitive products can be retained to a higher degree than by convective hot air drying. Therefore, it can be seen as an alternative to the drying of heat-sensitive products. From the final product quality perspective, it is observed that this method is superior to the separate use of IR and hot air drying. Also, energy and operating cost of the combined drying mode for several food and agricultural products is lower than convective drying systems, and optimal operating conditions for combined IR and hot air drying have been demonstrated by several researchers, the aim being to improve product quality and reduce operating and energy cost.

6.1 INTRODUCTION

The process of drying has been defined differently in many sources and also in different chapters of this book. Essentially, it is the removal of water from a material. The drying process is especially important in the food industry. Drying helps to prolong the shelf-life of food and agricultural products, makes the product smaller and lighter, and, since it does not require any cooling, makes storage and transportation processes easier while dramatically decreasing related costs as well. Today's limit-less market; the equipment required to ensure the necessary transportation capacity, product safety, and product quality; and the necessity of providing conditions suit-able for the transportation of fresh products all emphasize the need for drying for long-distance transportation of agricultural and food products.

Parallel to the change in lifestyles throughout the world, there have been changes in eating habits as well, and this change still continues. The need for more products in powder form, such as pharmaceuticals and food products, is still growing, as is the need for drying other essential products such as wood, cloth, and paper (Dhib, 2007). The changes in ready-to-serve food habits increase the demand for dried food and agricultural products that make up the ingredients used in the preparation of such foods. Therefore, there is a serious ongoing competition in agricultural products and food drying markets throughout the world.

Even though drying of agricultural products and foods—that is, the drying of material of biological origin—is known to be the oldest preservation method and is considered to be simple, it is actually a fairly complex method (involving heat and mass transfer, chemical reactions, changes in quality, dependence on material, the drying environment, etc.) and continues advancing in terms of drying technique and dryer type.

In many countries (especially in countries that have enough hours of sunshine), sun drying is applied in the open air, and different processing periods are defined based on the type of fruit and vegetable (3–12 days). This classic method entails some basic disadvantages, such as the nonuniformity of the products in the drying area, the need for large areas for drying, the difficulty of controlling of these large areas, the long drying time, high labor costs, undesired climate effects, and the fact that both the area and the products cannot be protected from environmental pollution. In addi-tion, the chemical structure, color, and hygiene of the dried good may be negatively affected by the different wavelengths sunlight has (Oztekin et al., 1999; Ertekin and Yaldiz, 2004; Doymaz, 2007). Artificial drying methods have been developed due to these undesired results. Hot air drying and dryers have been developed, and they have become standard by playing an active role in industrial drying.

In the convective hot air drying method, costs increase due to the following causes: the main energy source is mostly fossil fuel; energy efficiency is low due to indirect heat transfer for the conditioning (adjustment of temperature and humidity) of drying air; drying time is long; high inputs such as labor; etc. Moreover, drying systems that use fossil-based energy sources cause environmental pollution and may contribute to global warming.

In addition to the drawbacks stated above, the main reason the convective hot air drying process is slow is the low heat transfer characteristics of food and agricultural products. Leading problems of convective drying methods are the long drying times, enzymatic browning, weak drying characteristic, shrinkage, and nutrition value loss (Hebbar et al., 2004).

Fruits and vegetables play an important role in providing the protein and vitamin requirements vital for human nourishment. Generally, they are consumed both fresh and dried. In the processing of fruits and vegetables, drying is an important and sensitive process in terms of quality. Drying is also a processing stage that requires extensive energy consumption (Strumillo and Kudra, 1986; Itaya and Mori, 2006; Ratti and Mujumdar, 2006), which is why new drying systems should be developed, new dryers should be designed, the usage of new energy sources should be researched, energy inputs should be decreased, and methods that do not pollute the environment should be researched.

As stated above, one of the methods to overcome the undesired properties of convective drying is to use convective drying in combination with other drying methods. Among these are combined IR and hot air drying methods. The advantage of IR technology is that it can be combined with a convective-type mode (Adonis and Khan, 2004).

As was discussed in the previous sections of this book, thanks to some of its advantages, IR heat sources have been used for drying by some researchers (Strumillo and Kudra, 1986; Lewis, 1996; Fasina, 2003; Hebbar et al., 2004; Krishnamurthy et al., 2008). It has been found in recent studies that IR drying technology has higher energy efficiency, shorter drying time, and a better final product quality when compared with convective drying methods (Fasina, 2003; Hebbar et al., 2004; Kumar et al., 2005; Wang and Sheng, 2006). IR drying may be used in rural areas for low capacities and low-cost applications (Hebbar et al., 2004), as it has some advantages such as low investment cost, easy assembly, high rates of heating and drying, uniform temperature of the product during drying, simple equipment, high process control, and clean working conditions (Sandu, 1986; Sakai and Hanzawa, 1994; Chua and Chou, 2003). IR drying has started to be taken seriously in recent years, and many experimental studies on thin-layer drying of various food products via IR sources have been conducted; on the following: onion (Mongpraneet et al., 2002); carrot and garlic (Baysal et al., 2003); apple slices (Nowak and Lewicki, 2004); paddy (Das et al., 2004a, 2004b); carrot and potato (Hebbar et al., 2004); onion slices (Jain and Pathare, 2004); rough rice (Amaratunga et al., 2005); pear, carrot, sweet corn (Pan et al., 2005); sweet potato (Lin et al., 2005); apple slices (Togrul, 2005); onion slices (Sharma et al., 2005a, 2005b; Kumar et al., 2005; Kumar et al., 2006; Pathare and Sharma, 2006); peach (Wang and Sheng, 2006); banana slices (Nimmol et al., 2007); sweet potato slices (Lin et al., 2007); red pepper (Nasiroğlu and Kocabiyik, 2009); and carrot (Kocabiyik and Tezer, 2009). Information related to the mechanism and technique of IR drying may be found in Chapter 5. In this chapter, mainly the processing characteristics and product quality in combined IR and hot air drying are described. The effects of simultaneous and intermittent usage of IR and hot air drying are discussed.

6.2 DRYING TIME, DRYING CURVE, AND DRYING RATE

Today, convective hot air drying technology using gas fuel and an electrical fan system is generally used to dry grain products, fruits, vegetables, and other food and agricultural products. The long drying time is one of the biggest disadvantages of this method. In order to increase drying rate and production capacity and to decrease

TABLE 6.1

Drying Conditions versus Drying Times (Minutes) Observed

Air Temperature (°C)	Air Velocity (m/s)	Infrared Power		
		300 W	400 W	500 W
		Drying Time (Min)		
35	1.00	510	420	330
	1.25	510	420	330
	1.50	540	480	360
40	1.00	480	420	270
	1.25	510	420	300
	1.50	510	390	330
45	1.00	480	360	240
	1.25	480	360	300
	1.50	540	390	300

Source: Data from Sharma et al. 2005a. *Journal of Food Engineering* 71:282–286. With permission.

the drying time, the temperature and velocity of the drying air have to be increased. However, applications that increase temperature and air velocity increase the energy consumed in the drying process and may have some negative effects on the final product quality. Heat-sensitive products such as vegetables, fruits, etc., may lose more quality due to high temperature and long drying times. The falling rate period, where most of the product quality takes place, is one of the general properties of hot air drying systems and is considerably long. In order to shorten the long falling rate period of convective hot air drying and total drying time, making use of short drying time properties of IR drying and equipping convective hot air dryers with IR dryers are discussed.

In a combined IR and hot air system, the IR power or radiation intensity, air temperature, and air velocity affect the drying time (Table 6.1; Kumar et al., 2005, 2006; Sharma et al., 2005b; Kocabiyik and Tezer, 2009; Nasiroğlu and Kocabiyik, 2009). Sharma et al. (2005b) have stated that in thin-layer drying of onion slices via IR and convective hot air dryer when the IR power increased from 300 to 500 W, air temperature increased from 35°C to 45°C, air velocity increased from 1 to 1.5 m/s, and drying time decreased by a factor of 2.5. Hebbar et al. (2004) have determined that for the drying of potato and carrot via hot air, IR, and combined (IR and hot air) systems, the drying time for both products obtained by the combined system decreased about 48% in comparison with the hot air system, the longest drying time was obtained for the hot air system, and the drying time of the IR system was intermediate. Similarly, Kumar et al. (2005) have found that for onion slices, IR–hot air combination resulted in shorter drying times. In their research, corn was harvested at three different moisture contents (24%, 16%, and 15%) and dried under hot air, IR, and combined IR and hot air systems; the drying time required for the hot air

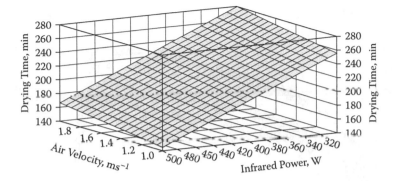

FIGURE 6.1 Variation of drying time with infrared power and inlet air velocity. (From Kocabiyik, H., and D. Tezer. 2009. *International Journal of Food Science and Technology* 44: 953–959. Reprinted with permission of John Wiley & Sons.)

system was determined to be about 255%, 360%, and 233% longer when compared with combined IR and hot air system (Yılmaz, 2008).

At a constant air temperature, the increase in IR power or IR radiation intensity and the decrease in air velocity shorten the drying time. The increase of IR power, and consequently of IR radiation, results in an increase in the IR energy penetrating the product to be dried. Thus, it increases the temperature and vapor pressure of water inside the product. Consequently, the drying rate increases, resulting in a shortening of drying time. Unlike convective hot air drying, the increase in air velocity at a constant IR radiation intensity and air temperature increases drying time. The reason for this is that increasing air velocity increases airflow, resulting in cooling at the surface of the product to be dried (Afzal and Abe, 2000; Kocabiyik and Tezer, 2009; Table 6.1 and Figure 6.1).

In combined IR and hot air drying, IR power or IR radiation intensity affects the drying curve and the change of drying rate just as it affects drying time (Mongpraneet et al., 2002; Sharma et al., 2005b; Wang and Sheng, 2006; Das et al., 2009; Kocabiyik and Tezer, 2009). At the start of drying, the product to be dried rapidly loses moisture and, as drying proceeds, the drying rate decreases. In combined IR and hot air drying, the drying rate may typically be divided into three regions (Figure 6.2) (Wang and Sheng, 2006; Kocabiyik and Tezer, 2009). At the beginning of drying, the temperature of the product starts to increase as IR radiation is absorbed by it, with the result that the water inside the product reaches boiling point and the drying rate continuously increases. This period is called the *heating up* or *rising period*. In this period, the drying time is considerably shorter in comparison to the total drying time. After this period, the drying rate tends to decrease, resulting in two falling rate periods. However, Mongpraneet et al. (2002) have determined a fairly short constant rate period. In the first falling rate period, the temperature of the product becomes constant due to the heat given by hot air and the absorption of the energy coming from the IR heater. The heat given to the product starts vaporizing the water inside the product and is used as the latent heat of vaporization of water. In this period, a relatively greater heat mass transfer takes place, and the drying rate tends to stay

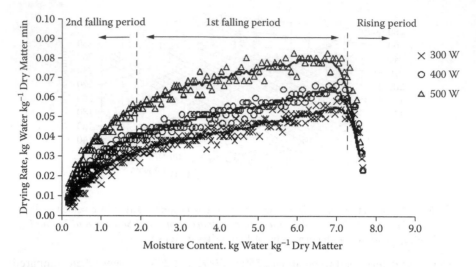

FIGURE 6.2 Drying rate versus moisture content of carrot slices at different infrared powers at air velocity 1.0 m/s. (From Kocabiyik, H., and D. Tezer. 2009. *International Journal of Food Science and Technology* 44: 953–959. Reprinted with permission of John Wiley & Sons.)

close to a constant drying rate. It is the period with the longest drying time and also the period during which the highest amount of water is removed. As a result of the decrease of the free water inside the product and the fact that bound water plays an active role, the drying rate dramatically drops. This period is called the *second falling rate period*. Since the water inside the product being dried decreases, if the same amount of heat is supplied by hot air, the surface temperature of the product rises until it reaches the dry-bulb temperature of the drying air. Most heat damage to food can therefore occur in the falling rate period, and the air temperature is controlled to balance the rate of drying and extent of heat damage (Fellows, 2007). Since the heat given by the IR heater is added to the heat given by hot air, the temperature of the product increases further, thereby affecting the product quality still further. The increase of IR radiation intensity increases drying rate and, as a result, decreases the drying time. The increase of IR radiation intensity increases product temperature and may result in loss of quality of the dried product. Color change, nutrient loss, taste anomalies, and vitamin losses may occur, especially in products sensitive to heat such as fruits and vegetables. These drawbacks can be overcome by using an intermittent IR heater (Afzal, 2003).

6.3   DRYING ENERGY

In convective hot air drying, the low energy efficiency and long drying time result in increase of the required total energy for drying. Sakai and Hanzawa (1994) stated that development of a continuous-drying apparatus equipped with far-IR (FIR) heaters, near-IR (NIR) heaters, and hot air blast can reduce the economic costs, drying time, and operating temperature. However, vegetable size should be restricted to no

TABLE 6.2
Specific Energy Consumption during Hot Air, Infrared, and Combined Drying of Potato and Carrot

Mode of Drying	Specific Energy Consumption (MJ/kg of Water Evaporation)	
	Potato	Carrot
Hot air	17.17	16.15
Infrared	7.60	7.15
Combined infrared and hot air	6.43	6.04

Source: Data from Hebbar et al. 2004. *Journal of Food Engineering* 65:557–563. With permission.

more than 5 mm in thickness to improve drying efficiency. As a result of obtaining a short drying time with IR and convective hot air combination, an advantage is obtained for the required specific energy needed to remove unit amount of water from the product. Hebbar et al. (2004) have determined that the specific energy consumption of IR and convective hot air drying of potato and carrot is about 63% less compared to convective hot air drying, and that the heat utilization efficiency of the dryer is 38.5% for potato and 38.9% for carrot (Table 6.2). In parallel to this, when this is compared with the drying process for barley performed with hot air at 70°C, it has been determined that the IR–hot air drying system decreased the total required energy by about 245% (Afzal et al., 1999).

After performing the drying of corn with three different moisture contents (24%, 16%, 15%) to 13% moisture content by convective hot air, IR, and combined IR and hot air systems, Yılmaz (2008) found that for first harvest periods where the initial moisture content of the product is high, specific energy consumption is 2.5 times less for combined system drying when compared with hot air drying; however, for final harvest periods where the initial moisture content of the product approaches that of the dried product, there was no significant difference between combined system and hot air system drying in terms of specific energy.

In the drying of paddy by using an IR grain dryer setup with vibrating tray, the specific energy consumption value ranged between 14.7 and 73.4 MJ/kg and decreased with increase in radiation intensity (5514, 4520, 3510, and 2520 W/m^2) and decrease in product depth (3, 6, 12, and 25 mm; Das et al., 2004b). In conditions where the air temperature (35°C, 40°C, and 45°C) and air velocity (1.0, 1.25, and 1.5 m/s) are constant, an increase in the effective moisture diffusion has been observed with increase of IR radiation intensity (26.5, 35.3, and 44.2 kW/m^2), and the activation energy ranged between 5.06 and 10.63 kJ/mol (Pathare and Sharma, 2006). Specific energy consumption values for red pepper, apple, and leek slices at a temperature of 30°C, drying air velocity range of 1.0–2.0 m/s, and at 300, 400, and 500 W IR lamp powers have been determined to range between 16.62 and 27.31 MJ/kg of water, 17.02 and 21.27 MJ/kg of water, 17.65 and 19.51 MJ/kg of water for red pepper; 13.67 and 19.00 MJ/kg of water, 16.42 and 24.21 MJ/kg of water, 15.81 and 19.50 MJ/kg of water for apple; and 19.18 and 19.15 MJ/kg of water, 15.01 and 22.56 MJ/kg of water, and 18.02 and 25.84 MJ/kg of water for leek (Nasiroglu 2007).

6.4 PRODUCT QUALITY

One of the most important indicators of the success of drying systems is the retention of the quality of dried food and agricultural products. Whichever drying system is used, some changes take place in the physical, chemical, and functional properties of the food or agricultural product. The quality of the final product after the drying process is desired to be similar to the quality before drying. The quality of many food products can be negatively affected by drying processes performed above room temperature. The heat added to the product or the time it was exposed to heat increases the rate of nutrient losses (Chua and Chou, 2005). For example, vitamin C is unstable and also sensitive to heat and oxidation (Fellows, 2007), and the degradation rate of ascorbic acid is dependent on environmental conditions such as water activity and temperature (Singh and Lund, 1984). Sokhansanj and Jayas (1995) have stated that during drying, ascorbic acid loss is between 10% and 50%. Drying time and the drying temperature applied to the product are important factors for quality loss. Since evaporative cooling takes place during the evaporation of the moisture on the product surface during the constant rate drying period, the nutrient losses of the product may be less. Quality losses of dried products increase due to the extension of the falling rate period during drying, vanishing of the evaporative cooling effect, and the long time the product is subjected to heat. Heat-sensitive food and agricultural products lose more quality. Generally, the drying time is fairly long in convective hot air drying, and since the falling rate period comprises most of the total drying time, it has an important effect on increasing quality losses. One of the methods used for decreasing drying time and improving quality is to supply heat by IR radiation (Timoumi et al., 2007). For example, IR radiation drying has improved the quality of spices (Paakkonen et al., 1999). Color loss can be properly prevented by shortening the drying time significantly via intermittent IR drying (Chua et al., 2004). Total color change and protein, carotenoid, and phenolic acid content have been similar for all drying systems for corn dried under hot air, IR, and combined IR and hot air system, while drying time and specific energy consumption are significantly decreased with the IR heater (Yılmaz, 2008).

Afzal et al. (1999) discovered that the drying process implemented via combined IR and hot air has a greater synergic effect compared to drying implemented via IR and hot air alone, and they also found that energy conservation and improvement in quality were obtained.

The rehydration capacity of the dried product may be related to the structural deformation of the product that takes place during drying. In hot air, case hardening and more shrinkage may be observed in the dried products since the drying process starts from the surface. The pore structure of the dry product breaks down due to shrinkage and case-hardening effects. Thus, rehydration capacity, which is one of the quality parameters of the dried product, may be negatively affected. Quality parameters such as color and case hardening of potato and carrot dried under combined mode have given better results in comparison to those of potato and carrot dried

TABLE 6.3
Quality of Onion Slices Dried under Selected Drying Processes

Mode of Drying	Drying Temperature (°C)	Slice Thickness (mm)	Air Temperature (°C)	Air Velocity (m/s)	Drying Time (mm)	Pyruvic Acid Content (μmol/g)	Total Color Change	Browning Index
Combined	60	2	40	2	220	16.95	20.63	13.74
Hot air	60	2	60	2	340	10.96	26.60	18.99
Infrared	60	2	—	—	280	9.83	30.62	15.85

Source: Data from Kumar et al. 2005. *Journal of Food Processing and Preservation* 29:132–150. With permission.

under hot air and only IR (Hebbar et al., 2004). Kumar et al. (2005) stated that the rehydration capacity of onion slices dried under combined IR and hot air method is better than that of onion slices dried under only hot air and IR. In the same study, a combined drying system yielded good results for quality parameters such as pyruvic acid content, total color change, and browning index (Table 6.3).

The change in IR radiation intensity can affect the rehydration capacity and shrinkage properties of the dried product. Mongpraneet et al. (2002) stated that in the FIR convection drying system where 40°C input air temperature and 0.54 m/s air velocity are used, the rehydration capacity of the Welsh onion increased while its shrinkage ratio decreased.

6.5 INTERMITTENT AND CONTINUOUS USES OF IR AND CONVECTIVE HEATING

The increase of IR radiation intensity causes an increase in the drying rate, thus shortening the drying time. However, since higher radiation intensity increases the temperature of the product, it can have negative effects on product quality (Abe and Afzal, 1997). When a biological material is subject to IR radiation for a long period of time, swelling takes place, ultimately resulting in the fracturing of the material (Jones, 1992). Fasina et al. (1997, 1999, 2001) stated that IR heating changes the physical, mechanical, chemical, and functional properties of barley and also that heating legume seeds up to 140°C via IR results in cracks on the surface. In order to overcome these negative effects, IR heating can be done intermittently during drying via combined IR and hot air system. Zbicinski et al. (1992) suggested intermittent irradiation combined with convective air drying for heat-sensitive materials. Also, Chua et al. (2004) stated that to dry heat-sensitive materials, a combined radiant-convective drying method or intermittent drying mode may be applied. Intermittent IR heating has the potential to decrease energy requirement, increase the quality of heat-sensitive product, and increase the surface temperature (Afzal, 2003). Controlling of timing of IR heaters (on/off) with appropriate temperature setting can reduced the product color degradation while shortening drying time and thus lessen radiative energy used for drying (Chua et al., 2004). Intermittent IR and continuous convective heating of thick porous materials have resulted in fine surface quality and energy efficiency (Dostie et al. 1989). In the FIR drying of barley performed by Afzal (2003), these results have been obtained: when compared with the simultaneous usage of IR heating, the total drying time has increased with less total energy input; higher IR heat intensity with intermittent application was the most energy-efficient method; discoloration was observed at higher intensities under continuous heating; however, during intermittent operation, no breakage or discoloration occurred; duration of exposure in intermittent heating had an effect on germination (Table 6.4).

6.6   MODELING STUDIES

In order to successfully transfer experimentally studied drying techniques into application and the drying system to industrial scale, their agreement with previously determined drying models may be further examined. Mathematical modeling of thin-layer drying is important for optimum management of operating parameters and prediction of performance of the drying system (Jain and Pathare, 2004). Numerous mathematical (empirical and semi-empirical) models have been proposed to define the drying behavior of food and agricultural products. The drying curves fitted for the models are often used to explain the mass transfer in thin layer drying, and empirical equations in drying simulations are easy to apply and have been widely used (Cenkowski et al., 2007; Laohavanich and Wongpichet, 2008).

Experimental studies have been made of drying various food and agricultural products via combined IR and hot air drying. The agreement of obtained drying kinetics and data with different simulation models developed by various researchers has been analyzed, and new model coefficients have been determined. As a result of these studies, natural or forced convective systems have been found to be in accordance with most of the models used in simulations. Many of the adapted model equations have had high correlation coefficients and low chi-square values. Models giving the best results have been determined (Table 6.5).

Even though the use of IR radiation for drying was defined a long time ago, studies related to drying have gained importance in the last 20 years. The good results obtained via simulation models in combined IR and hot air system drying systems will be an important step in the transfer of this method to industry and in the design of new dryers and prototypes.

6.7   OPTIMIZATION OF COMBINED IR AND HOT AIR DRYING

In the combined IR and hot air drying system, process variables such as IR radiation intensity, the temperature of inlet air, air velocity, and form of product affect the drying time, drying rate, the required energy for drying or specific energy consumption and various quality parameters of the final product of agricultural and food products. Even though drying time shortens with increase of IR power or IR radiation intensity at constant temperature, the quality features of the product may be negatively affected and energy cost may increase. Thus, as the quality parameters of the product such as color, nutritional value, vitamin content, taste, aroma, rehydration, etc.,

TABLE 6.4

Effects of Intermittent Infrared Drying on Drying Time, Specific Energy Consumption, and Quality of Barley

Duty Cycle On/Off	Infrared Heat Flux (W/cm²)	Inlet Air Temperature (°C)	Operation Time (h)	Moisture Reduction/ Unit Operating Time (%/h)	Specific Energy Consumption (MJ/kg of Water)	Germination Mean (%)	Bulk Density Mean (g/cm³)
5/5	0.333	30	1.67	4.55	29.957	97	0.804
	0.500	30	0.75	9.33	14.655	96	0.809
10/10	0.333	30	1.73	4.62	31.595	97	0.800
	0.500	30	0.80	8.75	14.856	89	0.813
20/20	0.333	30	1.60	4.38	32.471	97	0.815
	0.500	30	0.83	8.55	17.298	85	0.815
5/15	0.333	30	2.97	2.29	34.253	97	0.806
	0.500	30	1.50	4.53	17.672	95	0.815
Continuous heating	0.333	30	0.92	8.15	29.295	98	0.809
	0.500	30	0.52	14.81	16.508	61	0.806

Source: Adapted from Afzal, T.M. 2003. Intermittent far infrared radiation drying, ASAE Paper No: 036201. With permission.

TABLE 6.5

Selected Modeling Works on Combined Infrared and Hot Air Drying of Selected Agricultural and Food Products

Evaluated Models	Products	Suitable Model and Equation	References
Exponential Page Diffusion Approximation of diffusion	Rough rice	Page model $MR = \exp(-kt^n)$	Abe and Afzal (1997)
Page	Barley	Page model $MR = \exp(-kt^n)$	Afzal and Abe (2000)
Exponential Page Approximation of diffusion	Onion slices	Page model $MR = \exp(-kt^n)$	Wang (2002)
Newton Page Modified Page Henderson and Pabis Asymptotic (logarithmic) Logistic Two-term exponential Geometric Wang and Singh	Onion slices	Asymptotic (logarithmic) $MR = a_0 + a\exp(-kt)$	Jain and Pathare (2004)
Page	Paddy	Page model $MR = \exp(-kt^n)$	Das et al. (2004a)
Newton Page Modified Page Henderson and Pabis Logarithmic Two-term exponential Wang and Singh	Onion slices	Page model $MR = \exp(-kt^n)$	Sharma et al. (2005a)
Page Modified Page Fick's Exponential	Onion slices	Modified Page model $MR = \exp[-(kt)^n]$	Kumar et al. (2006)
Simple exponential Biparametric exponential Page Two-term exponential	Paddy	Page model $MR = \exp(-kt^n)$	Das et al. (2009)

are maximized, operation parameters such as drying time, energy requirement, capital, labor, etc., should be minimized. Therefore, determination of optimal operating conditions for combined IR and hot air drying is very important.

The optimization of thermal conditions for drying of porous media is an active field of development and research in industry (Dutournie et al., 2006). Several researchers have specifically aimed to obtain optimal conditions of a dryer in combined IR and hot air drying.

Wang and Sheng (2006) have determined the highest sensory quality and lower energy consumption rate for peach to be at 2.891 and 2.860 kW/kg IR drying points, respectively. During the drying of onion slices under IR–hot air combination, the optimum drying conditions at which their color and pyruvic acid content is best conserved is at 2 m/s air velocity, 40°C air temperature, 60°C drying temperature, and an onion slice thickness of 2 mm (Kumar et al., 2005). As a result of their studies for carrot, Kocabiyik and Tezer (2009) have maximized quality parameters of carrots such as rehydration and shrinkage while minimizing operational parameters, drying time, and specific energy, and they have stated the optimum drying condition to be 409.12 W IR power and 1.0 m/s air velocity. Chua et al. (2004) have aimed to study the possibility of minimizing overall color degradation and reducing drying time of agroproducts (carrot, potato, and banana) through suitable osmotic treatment and appropriate intermittence period and intensity level of IR radiation heat input, and determined that with appropriate control and regulation of operational variables, it is possible to produce premium quality agroproducts while maintaining high drying rates.

There are a numbers of operational parameters in combined drying systems, and all these parameters are very important and have different effects. Due to the number of operational parameters, experiments and the numerical program used to obtain optimal operational conditions in the combined IR dryer and hot air dryer are very complex and time consuming. The most economical and accurate method for performing process optimization must be selected. In this regard, numerical programs have been previously developed and experimentally validated to model the thermal and hygroscopic behavior of the product. To obtain a behavioral reduced model that allows the operating optimums to be calculated quickly, design of experimental methodology that is based on using response surface methodology can be used (Dutournie et al., 2006).

6.8 CONCLUSION

When the combined IR and hot air drying system is compared to the usage of IR and hot air drying separately, it is seen that it has some advantages. Heat and mass transfer may take place in a better way and, consequently, it is observed that drying time dramatically decreases, energy efficiency increases, and specific energy consumption decreases. Therefore, it is hoped that some savings in operating cost may be attained. From the final product quality perspective, it is observed that this method is superior to the separate use of IR and hot air drying. When compared with convective hot air drying, lower temperature drying air can be used in combined IR and hot air drying. Therefore, it can be seen as an alternative to the drying of heat-sensitive products.

Even though it was determined a long time ago that IR radiation can be used for drying, studies related to IR radiation drying have gained importance in the last 20 years. A vast majority of the studies performed are experimental, and their number is quite limited. However, this combined drying system has been adapted for industrial use, and there are samples for commercial applications today. The fact that better results are obtained from simulation models in drying studies via combined IR and hot air system may be an important step in the transfer of this method to industry and in the design of new dryers and prototypes.

Experimental studies should be conducted in the field of combined IR and hot air drying in order to judge the effects on drying kinetics, drying time, drying energy, operating cost, and final product quality of pretreatments that might be applied to the product to be dried. Studies should be conducted using different products in order to understand the effects of the intermittent use of IR heating and hot air heating methods.

REFERENCES

Abe, T., and T.M. Afzal. 1997. Thin-layer infrared radiation drying of rough rice. *Journal of Agricultural Engineering Research* 67:289–297.

Adonis, M., and M.T.E. Khan. 2004. Combined convective and infrared drying model for food applications. *IEEE Africon* 2:1049–1052.

Afzal, T.M. 2003. Intermittent far infrared radiation drying. ASAE Paper No: 036201.

Afzal, T.M., and T. Abe. 2000. Simulation of moisture changes in barley during far infrared radiation drying. *Computers and Electronics in Agriculture* 26 (2):137–145.

Afzal, T.M., T. Abe, and Y. Hikida. 1999. Energy and quality aspects during combined FIR-convection drying of barley. *Journal of Food Engineering* 42:177–182.

Amaratunga, K.S.P., Z. Pan, X. Zheng, and J.F. Thompson. 2005. Comparison of drying characteristics and quality of rough rice dried with infrared and heated air. ASAE Paper No.056005.

Baysal, T., F. Icier, S. Ersus, and H. Yildiz. 2003. Effects of microwave and infrared drying on the quality of carrot and garlic. *European Food Research and Technology* 218:68–73.

Cenkowski, S., S.D. Arntfield, and M.G. Scanlon. 2007. Far infrared dehydration and processing. In *Food Drying Science and Technology: Microbiology, Chemistry, Application*, Ed. Y.H. Hui, C. Clary, M.M. Farid, O.O. Fasina, A. Noomhorm, J. Welti-Chanes, 157–201. Lancaster, PA: DEStech Publications.

Chua, K.J., and S.K. Chou. 2003. Low-cost drying methods for developing countries. *Trends in Food Science and Technology* 14:519–528.

Chua, K.J., and S.K. Chou. 2005. New hybrid drying technologies. In *Emerging Technologies for Food Processing*, Ed. D. W. Sun, 535–551. London: Elsevier Academic Press.

Chua, K.J., S.K. Chou, A.S. Mujumdar, J.C. Ho, and C.K. Hon. 2004. Radiant-convective drying of osmotic treated agro-products: Effects on drying kinetics and product quality. *Food Control* 15:145–158.

Das, I., S.K. Das, and S. Bal. 2004a. Drying performance of a batch type vibration aided infrared dryer. *Journal of Food Engineering* 64:129–133.

Das, I., S.K. Das, and S. Bal. 2004b. Specific energy and quality aspects of infrared (IR) dried parboiled rice. *Journal of Food Engineering* 62:9–14.

Das, I., S.K. Das, and S. Bal. 2009. Drying kinetics of high moisture paddy undergoing vibration-assisted infrared (IR) drying. *Journal of Food Engineering* 95:166–171.

Dhib, R. 2007. Infrared drying: From process modeling to advanced process control. *Drying Technology* 25:97–105.

Dostie, M., J.N. Seguin, D. Maure, Q.A. Ton-That, and R. Chatingy. 1989. Preliminary measurement on the drying of thick porous materials by combinations of intermittent infrared and continuous convection heating. In *Drying'89*, Eds. S.A. Mujumdar and M.A. Roques. New York: Hemisphere Press.

Doymaz, I. 2007. Air-drying characteristics of tomatoes. *Journal of Food Engineering* 78:1291–1297.

Dutournie, P., P. Salagnac, and P. Glouannec. 2006. Optimization of radiant drying of a porous medium by design of experiment methodology. *Drying Technology* 24:953–963.

Ertekin, C., and O. Yaldiz, 2004. Drying of eggplant and selection of suitable thin-layer drying model. *Journal of Food Engineering* 63:349–359.

Fasina, O. 2003. Infrared heating of food and agricultural materials. ASAE Paper No:036219.

Fasina, O., B. Tyler, and M. Pickard. 1997. Infrared heating of legume seeds effect on physical and mechanical properties. ASAE Paper No: 976013.

Fasina, O., R.T. Tyler, M. Pickard, and G.H. Zheng. 1999. Infrared heating of hulless and pearled barley. *Journal of Food Processing and Preservation* 23:135–151.

Fasina, O.O, R.T. Tyler, M. Pickard, G.H. Zheng, and N. Wang. 2001. Effect of infrared heating on the properties of legume seeds. *International Journal of Food Science and Technology* 36:79–90

Fellows, J.P. 2007. *Food Processing Technology Principles and Practice.* Cambridge: Woodhead Publishing Limited.

Hebbar, H.U., K.H. Viswanathan, and M.N. Ramesh. 2004. Development of combined infrared and hot air dryer for vegetables. *Journal of Food Engineering* 65:557–563.

Itaya, Y., and S. Mori. 2006. Recent R&D on drying technology in Japan. *Drying Technology* 24:1187–1196.

Jain, D., and P.B. Pathare. 2004. Selection and evaluation of thin layer drying models for infrared radiative and convective drying of onion slices. *Biosystems Engineering* 89: 289–296.

Jones, W. 1992. *A Place in the Line Micronizer.* Special Report. Micronizing Company Ltd.

Kocabiyik, H., and D. Tezer. 2009. Drying of carrot slices using infrared radiation. *International Journal of Food Science and Technology* 44:953–959.

Krishnamurthy, K., H.K. Khurana, S. Jun, J. Irudayaraj, and A. Demirci. 2008. Infrared heating in food processing: An overview. *Comprehensive Reviews in Food Science and Food Safety* 7(1):2–13.

Kumar, D.G.P., H.U. Hebbar, and M.N. Ramesh. 2006. Suitability of thin layer models for infrared-hot air drying of onion slices. *LWT—Food Science and Technology* 39:700–705.

Kumar, D.G.P., H.U. Hebbar, D. Sukumar, and M.N. Ramesh. 2005. Infrared and hot-air drying of onions. *Journal of Food Processing and Preservation* 29:132–150.

Laohavanich, J., and S. Wongpichet. 2008. Thin layer drying model for gas-fired infrared drying of paddy. *Songklanakarin Journal of Science and Technology* 30:343–348.

Lewis, M.J. 1996. *Physical Properties of Food and Food Processing Systems.* Cambridge: Woodhead Publishing Limited.

Lin, Y.P., T.Y. Lee, J.H. Tsen, and V.A.E. King. 2007. Dehydration of yam slices using FIR-assisted freeze drying. *Journal of Food Engineering* 79:1295–1301.

Lin, Y.P., J.H. Tsen, and V.A.E. King. 2005. Effects of far-infrared radiation on the freeze-drying of sweet potato. *Journal of Food Engineering* 68:249–255.

Mongpraneet, S., T. Abe, and T. Tsurusaki. 2002. Far infrared-vacuum and -convection drying of welsh onion. *Transaction of the ASAE* 45(5):1529–1535.

Nasiroğlu, S. 2007. Using of infrared drying technique in red pepper, apple and leek drying. Masters thesis, Canakkale Onsekiz Mart University.

Nasiroğlu, S., and H. Kocabiyik. 2009. Thin-layer infrared radiation drying of red pepper slices. *Journal of Food Process Engineering* 32(1):1–16.

Nimmol, C., S. Devahastin, T. Swasdisevi, and S. Soponronnarit. 2007. Drying of banana slices using combined low-pressure superheated steam and far-infrared radiation. *Journal of Food Engineering* 81:624–633.

Nowak, D., and Lewicki, P.P. 2004. Infrared drying of apple slices. *Innovative Food Science and Emerging Technologies* 5:353–360.

Oztekin, S., A. Bascetincelik, and Y. Soysal. 1999. Crop drying programme in Turkey. *Renewable Energy* 16:789–794.

Paakkonen, K., J. Havento, B. Galambosi, and M. Pyykkonen. 1999. Infrared drying of herb. *Agricultural and Food Science in Finland* 8:19–27.

Pan, Z., D.A. Olson, K.S.P. Amaratunga, C.W. Olsen, Y. Zhu, and T.H. Mchugh. 2005. Feasibility of using infrared heating for blanching and dehydration of fruits and vegetables. ASAE Paper No.056086.

Pathare, P.B., and G.P. Sharma. 2006. Effective moisture diffusivity of onion slices undergoing infrared convective drying. *Biosystems Engineering* 93(3):285–291.

Ratti, C., and A.S. Mujumdar. 2006. Infrared drying. In *Handbook of Industrial Drying*, 3rd enhanced ed. Ed. A.S. Mujumdar, 423–438. New York: Taylor & Francis.

Sakai, N., and T. Hanzawa. 1994. Applications and advances in far-infrared heating in Japan. *Trends in Food Science and Technology* 5:357–62.

Sandu, C. 1986. Infrared radiative drying in food engineering: A process analysis. *Biotechnology Process* 2:109–119.

Sharma, G.P., R.C. Verma, and P.B. Pathare. 2005a. Mathematical modeling of infrared radiation thin layer drying of onion slices. *Journal of Food Engineering* 71:282–286.

Sharma, G.P., R.C. Verma, and P.B. Pathare. 2005b. Thin-layer infrared radiation drying of onion slices. *Journal of Food Engineering* 67:361–366.

Singh, R.K., and D.B. Lund. 1984. Kinetics of ascorbic acid degradation in stored intermediate moisture apple. In *Proceedings of the 3rd International Congress on Engineering and Food Engineering Sciences in the Food Industry*, Ed. B.M. Mckenna, Amsterdam.

Sokhansanj, S., and D.S. Jayas. 1995. Drying of foodstuffs. In *Handbook of Industrial Drying*, Ed. A.S. Mujumdar, 567–588. New York: Marcel Dekker.

Strumillo, C., and F.T. Kudra. 1986. *Drying: Principles, Applications and Design*. New York: Gordon and Breach Science Publishers.

Timoumi, S., D. Mihoubi, and F. Zagrouba. 2007. Shrinkage, vitamin C degradation and aroma losses during infra-red drying of apple slices. *LWT—Food Science and Technology* 40:1648–1654.

Togrul, H. 2005. Simple modeling of infrared drying of fresh apple slices. *Journal of Food Engineering* 71:311–323.

Wang, J. 2002. A single-layer model for far-infrared radiation drying of onion slices. *Drying Technology* 20:1941–1953.

Wang, J., and K. Sheng. 2006. Far infrared and microwave drying of peach. *LWT—Food Science and Technology* 39:247–255.

Yilmaz, N. 2008. Effects of different drying techniques on physical, chemical and toxicologic properties of maize. Masters thesis, Canakkale Onsekiz Mart University.

Zbicinski, J., A. Jakobsen, and J.L. Driscoll. 1992. Application of infra-red radiation for drying of particulate material. In *Drying'92*, Ed. A.S. Mujumdar. Amsterdam: Elsevier Science Publisher.

7 Combined Infrared Radiation and Freeze-Drying

Griffiths Gregory Atungulu and Zhongli Pan

CONTENTS

7.1 Introduction 117
7.2 Principles of IR and Freeze-Drying 118
7.3 Characteristics of Combined IR and Freeze-Drying................... 119
    7.3.1 Drying Rate 119
    7.3.2 Quality Attributes of Combined IR Radiation and Freeze-Dried Products 122
        7.3.2.1 Rehydration Potential... 122
        7.3.2.2 Texture Modification.. 124
        7.3.2.3 Retention of Color and Chemical Components 130
    7.3.3 Dehydration Cost 133
7.4 Equipment.. 133
    7.4.1 Commercial-Scale Combined IR and Freeze Dryers 133
    7.4.2 Laboratory-Scale Combined Infrared and Freeze Dryers............... 133
7.5 Modeling Combined IR and Freeze-Drying............................... 134
7.6 Commercial Potential 136
7.7 Conclusion 136
Acknowledgment 138
Nomenclature 138
References.. 139

7.1 INTRODUCTION

The demand for improvement in quality of agricultural products and more economic processing operations has brought much attention to the replacement of conventional operations with a number of novel food processing and preservation methods and equipment. Combining infrared (IR) radiation and freeze-drying either simultaneously or sequentially are relatively new methods that have shown great potential for industrial application. Especially, sequential IR radiation and freeze-drying (SIRFD) have become the most popular of the emerging novel combinations. SIRFD is a two-step drying process of IR predehydration followed by freeze-drying. Traditionally, hot air or freeze-drying has been used in the food industry to dry fruits and vegetables.

However, hot air drying is a time- and energy-consuming process. Its low heat transfer to the product results in low energy efficiency. The long drying process may cause undesirable changes in the product, resulting in poor product quality (Nowak and Lewicki, 2004).

Typically, freeze-drying minimizes the negative impacts of drying on the final product quality and could produce the highest-quality food product among the drying methods. The prominent factors are the low temperature and the product structural maintenance during sublimation (Singh and Heldman, 1993). The rigidity in structure prevents the solid matrix in food from collapsing after drying, which results in a porous structure in the dried product (Mujumdar, 1995). Freeze-drying results in less shrinkage in food (Shishehgarha et al., 2002). There is also good retention of aroma, flavor, and nutrients in the finished product. The texture could be crispy, which is desirable for many food applications. Freeze-dried food can be stored at ambient temperature (Baker, 1997), and the porous structure of the product facilitates rapid and almost complete rehydration when water is added to the substance at a later time (Mujumdar, 1995). However, freeze-drying is an expensive process of dehydration for foods because of high capital and operating costs, and it is also time consuming because of its slow drying rate. The freeze-drying process usually has high capital costs of equipment and high energy costs. On an industrial scale, the operational cost of freeze-drying processes is on the order of four to five times higher than that of the spray-drying technique and eight to ten times higher than that of the single-stage evaporator (Flink, 1977). Therefore, the freeze-drying process is usually used only for high-value products to contain high manufacturing costs.

Combining IR and freeze-drying is gaining interest, especially for snacks or add-ins in breakfast cereals, as an application that can enhance improved drying rates while at the same time maintaining desirable food characteristics. Recent research reports indicate that some progress has been made in the use of SIRFD to produce high-quality and crispy dried fruits and vegetables with reduced drying time and improved energy efficiency. Lin et al. (2007) studied yam dehydration using freeze-drying with far-IR radiation with a three-level, three-factor design and aimed at determining the optimum combination of drying temperature, distance between sample and far-IR heater, and thickness that can lead to the best results of drying time, rehydration ratio, and total color difference.

7.2 PRINCIPLES OF IR AND FREEZE-DRYING

IR radiation is energy in the form of electromagnetic waves or electromagnetic radiation. IR can be transferred from the heating element to the product surface without heating the surrounding air; therefore, the energy transfer is highly efficient (Jones, 1992). Radiation heat transfer can occur between two bodies separated by a medium colder than both bodies (Cengel, 1998). The wavelength of IR falls in the spectrum of 0.76 to 1000 μm and can be typically categorized into near-IR (NIR) radiation (0.76–2 μm), medium-IR (MIR) radiation (2–4 μm), and far-IR (FIR) radiation (4–1000 μm). For agricultural food product processing, the high temperatures corresponding to NIR could cause product discoloration and quality deterioration and, therefore, temperature needs to be carefully controlled if NIR is used. FIR is associated with low temperature and energy emission and has found considerable application in food processing.

In the regular freeze-drying process, liquid in a frozen foodstuff is removed by sublimation, generally under reduced pressure. The reduced or low partial pressure usually means below about 607.95 Pa (0.006 atm), which is the triple point of water. In industrial applications, the low temperature is created by refrigeration, and the resulting low partial pressure of water is accommodated by carrying out the process under vacuum. The energy for sublimation is provided by electrical or steam heating of trays in contact with the frozen product. The freeze-drying process involves three stages:

a. Freezing stage
b. Primary drying stage
c. Secondary drying stage

In the freezing stage, the foodstuff to be processed is cooled down to a temperature where all the material is in a frozen state; and in the primary drying stage, the frozen liquid is removed by sublimation. As the frozen liquid sublimes, the sublimation interface, which starts at the outside surface, recedes, and a porous shell of dried material remains. The heat for the latent heat of sublimation can be conducted through the layer of dried material and through the frozen layer. The vaporized frozen liquid vapor is transported through the porous layer of dried material. During the primary drying stage, some of the sorbed water (nonfrozen water) in the dried layer may be desorbed. The desorption process in the dried layer could affect the amount of heat that arrives at the sublimation interface and, therefore, it could affect the velocity of the moving sublimation front. The time at which there is no more frozen layer is taken to represent the end of the primary drying stage. The secondary drying stage involves the removal of water that did not freeze (bound water). At the secondary stage, the desorbed water vapor is transported through the pores of the material being dried (Mujumdar, 1995).

7.3   CHARACTERISTICS OF COMBINED IR AND FREEZE-DRYING

There are three major aspects of the dehydration process that can be used to compare different techniques: the rate of dehydration, the quality and characteristics of the final product, and the economic and energy cost of the process. Some of the limitations of conventional dehydration technologies, such as air drying or freeze-drying, in these areas can be addressed by combined IR and freeze-drying.

7.3.1   DRYING RATE

The combination of IR heating and freeze-drying allows solid food pieces to be dried more rapidly than by freeze-drying alone, while maintaining the product quality relatively high. A comparison of the drying curves of replicate samples of 4-mm-thick strawberry slices dehydrated using SIRFD at a radiation intensity of 5000 W/m² illustrates that rapid dehydration rates can be achieved during freeze-drying (Figure 7.1). Freeze-drying tests showed that drying time can be reduced

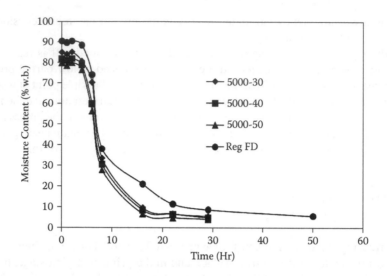

FIGURE 7.1 Dehydration curves of 4-mm-thick strawberry slices dehydrated using sequential infrared radiation and freeze-drying method and freeze-drying alone. In the legend, the first number indicates the radiation intensity (W/m²) and the second number indicates the weight reduction in the infrared predehydration step (%). (From Shih et al. 2008. *Transactions of the ASABE*, 51(1), 205–216. Copyright 2008. Reprinted with permission from American Society of Agricultural and Biological Engineers.)

when products are predehydrated by IR predehydration treatment as compared to freeze-drying alone. The regular freeze-drying sample took 50 h to achieve a final moisture content of 5% (w.b.) in strawberry slices, whereas it took about 29 h with the SIRFD method. Without predehydration, it takes a longer time to dry the slices, while the SIRFD process could save about 42% of processing time. The SIRFD samples display the same typical curve as that for the freeze-drying process curve, consisting of (a) the freezing stage, (b) the primary drying stage, and (c) the secondary drying stage. In the freezing stage, the foodstuff to be processed is cooled down to a temperature at which all the material is in a frozen state. During the freezing stage, moisture change is nearly negligible. Drying of foodstuffs takes place at the primary drying stage when the drying chamber is evacuated and its pressure is reduced to a value that would allow the sublimation of frozen water to occur. In general, high radiation intensity and weight reduction in predrying result in less needed freeze-drying time to achieve a specific moisture content. The relationships between the moisture ratio and time for different weight reduction (WR) during the freeze-drying process are shown in Figure 7.2. A 50% WR sample was freeze-dried faster than 30% or 40% WR samples. The final moisture contents of the samples with different weight reductions were significantly different, although they were very close at the end of the drying process.

An experimental dryer was developed by Lin et al. (2005) to determine the drying characteristics of sweet potato during freeze-drying with far-infrared radiation. The experimental drying time of sweet potato cubes dehydrated by three drying methods, i.e., air-drying, freeze-drying, and freeze-drying with far-infrared radiation,

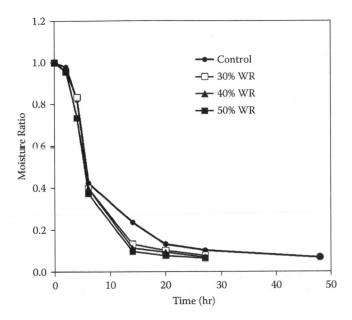

FIGURE 7.2 Moisture ratios at different weight reductions during the freeze-drying process for 4-mm-thick strawberry slices without considering the freezing step. (From Shih et al. 2008. *Transactions of the ASABE*, 51(1), 205–216. Copyright 2008. Reprinted with permission of American Society of Agricultural and Biological Engineers.)

were compared, and freeze-drying with far-infrared radiation was found to be able to reduce the drying time of sweet potato.

The experimental drying time of sweet potato cubes with various sizes dehydrated by using different methods is shown in Table 7.1. Simultaneous combined FIR–freeze-drying uses nearly half the drying time required by regular freeze-drying. The size of product is found to be an important influencing factor for drying

TABLE 7.1

The Experimental Drying Time (h) of Sweet Potato Cubes Dried by Various Drying Methods

Methods	Different Lengths of Potato Cubes (mm)		
	10	17.5	25
Freeze-drying with far-infrared	5.167	13.382	25.891
Freeze-drying (*p* = 2 mmHg)	12.353	23.063	30.936
Air drying (50°C)	9.966	24.105	49.147

Source: Lin et al. 2005. *Journal of Food Engineering*, 68(2), 249–255. Copyright 2005. Reprinted with permission of Elsevier.

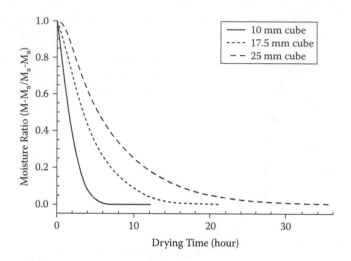

FIGURE 7.3 The change of moisture ratio of sweet potato cubes during freeze-drying with far-infrared treatment. (From Lin et al. 2005. *Journal of Food Engineering*, 68(2), 249–255. Copyright 2005. Reprinted with permission of Elsevier.)

time. Since the penetration of FIR radiation decreased with the thickness of sample, the effect of drying time reduction is not significant for 25 mm in the case of sweet potato cube. The air-drying method needed a longer drying time than FIR combined with freeze-drying. The change of moisture ratio of sweet potato cubes during freeze-drying with far-infrared treatment is shown in Figure 7.3. The utilization of FIR combined with freeze-drying provides an improvement of the drying rate.

7.3.2 QUALITY ATTRIBUTES OF COMBINED IR RADIATION AND FREEZE-DRIED PRODUCTS

7.3.2.1 Rehydration Potential

Many dried foods are rehydrated before consumption and, therefore, rehydration behavior is a critical functional property of these products. Although the ideal rehydration requirement differs from product to product, in many instances, a rapid rate and complete hydration are desirable. Freeze-drying is generally acknowledged to allow the greatest rehydration potential of any food-drying technique, both in terms of rate and amount of water uptake (Barbosa-Canovas and Vega-Marcado, 1996). Because water is removed by sublimation while tissues are at a low temperature, the original cellular structure is largely retained. Rehydration capability (rehydration ratio) can be used to reveal the physical and chemical changes caused by drying and treatments preceding dehydration (Lewicki, 2006; Feng and Tang, 1998). Rehydration of dried plant tissues is composed of three simultaneous processes: the imbibition of water into the dried material, the swelling, and, finally, the leaching of soluble solids from the dried material (McMinn and Magee, 1997). It is generally accepted that the degree of rehydration is dependent on the degree of cellular and structural disruption, and the rehydration ratio is used as a metric. The rehydration ratio is the ratio of the drained weight of the rehydrated sample to the weight of the dry sample used for rehydration.

The rehydration rate of the SIRFD samples is slower than that of the regular freeze-dried products but much faster than that of the sequential hot air– and freeze-dried (SHAFD) products. For the purpose of producing cereal add-ins; for instance, the SIRFD method can be recommended to produce products that do not become soggy as quickly as in regular freeze-drying. The differences in rehydration capacity of products dried by different methods under different conditions are shown in Figure 7.4. The SIRFD samples generally exhibit rehydration properties intermediate

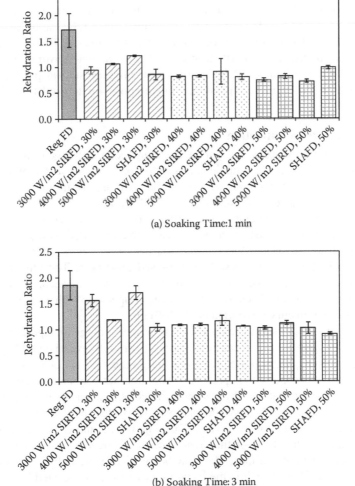

(a) Soaking Time: 1 min

(b) Soaking Time: 3 min

FIGURE 7.4 Rehydration ratios of 4-mm-thick strawberry slices dried with different drying methods after (a) 1-min and (b) 3-min soaking time. (From Shih et al. 2008. *Transactions of the ASABE*, 51(1), 205–216. Copyright 2008. Reprinted with permission of American Society of Agricultural and Biological Engineers.)

to those of freeze-dried and SHAFD products. This phenomenon could be explained by crust formation on the surfaces of the SIRFD sample, which could slow down the liquid (rehydrant) penetration into the dried sample during the rehydration process. On the other hand, the more porous structure in the regular FD sample facilitates rapid rehydration.

In a review by Sakai and Hanzawa (1994), the rehydration of Welsh onions dried with FIR radiation under vacuum was greater than that dried with hot air. Similar results were also reported by Kumar et al. (2005) for IR and hot air drying of onions. The rehydration ratio of the SIRFD samples decreases as the IR-drying weight reduction increases. This is expected because more shrinkage occurs with higher weight reduction and, as a result, products do not easily soak up the rehydrant. In practice, product slices processed with the SIRFD method rehydrate much faster than those produced by SHAFD but have slightly slower rehydration than product dried without predehydration. In comparison to hot air drying, Baysal et al. (2003) observed that microwave and IR drying resulted in higher rehydration capability for carrot and garlic samples. Rehydration of some tissues, such as potato, is known to be hindered by drying temperatures above approximately 60°C (Talburt and Smith, 1967).

7.3.2.2 Texture Modification

Texture is an important sensory attribute for many foods, and loss of the desired texture leads to a loss in product quality and a reduction in shelf-life. Combined IR and freeze-drying can be used to create a desirable crisp texture in foods that are consumed in the dehydrated form, such as snack food or crackers. A crisp food should be firm and snap easily when deformed, emitting a crunchy sound. A crisp texture is the result of a tissue that consists of cells or cavities that are filled with air, surrounded by a brittle structural phase (Vickers and Bourne, 1976). Factors that maximize the in situ vaporization of water result in increased crispiness. The physicochemical characteristics of the sample also affect textural parameters. Mechanical tests such as compression tests have been used to correlate crispness to a physical parameter in a force-deformation curve (Krokida et al., 2001a,b).

Drying methods can significantly affect the crispness of the final products. During drying of plant material, water can be transported through several possible pathways. In the first pathway, water passes from one cell to the next via cytoplasmic strands (plasmodesmata). In the second pathway, water alternatively enters and leaves successive cells along its pathway by passing through plasmalemma membrane boundaries. However, the most important pathway for water movement through plant tissues is through the cell wall (Tyree, 1970). Figures 7.5 and 7.6 show the scanning electron micrograph (SEM) of the cross section of strawberry and banana slices under different drying methods. The strawberry slices were dehydrated by regular FD, SIRFD, and SHAFD, while the bananas slices were dehydrated by regular FD and SIRFD with IR predehydration, resulting in 20% or 40% weight reduction. The regular FD sample shows uniform and small porous structure and little or no damage or disruption of cell walls at the heated surface. On the other hand, the SIRFD sample shows that cells at the surface collapse, forming a dense layer or crust on the surfaces of product. The rapid increase of surface temperature during IR heating

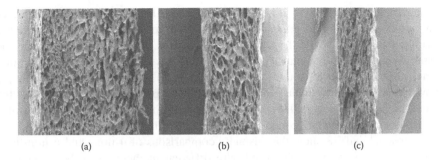

FIGURE 7.5 SEM of cross section of strawberry slices dried under different drying methods. (a) Regular FD, (b) SIRFD, (c) SHAFD. (From Shih et al. 2008. *Transactions of the ASABE*, 51(1), 205–216. Copyright 2008. Reprinted with permission of American Society of Agricultural and Biological Engineers.)

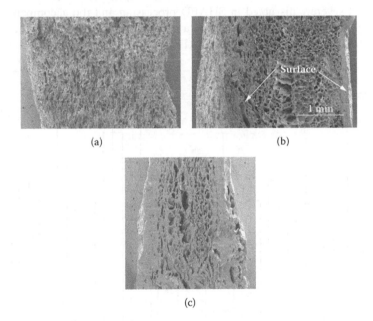

FIGURE 7.6 SEM of cross section of dried banana slices with no acid treatment under different drying methods. (a) Regular FD; (b) IR predehydration, 20% weight reduction; (c) IR predehydration, 40% weight reduction. (From Pan et al. 2008. *LWT—Food Science and Technology*, 41 (10), 1944–1951. Copyright 2008. Reprinted with permission of Elsevier.)

results in surface collapse at a rate comparable or higher than the rate at which moisture moves from the interior to the surface. Large pores exist in the center region of the strawberry and banana slices, which could be due to water vapor created during IR drying. Because the surface temperature of the product in IR heating rises quickly, although it may not change much at the center, the water vapor forces the expansion of the cellular walls, and this causes the development of large pores within the material (Jamradloedluk et al., 2007). Unlike in IR heating, the temperature of samples dried with hot air gradually increases from the ambient temperature to the drying temperature. While the temperature increases slowly, some moisture in the materials is released, thus making the vapor pressure driven by internal evaporation of moisture not as much as in the case of IR drying. Therefore, the SHAFD sample shows severe structural damage of the cell walls. In particular, the cell wall of the center region completely collapses. The long drying time during hot air drying also contributes to the structure collapse in the SHAFD samples.

Figures 7.7 and 7.8 show the crispness comparisons of 4-mm- and 6-mm-thick strawberry slices, respectively, dried with different methods. Strawberry samples processed with SIRFD had higher crispness than those processed with either the regular FD or SHAFD method. The crispness of the final product mainly relates to crust/dense layer formation and structural change during the drying process. Because the SIRFD samples form modest crust and large pores, the resultant product exhibits higher crispness than SHAFD products. The integrity of the cells of biological material get severely disrupted in SHAFD processing and the accompanying loss of binding forces between cells results in a less crisp product (Alvarez et al., 1995). The radiation intensity applied during SIRFD processing affects the crispness of the

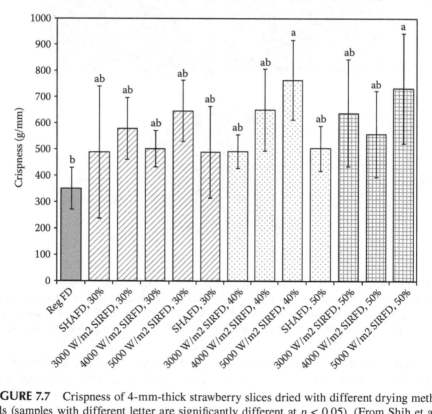

FIGURE 7.7 Crispness of 4-mm-thick strawberry slices dried with different drying methods (samples with different letter are significantly different at $p < 0.05$). (From Shih et al. 2008. *Transactions of the ASABE*, 51(1), 205–216. Copyright 2008. Reprinted with permission of American Society of Agricultural and Biological Engineers.)

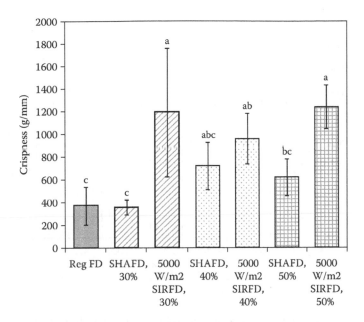

FIGURE 7.8 Crispness comparisons of 6-mm-thick strawberry slices dried with different methods. (From Shih, Y. C. 2006. Strawberry dehydration using sequential infrared radiation and freeze-drying method. Dissertation, Biological and Agricultural Engineering, University of California, Davis.)

product. Increased temperature at higher radiation intensity causes quicker moisture removal in the products. Crispy product texture can be obtained with higher radiation intensity. The SIRFD samples were crisper than the regular FD and SHAFD samples with the use of 5000 W/m^2 radiation intensity.

Figure 7.9 shows the firmness results of the strawberry samples dried with different drying methods. Statistical analysis indicated no significant differences between products obtained using SIRFD and SHAFD and the regular FD product. However, SIRFD samples subjected to 5000 W/m^2 radiation intensity had a higher firmness value compared to the SHAFD samples.

Water removal affects many aspects of cell structure. Hot air drying, which is the most economical and widely used method for dehydrating vegetable and fruit pieces, has the problem of irreversible damage to the texture, which leads to shrinkage, slow cooking, and incomplete rehydration (Mujumdar, 1995). The possible causes of these texture deteriorations are loss of differential permeability in the protoplasmic membrane, loss of turgor pressure in the cell, protein denaturation, starch crystallinity, and hydrogen bonding of macromolecules. Texture of air-dried vegetables deteriorates during storage if the product is exposed to high temperature or if it is inadequately dehydrated (Richardson, 2001).

In studies on air-dried carrots, parsnips, and turnips, Pedlington and Ward (1965) observed several changes, including a loss in the selective permeability of cytoplasmic membranes of cells responsible for maintaining turgidity and crisp texture of vegetables. They also found that loss of water resulted in rigidity of cell walls and

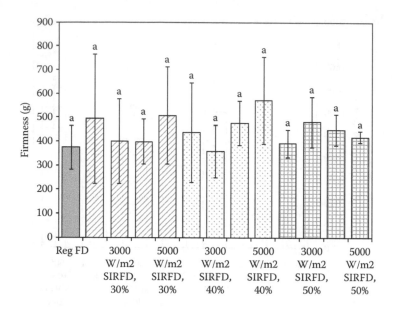

FIGURE 7.9 Firmness comparisons of 4-mm-thick strawberry slices dried with different methods. (From Shih, Y. C. 2006. Strawberry dehydration using sequential infrared radiation and freeze-drying method. Dissertation, Biological and Agricultural Engineering, University of California, Davis.)

leads to slow collapse by the stresses set up by shrinkage of neighboring cells. SIRFD samples tend to have slightly higher shrinkage in thickness than samples dried with freeze-drying. However, they have much less shrinkage than samples dried with hot air drying (Figure 7.10). In case of regular FD, structural rigidity of the dehydrated product is created during the freezing stage of the freeze-drying process. This structural rigidity prevents the collapse of the solid matrix during drying (Mujumdar, 1995). Shrinkage due to SIRFD is related to weight reduction level in predehydration. Ketelaars et al. (1992) found that shrinkage during drying is attributed to moisture removal and stresses developed in the cell structure during drying. Product shrinkage also depends on radiation intensity. A decrease in thickness shrinkage occurs as radiation intensity increases. The shrinkage results showed that for samples with 50% weight reduction level, the SHAFD sample had higher thickness shrinkage compared to that for the SIRFD sample dried at 5000 W/m². This might be due to the longer drying time in the SHAFD method, causing more cells to collapse. To have a product with less thickness shrinkage, it is recommended that higher radiation intensity, such as 5000 W/m², be used during strawberry predehydration, to take advantage of higher drying rate than observed in freeze-drying alone.

Pretreatment with dipping solutions containing 10 g/L ascorbic acid and 10 g/L citric acid before the IR predehydration in the SIRFD method affects the thickness shrinkage of the banana products (Figure 7.11). Acid dipping has a significant effect ($p < 0.05$) on the shrinkage of the final product. A dramatic decrease in thickness shrinkage occurs for acid-dipped slices compared to undipped samples, which could be due to modification of surface properties of banana slices caused by the dipping

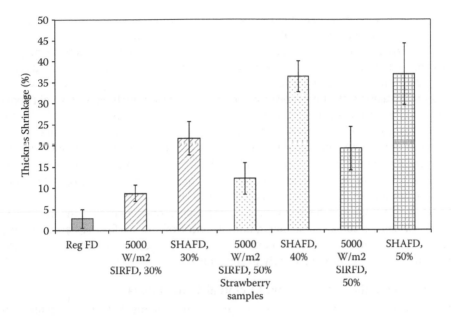

FIGURE 7.10   Thickness shrinkage of 6-mm-thickness dehydrated strawberry slices dried under different methods and conditions. (From Shih et al. 2008. *Transactions of the ASABE*, 51(1), 205–216. Copyright 2008. With permission from American Society of Agricultural and Biological Engineers.)

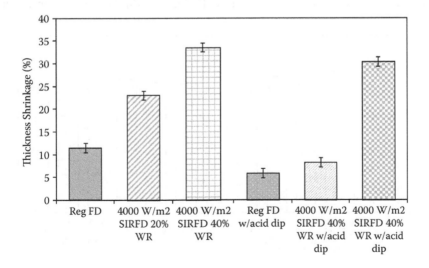

FIGURE 7.11   Thickness shrinkage of dehydrated banana slices under different treatments. (From Pan et al. 2008. *LWT—Food Science and Technology*, 41(10), 1944–1951. Copyright 2008. Reprinted with permission from Elsevier.)

treatment. For example, the thickness shrinkage decreased from 23.1% of undipped sample to 8.3% of dipped samples with 20% weight reduction.

Depending on the products, their structure and chemical modification during processing influences the drying and quality characteristics. Banana samples treated with 20% and 40% weight reductions and with no acid-dipping treatment took 23 and 38 h to achieve the final moisture content of 5 g moisture/100 g wet weight compared to 21.1 h for the non-pre-dehydrated (regular FD) sample (Pan et al., 2008). These results were contrary to what had been observed with strawberry slices studies (Shih et al., 2008) and could be attributed to shrinkage or crust formation during the predehydration of banana samples based on the SEM micrographs of the dried banana. Crust formation is more apparent in case of higher weight reduction and results in reduced drying rate during freeze-drying. Acid-dipping treatment prior to IR drying improves the drying rate during freeze-drying because the acid treatment washes out some starch, sugar, and protein from the surface of banana slices, resulting in formation of more porous surfaces. This effect of improved drying rate is mainly observed during the early drying stage.

7.3.2.3   Retention of Color and Chemical Components

Flavor, color, nutrient, or other biologically active chemicals that are sensitive to thermal or oxidative degradation are typically affected by drying method. Very little work has been done to reveal the effects of SIRFD on the retention of chemical components of dehydrated food products. Available information in the literature mainly addresses color changes. SIRFD produces samples with decreased L values, as weight reduction in the IR predehydration increases under the same radiation intensity (Figure 7.12). Li and Ma (2003) demonstrated increased strawberries' brightness/whiteness after freeze-drying. IR predehydration results in a significant improvement in redness compared to SHAFD (Figure 7.13). The decrease in a values

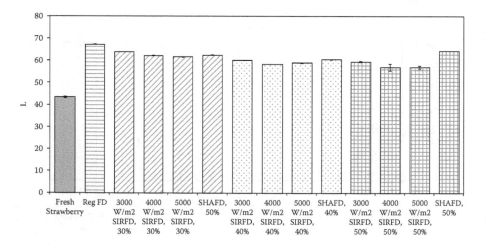

FIGURE 7.12 L color value of 4-mm-thick strawberry samples with different drying methods. (From Shih et al. 2008. *Transactions of the ASABE*, 51(1), 205–216. Copyright 2008. Reprinted with permission of American Society of Agricultural and Biological Engineers.)

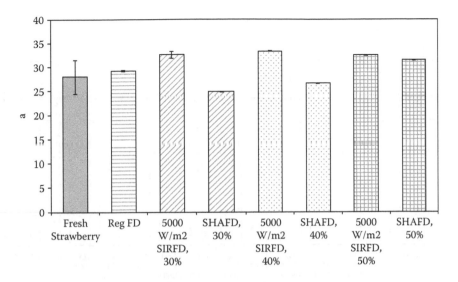

FIGURE 7.13   *a* Color values of 6-mm-thick strawberry samples dried with different drying methods. (From Shih et al. 2008. *Transactions of the ASABE,* 51(1), 205–216. Copyright 2008. Reprinted with permission of American Society of Agricultural and Biological Engineers.)

for SHAFD may be attributed to destruction of anthocyanin (Maskan et al., 2002a,b) and structural changes in the product tissues that occur during heating and drying. Phenomenological red color increase is due to the water reduction effect and the associated concentration increase of the red pigments (anthocyanins) in the dried product (Hammami and René, 1997). Higher drying temperatures lead to greater *a* values because of accelerated nonenzymatic browning reaction (Jamradloedluk et al., 2007). In IR predrying, the product temperature of the thin-sliced product increases rapidly than in hot air drying. Therefore, SIRFD strawberry would obviously be more reddish than that dried by SHAFD. Statistically, radiation intensity and weight reduction in the pretreatment step have a significant effect ($p < 0.05$) on color change in *L* and *a* color values for the SIRFD process. The weight reduction factor has the most influence on the *L* and *a* values. Baysal et al. (2003) found that hue angles were not significantly different among raw, hot air–, microwave-, and IR-dried samples in the study of drying carrot and garlic. For the SIRFD samples, hue angle values change with radiation intensity and weight reduction (Figure 7.14). When the radiation intensity is low, increased weight reduction level increases hue angle value. However, when radiation intensity is high, increased weight reduction decreases the hue angle value of the strawberry slices. Figure 7.15 shows the effect of (a) regular freeze-drying, (b) SIRFD, and (c) SHAFD on the appearance of dried strawberries. From visual observation, the SIRFD samples have a darker red color tone that is more desirable than the light pinkish color tone of the product in the regular FD and the light orange-reddish color tone of the product in the SHAFD method (Figure 7.15).

Very little work has been done to reveal the effects of SIRFD on the retention of chemical components. Biologically active components present in foods may benefit

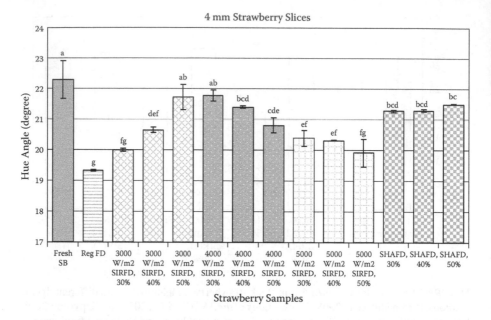

FIGURE 7.14 Hue angle of 4-mm-thick strawberry sample for all drying tests. *Note:* Samples with different letters are significantly different at $p < 0.05$. (From Shih et al. 2008. *Transactions of the ASABE*, 51(1), 205–216. Copyright 2008. Reprinted with permission from American Society of Agricultural and Biological Engineers.)

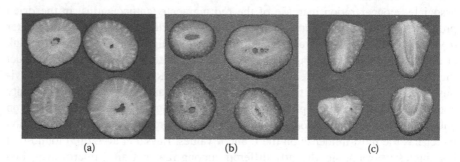

FIGURE 7.15 Effect of drying method on appearance of dried strawberries. (a) Regular FD strawberry samples, (b) SIRFD strawberry samples, (c) SHAFD strawberry samples. (From Shih et al. 2008. *Transactions of the ASABE*, 51(1), 205–216. Copyright 2008. Reprinted with permission of American Society of Agricultural and Biological Engineers.)

from the synergistic fast processing rates and low temperature accorded by combining IR predehydration and freeze-drying respectively in the SIRFD process. As scientific studies confirm links between enhanced health effects and consumption of specific food components, the market potential for functional foods will continue to grow (Gray et al., 2003). Therefore, the enhanced retention of functional components in food material as a result of SIRFD could be a significant advantage to food manufacturers.

7.3.3 DEHYDRATION COST

The cost of a dehydration process is a function of capital costs, labor cost, energy costs, and energy efficiency. Although few data are available on combined IR and freeze-drying operations, operational costs and energy costs of combined IR and freeze-drying are expected to be slightly more than natural gas-fueled convection hot air drying but less than regular freeze-drying. The new alternative for high-energy-efficient drying by using SIRFD uses IR heating during the predrying stage to remove a significant amount of moisture in fruits and vegetables before freeze-drying is used in the final stage of the drying process. Because IR heating is a very energy-efficient drying method with significantly short drying time, the energy savings of the SIRFD should be significant compared with the current drying practices.

7.4 EQUIPMENT

7.4.1 COMMERCIAL-SCALE COMBINED IR AND FREEZE DRYERS

Commercial-scale equipment that combines catalytic IR (CIR) energy with a unique conveying system has been built by Catalytic Drying Technologies LLC. The system is designed to efficiently and effectively dehydrate herbs, vegetables, and smaller fruits and fruit pieces. The trial has shown remarkable improvement in quality and drying efficiency when applied to a variety of conventionally dried products. In this technological advancement, significant energy reduction is achieved by predrying high-moisture products prior to their entering the freeze dryer. The CIR energy is generated by catalyzing natural gas or propane with a proprietary, enhanced platinum catalyst. Natural gas, when combined with air across the platinum catalyst, reacts by oxidation reduction to yield IR energy and small amounts of CO_2 and water vapor. The peak wavelengths, ranging from approximately 3 to 7 μm, are achieved by controlling emitter temperature. Water absorbs energy very efficiently at 3, 4.5, and 6 μm. By directing this energy spectrum at water-containing products, evaporative energy is targeted to the water within the product.

7.4.2 LABORATORY-SCALE COMBINED INFRARED AND FREEZE DRYERS

There are several lab-scale combined IR and freeze-drying approaches described in published reports. One is a simple system that was described by Lin et al. (2005) and consists of a vacuum chamber, cold trap condenser, and vacuum pump. The lab-scale experimental FIR freeze dryer is constructed by combining a freeze dryer (Model FDU-540, Eyela Co., Japan) and an FIR ceramic radiator (Figure 7.16). The FIR ceramic radiator emits thermal radiation in the wavelength range of 4–50 μm. Wire-equipped trays that contain the materials to be dried are fitted in the interior of a stainless steel drying chamber. The trays are placed under FIR ceramic radiator, and the freeze-drying operation is accomplished under vacuum at an absolute pressure of 0.3 mmHg.

Other equipment for combining IR and freeze-drying includes the SIRFD approach, as mentioned earlier, in which products are dried with IR heating followed

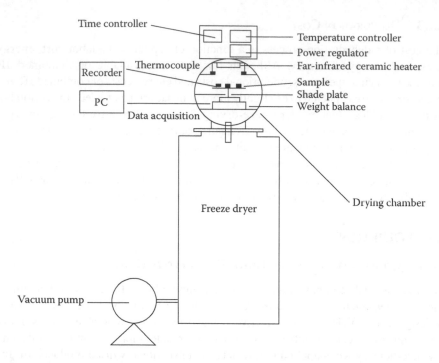

Time controller

Temperature controller

Power regulator

Thermocouple

Far-infrared ceramic heater

Recorder

Sample

PC

Shade plate

Weight balance

Data acquisition

Freeze dryer

Drying chamber

Vacuum pump

FIGURE 7.16 Schematic view of experimental freeze-dryer with far-infrared radiation. (From Lin et al. 2005. *Journal of Food Engineering*, 68(2), 249–255. Copyright 2005. Reprinted with permission of Elsevier.)

by freeze-drying. The IR dryer can be used to target various weight reductions during the predehydration step before introducing the samples into the freeze dryer.

7.5 MODELING COMBINED IR AND FREEZE-DRYING

Models describing heat and moisture transfer during SIRFD can provide valuable information to aid process optimization and equipment design. While there have been several approaches to modeling freeze-drying and IR heating independently, relatively little work has been carried out for combined IR heating and freeze-drying. Lin et al. (2005) used different models such as exponential model, Page model, diffusion model, and an approximation of the diffusion model to describe the behavior of freeze-drying combined with FIR radiation during the drying process of sweet potato. Drying data obtained were fitted to the four drying models and nonlinear regression analysis in the SPSS software was used to estimate the parameters of those four models. Table 7.2 shows the coefficients of the models obtained in the combined freeze-drying of sweet potato cubes with FIR heating. The coefficients of determination (R^2) for exponential, Page, and approximate diffusion model were all above 0.98, and that for the diffusion model was lower but still above 0.92. The largest coefficient of determination (R^2) could be found in Page and the approximate diffusion model. The larger values of coefficients k, K, and B indicated that less drying time was used. Constant N values could be obtained in the Page

TABLE 7.2
Coefficients of Various Models Obtained in the Freeze-Drying of Sweet Potato Cubes with Far-Infrared

Length of Cube mm	Exponential Model ($MR = \exp(-kt)$)		Page Model ($MR = \exp(-Kt^N)$)			Diffusion Model ($MR = (8/\pi^2) \times \exp[(-\pi^2/L^2) \times Dt]$)		Approximation of the Diffusion Model ($MR = A \times \exp(-Bt)$)		
	k	R^2	K	N	R^2	(D(m²/h) $\times 10^6$)	R^2	A	B	R^2
10	0.544	0.9861	0.430	1.286	0.9995	4.606	0.9338	1.109	0.597	0.9935
17.5	0.225	0.9872	0.151	1.237	0.9984	5.815	0.9281	1.112	0.249	0.9958
25	0.135	0.9910	0.092	1.176	0.9980	7.065	0.9344	1.098	0.148	0.9978

Source: Lin et al. 2005. *Journal of Food Engineering*, 68(2), 249–255. Copyright 2005. Reprinted with permission of Elsevier.

model. Additional statistical analysis to examine the accuracy of those four drying models using the residual plots (Figure 7.17) and root mean squared error (RMSE) value revealed the difference between experimental results and predicted values, and showed that the Page model gave the best results because of the smaller residual as well as RMSE. The Page model described the FIR freeze-drying characteristics of sweet potato properly. Abe and Afzal (1997) used various models similar to the experiments of Lin et al. (2005) to model the thin-layer IR radiation drying of rough rice. Their results similarly showed that the Page model provides the highest prediction accuracy of the drying models.

The reported study showed that the Henderson and Pabis drying model can be an appropriate model for describing the SIRFD process for strawberry slices. Table 7.3 shows the coefficients of Page and Henderson and Pabis models obtained in the freeze-drying of control and SIRFD strawberry slices. Acceptable correlation coefficient values (R^2) were obtained for both models for all weight reduction levels during the IR predehydration process. The obtained R^2 values indicated that both models fit well with the experimental data. However, the Henderson and Pabis model performed better than the Page model when the moisture ratio was low. Although the Page model showed a higher R^2, the predicted data did not fit well with the experimental data at the middle and the end of the drying process. The Page model underpredicted the moisture content values of the IR treated samples and overpredicted the moisture content value of the control sample (Pan, 2006; Shi and Pan, 2006). Conversely, the Henderson and Pabis model gave better predicted data although the R^2 value was slightly lower than that shown in the Page model. The predicted data generally banded around the experimental value, which showed the suitability of the fitted model in describing the drying behavior of SIRFD strawberry slices in the freeze-drying process.

7.6 COMMERCIAL POTENTIAL

Combined IR freeze-drying is expected to provide substantial cost savings when compared to regular freeze-drying. As a novel processing technology, combined IR and freeze-drying has the potential not only in cost savings but also the capacity to create new products, or products with unique characteristics, that cannot be replicated effectively with other technologies. The development of the equipment, which is still at a very early stage, will allow the manufacturer to effectively utilize new technology to deliver to consumers desirable food characteristics, such as excellent texture, better nutrient retention, and more convenience than competing technologies, thereby allowing the technology to grow rapidly and become widespread.

7.7 CONCLUSION

The novel combined IR and freeze-drying technology has been shown to be much more rapid than regular freeze-drying. The resulting tissue structure of products processed by SIRFD tends to have higher crispness than those processed with either regular FD or SHAFD method. SIRFD also produces products with rehydration

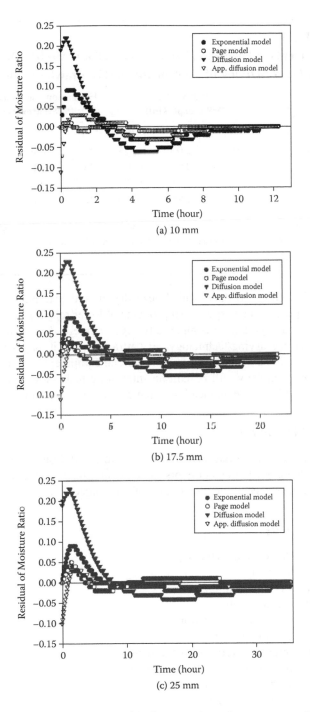

FIGURE 7.17 Plots of moisture ratio residuals versus time of sweet potato cubes with different sizes during freeze-drying with far-infrared treatment. (From Lin et al. 2005. *Journal of Food Engineering*, 68(2), 249–255. Copyright 2005. Reprinted with permission of Elsevier.)

TABLE 7.3
Coefficients of Page and Henderson and Pabis Models Obtained in the Freeze-Drying of Control and SIRFD Strawberry Slices

Weight Reduction (%)	Page Model (MR = exp(−ktn))			Henderson and Pabis Model (MR = a*exp(−kt))		
	k	n	R^2	a	k	R^2
Control	7.08E–2	1.000	0.828	1.076	0.105	0.952
30	9.545E–4	3.833	0.976	1.095	0.121	0.936
40	6.854E–3	2.737	0.981	1.088	0.130	0.955
50	5.912E–3	2.858	0.986	1.094	0.136	0.950

Source: Shih et al. 2008. *Transactions of the ASABE*, 51(1), 205–216. Copyright 2008. Reprinted with permission of American Society of Agricultural and Biological Engineers.

properties intermediate to those with freeze-drying and SHAFD. The production of some high-quality and crisper dried products of lower rehydration ratio with significantly reduced drying time compared to regular freeze-drying makes the combined IR and freeze-drying technology very attractive, and it may have widespread application in the industry. If, however, commercial equipment manufacturing is not addressed, the technology will remain an interesting technology for research purposes only but with limited application to the commercial food process.

ACKNOWLEDGMENT

The authors are extremely grateful to their co-workers Shih Connie and Dr. Elizabeth Atungulu, whose contributions of expertise and care, respectively, were invaluable throughout all the stages of the preparation of this chapter.

NOMENCLATURE

CDT   Catalytic drying technology
FD   Freeze-drying
FIR   Far-infrared radiation
IR   Infrared radiation
MIR   Medium-infrared radiation
NIR   Near-infrared radiation
SIRFD   Sequential infrared and freeze-drying
SHAFD   Sequential hot air and freeze-drying
SEM   Scanning electron microscopy
RMSE   Root mean squared error
w.b.   Wet basis
WR   Weight reduction

REFERENCES

Abe, T., and T. M. Afzal. 1997. Thin-layer infrared radiation drying of rough rice. *Journal of Agricultural Engineering Research*, 67, 289–297.

Alvarez, C. A., R. Aguerre, R. Gomez, S. Vidales, S. M. Alzamora, and L. N. Gerschenson. 1995. Air dehydration of strawberries: effects of blanching and osmotic pretreatments on the kinetics of moisture transport. *Journal of Food Engineering*, 25(2), 167–178.

Baker, C. G. J. 1997. *Industrial Drying of Foods*. New York: Chapman & Hall, pp. 229–263.

Barbosa-Canovas, B. V., and H. Vega-Mercado. 1996. In *Dehydration of Foods*. Freeze dehydration. New York: Chapman & Hall.

Baysal, T., F. Icier, S. Ersus, and H. Yildiz. 2003. Effects of microwave and infrared drying on the quality of carrot and garlic. *European Food Research and Technology*, 218, 68–73.

Cengel, Y. A. 1998. *Heat Transfer a Practical Approach*. Boston: McGraw-Hill.

Feng, H., and J. Tang. 1998. Microwave finish drying of diced apples in a spouted bed. *Journal of Food Science*, 63, 679–683.

Flink, J. 1977. Energy analysis in dehydration processes. *Food Technology*, 31, 77–84.

Hammami, C., and F. René. 1997. Determination of freeze-drying process variables for strawberries. *Journal of Food Engineering*, 32, 133–154.

Jamradloedluk, J., A. Nathakaranakule, S. Soponronnarit, and S. Prachayawarakorn. 2007. Influences of drying medium and temperature on drying kinetics and quality attributes of durian chip. *Journal of Food Engineering*, 78, 198–205.

Jones, P. 1992. Electromagnetic wave energy in drying processes. In Mujumdar, A.S. (Ed.), *Drying '92*. Elsevier Science Publisher B.V., pp. 114–136.

Ketelaars, A., W. Jomaa, J. Puigalli, and W. Coumans. 1992. Drying shrinkage and stress. In A. S. Mujumdar (Ed.), *Drying '92, Part A*, Amsterdam, The Netherlands: Elsevier, pp. 293–303.

Krokida, M. K., Z. B. Maroulis, and G. D. Saravacos. 2001a. The effect of the method of drying on the colour of dehydrated products. *International Journal of Food Science and Technology*, 36, 53–59.

Krokida, M. K., V. Oreopoulou, Z. B. Maroulis, and D. Marinos-Kouris. 2001b. Effect of pretreatment on viscoelastic behaviour of potato strips. *Journal of Food Engineering*, 50, 11–17.

Kumar, D. G. P., H. U. Hebbar, D. Sukumar, and M. N. Ramesh. 2005. Infrared and hot-air drying of onions. *Journal of Food Processing and Preservation*, 29, 132–150.

Lewicki, P. P. 2006. Design of hot air drying for better foods. *Food Science and Technology*, 17, 153–163.

Li, G., and Z. Ma. 2003. Vacuum freeze-drying process of strawberry. *Food and Machinery*, 3, 18–19.

Lin, Y., T. Lee, J. Tsen, and V. A. King. 2007. Dehydration of yam slices using FIR-assisted freeze drying. *Journal of Food Engineering*, 79(4), 1295–1301.

Lin, Y., J. Tsen, and V. An-Erl King. 2005. Effects of far-infrared radiation on the freeze-drying of sweet potato. *Journal of Food Engineering*, 68(2), 249–255.

Maskan, A., S. Kaya, and M. Maskan. 2002a. Effect of concentration and drying processes on color change of grape juice and leather (pestil). *Journal of Food Engineering*, 54, 75–80.

Maskan, A., S. Kaya, and M. Maskan. 2002b. Hot air and sun drying of grape leather (pestil). *Journal of Food Engineering*, 54, 81–88.

McMinn, W., and T. Magee. 1997. Physical characteristics of dehydrated potatoes: Part II. *Journal of Food Engineering*, 33(1), 49–55.

Mujumdar, A. S. 1995. *Handbook of Industrial Drying*, 2nd ed. New York: Marcel Dekker.

Nowak, D., and P. P. Lewicki. 2004. Infrared drying of apple slices. *Innovative Food Science and Emerging Technologies*. 5, 353–360.

Pedlington, S., and J. P. Ward. 1965. Histological examination of some air dried and freeze dried vegetables. *Proceedings of the First International Congress on Food Science and Technology*, 4, 55.

Pan, Z., Y. C. Shih, T. McHugh, and E. Hirschberg. 2008. Study of banana dehydration using sequential infrared radiation heating and freeze-drying. *LWT—Food Science and Technology,* 41(10), 1944–1951.

Richardson, P. 2001. *Thermal Technologies in Food Processing*. Boca Raton, FL: CRC Press.

Sakai, N., and T. Hanzawa. 1994. Applications and advances in far-infrared heating in Japan. *Trends in Food Science and Technology*, 5, 357–362.

Shih, C. 2006. Strawberry dehydration using sequential infrared radiation and freeze-drying method. Dissertation, Biological and Agricultural Engineering, University of California, Davis.

Shih, C., Z. Pan, T. McHugh, D. Wood, and E. Hirschberg. 2008. Sequential infrared radiation and freeze-drying method for producing crispy strawberries. *Transactions of the ASABE*, 51(1), 205–216.

Shishehgarha, F., J. Makhlouf, and C. Ratti. 2002. Freeze-drying characteristics of strawberries. *Drying Technology* 20(1), 131–145.

Singh, R. P., and D. R. Heldman. 1993. *Introduction to Food Engineering*, 2nd ed. San Diego: Academic Press.

Talburt, W. F., and O. Smith. 1967. *Potato Processing*. Westport, CT: AVI Publishing.

Tyree, M. T. 1970. The symplast concept: A general theory of symplastic transport according to the thermodynamics of irreversible processes. *Journal of Theoretical Biology*, 26, 181–214.

Vickers, Z. M., and M. C. Bourne. 1976. A psychoacoustical theory of crispness. *Journal of Food Sciences*, 4, 1158–1164.

8 Vacuum Infrared Drying

Chatchai Nimmol and Sakamon Devahastin

CONTENTS

8.1 Introduction 141
8.2 Vacuum IR Drying System.. 142
8.3 Vacuum IR Drying of Foods and Agricultural Materials 143
8.4 Advances in Vacuum IR Drying... 148
    8.4.1 Energy Consumption of the Process................................... 158
    8.4.2 Quality of Dried Products 159
8.5 Concluding Remarks 165
References.. 165

8.1 INTRODUCTION

During the past decade, infrared (IR) radiation has received more attention as a thermal energy source (or as an auxiliary energy source) for drying of many foods and agricultural materials. When IR radiation is applied to a drying process, the energy requirement of the process is generally reduced. This is because energy in the form of electromagnetic waves is absorbed directly by the product without loss to the environment, leading to considerable energy savings (Ginzburg, 1969; Sandu, 1986; Ratti and Mujumdar, 1995; Abe and Afzal, 1997, 1998; Afzal et al., 1999; Mongpraneet et al., 2002b). The dried product quality is also generally improved when compared with the quality of air-dried products. For example, lower losses of vitamin C, volatile components, and flavors (Niibori and Motoi, 1988), lower loss of β-carotene (Mogi et al., 1990), and better color retention (Itoh, 1986) have been reported.

Although IR radiation can accelerate a drying process, foods and agricultural materials, which are heat sensitive in nature, may be damaged or degraded if radiation intensity, which is the main factor influencing the product temperature, is not properly regulated. In order to speed up the drying process without affecting the dried product quality, the drying process should be performed at a lower temperature. Drying in vacuum (or vacuum drying) is one possible means to achieve this goal because in vacuum, water (moisture) evaporates at a lower temperature due to its lower boiling point; drying can thus take place at a lower temperature (lower than 100°C) compared with conventional atmospheric drying. Under such conditions, moisture migration within a material being dried also increases due to rapid increase in the internal vapor pressure, leading to a higher drying rate (Audebert and Temmar, 1997; Jomaa and Baixeras, 1997). Drying in vacuum may also lead to dried products with superior quality. This is again due to the lower drying temperature and lower level of oxidative reactions (Wu et al., 2007).

To combine the advantages of IR radiation and drying in vacuum, vacuum IR drying has recently been proposed and applied to an array of foods and agricultural materials with various degrees of success. In this chapter, information on vacuum infrared drying, namely, system configuration and its potential to dry various materials, especially heat-sensitive materials, is discussed. In addition, selected advances in vacuum IR drying, which is considered a novel technology, are also discussed.

8.2   VACUUM IR DRYING SYSTEM

A vacuum IR drying system generally consists of an insulated drying chamber, which is designed to withstand lower level of pressure; an IR radiator (or IR heater), which is used to supply thermal radiation to a drying product; a vacuum pump, which is used to maintain vacuum in the drying chamber; and a control system for the IR radiator. A typical schematic diagram of a vacuum IR dryer is given in Figure 8.1.

Typically, there are two methods that can be used to control the operation of an IR radiator. First, an IR radiator can be controlled through the surface temperature of a drying product or the temperature of a drying medium presented in a drying chamber (Nimmol et al., 2005, 2007a,b; Swasdisevi et al., 2007). This control strategy is generally applied to a product that may be damaged at higher temperature. Alternatively, an IR radiator can be controlled by adjusting the energy input to the radiator (Abe and Afzal, 1997, 1998; Afzal et al., 1999; Mongpraneet et al., 2002a,b). In this case, the output of the radiation intensity, which is the radiation power output per unit area, is regulated to the desired level.

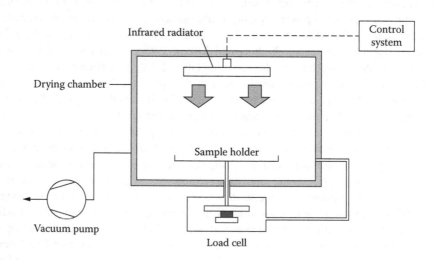

FIGURE 8.1   A schematic diagram of a vacuum IR dryer.

8.3   VACUUM IR DRYING OF FOODS
AND AGRICULTURAL MATERIALS

In this section, the focus is on the advantages of vacuum IR drying when applied to foods and agricultural materials. The effects of various parameters on the drying kinetics, energy consumption of the drying process, and dried product quality are briefly reviewed.

Nimmol et al. (2005) proposed a combined far-IR radiation and vacuum drying system (FIR-VACUUM) for heat sensitive materials; a carrot cube was used as test material. In their system, operation of a far-IR radiator was controlled by the surface temperature of the sample (at about 1 mm below the top surface of the sample). Drying experiments were conducted at the controlled surface temperatures of 60°C, 70°C, and 80°C and absolute chamber pressures of 7 and 10 kPa. Comparison was also made with similar sets of data obtained from low-pressure superheated steam drying (LPSSD) and vacuum drying experiments conducted in the same drying chamber (Devahastin et al., 2004).

Based on the experimental drying data, it was found that the drying time decreased with an increase in the controlled surface temperature and a decrease in the chamber pressure. In addition, the drying time of carrots subjected to FIR-VACUUM was less than that required by LPSSD and vacuum drying under all drying conditions although, in some cases, the differences were not much significant, especially when compared with the time required by vacuum drying. This is probably due to the use of rather thick carrot samples (10 × 10 × 10 mm). Since FIR could not penetrate very far though the sample, the effect of using FIR was not clear when applying it to a thick sample.

A comparison of the drying curves presented in Figure 8.2 reveals that carrots subjected to FIR-VACUUM at the same pressure lost their moisture at almost the same rates during the first 70 min of drying although their controlled surface temperatures were different. This is because carrots were still probably in the constant rate period and, hence, their surface temperature was always at a fixed value close to the boiling point of water at the corresponding dryer operating pressure (not at the boiling point, since far-IR radiator was present). Once the surface of carrots started to dry, the surface temperature of carrot began to rise, and the effect of the different levels of temperature became more obvious. During the later stage of drying, higher controlled surface temperatures resulted in higher rates of moisture reduction, as expected.

Figure 8.3 illustrates the temperature evolution of carrots subjected to FIR-VACUUM under different drying conditions. It can be seen from this figure that the temperature of the carrot cube, measured at its center, fell suddenly from its initial value within the first 5 min of the process. This is due to the rapid reduction of the chamber pressure, which led to some flash evaporation of the surface moisture. After this period, the temperature of carrots gradually rose to the predetermined controlled surface temperature.

It is also observed from Figure 8.3 that the rates of change of the carrot temperature within the first 90 min (in the case of drying at 10 kPa) and 75 min (in

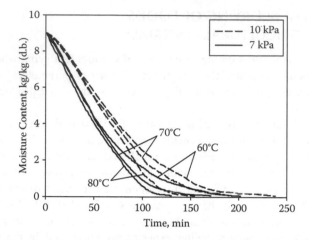

FIGURE 8.2 Drying curves of carrots undergoing FIR-VACUUM under different drying conditions. (From Nimmol et al. 2005. Drying kinetics and quality of carrot dried by far-infrared radiation under vacuum condition. Paper presented at the International Agricultural Engineering Conference, December 6–9, in Bangkok, Thailand. With permission.)

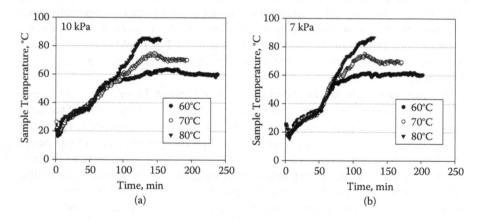

FIGURE 8.3 Temperature evolution of carrots undergoing FIR-VACUUM under different drying conditions. (a) Before; (b) after. (From Nimmol et al. 2005. Drying kinetics and quality of carrot dried by far-infrared radiation under vacuum condition. Paper presented at the International Agricultural Engineering Conference, December 6–9, in Bangkok, Thailand. With permission.)

the case of drying at 7 kPa) were nearly the same at any controlled surface temperature. After these initial periods, the temperature of the carrots rose to the predetermined controlled surface temperature. This is because the far-IR radiator was switched on continuously until the controlled surface temperature for each condition was reached. Therefore, the level of radiation experienced by the drying sample was more or less the same in all cases.

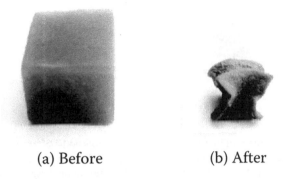

(a) Before   (b) After

FIGURE 8.4   Photographs of FIR-VACUUM-dried carrots both before and after drying. (From Nimmol et al. 2005. Drying kinetics and quality of carrot dried by far-infrared radiation under vacuum condition. Paper presented at the International Agricultural Engineering Conference, December 6–9, in Bangkok, Thailand. With permission.)

In terms of quality, it was observed that only with FIR-VACUUM was the shrinkage of carrots higher for longer drying duration, namely, at a higher operating pressure or at a lower controlled surface temperature. However, the effect of the drying pressure on the shrinkage of carrots at a higher controlled surface temperature was not significant. These results are similar to those reported by Mongpraneet et al. (2002b), who observed that the shrinkage of Welsh onion was higher at a lower level of radiation intensity.

Regarding the effects of different drying methods, it was found that the degrees of shrinkage of FIR-VACUUM, LPSSD, and vacuum-dried carrots were quite similar. However, the shrinkage pattern of carrots dried by FIR-VACUUM, which is a rapid drying process, was nonuniform when compared with those dried by LPSSD (see Figure 8.4). This is because the surface of the FIR-VACUUM-dried carrots became dry and rigid long before the center had dried out; the center dried and shrank much later than the outer surface did and pulled away from the rigid surface layers and caused nonuniform shrinkage (Devahastin et al., 2004).

In the case of rehydration ability, it was found that FIR-VACUUM-dried carrot had lower rehydration ability when dried for a shorter drying duration; for example, at a lower drying pressure or at a higher controlled surface temperature. This is because under these conditions, dense or rigid layers formed on the sample surface; these layers prevented adsorption of water during rehydration. Figure 8.5 shows the photographs of FIR-VACUUM-dried carrot cubes both before and after rehydration. In comparison with LPSSD and vacuum drying, the rehydration ability of FIR-VACUUM-dried carrot was much lower than that of LPSSD and vacuum-dried carrots obtained under similar conditions, especially under prolonged drying conditions. This is because, in the case of FIR-VACUUM, dense or rigid layers were formed rapidly compared with the cases of LPSSD and vacuum drying.

In terms of color, it was found that FIR-VACUUM-dried carrot was redder than the fresh one under all drying conditions. This is probably due to an increase in color

(a) Before   (b) After

FIGURE 8.5 Photographs of FIR-VACUUM-dried carrots both before and after rehydration. (From Nimmol et al. 2005. Drying kinetics and quality of carrot dried by far-infrared radiation under vacuum condition. Paper presented at the International Agricultural Engineering Conference, December 6–9, in Bangkok, Thailand. With permission.)

pigment concentration caused by the moisture removal. In the case of lightness, it was observed that FIR-VACUUM led to darker dried carrot when compared with the fresh one, under all drying conditions. However, the effects of the controlled surface temperature and the drying pressure on the redness and lightness of carrot were not significant in all cases. This is probably because the ranges of the controlled surface temperature and drying pressure tested were not large enough. The results obtained by FIR-VACUUM were significantly different from those of LPSSD and vacuum drying. LPSSD yielded dried carrots, which were redder and lighter than those obtained by FIR-VACUUM and vacuum drying.

Swasdisevi et al. (2007) dried banana slices with the same dryer setup as that used by Nimmol et al. (2005). Banana slices were dried at various absolute chamber pressures (5, 10, and 15 kPa), temperatures (50°C, 55°C, and 60°C), and sample thicknesses (2, 3, and 4 mm). Similar to the results reported by Nimmol et al. (2005), the rate of moisture reduction of bananas increased with an increase in the drying temperature and a decrease in the drying pressure. In addition, the rate of moisture reduction increased with a decrease in the sample thickness. This is because, in the case of the thinner sample, far-IR could easily penetrate into the sample, leading to a higher absorbed energy. The temperature evolution pattern of banana slices during FIR-VACUUM was also similar to that reported by Nimmol et al. (2005).

In terms of quality, it was observed that the lightness of the sample decreased with increasing surface temperature, while yellowness increased with increasing surface temperature due to enhanced browning during drying. However, pressure did not significantly affect the banana surface color. In the case of shrinkage, it was found that shrinkage increased with decreasing surface temperature. This is because drying at high temperatures caused the banana surface to become dry more rapidly. Since hardening occurred at the surface, it helped maintain the shape of the sample. In addition, shrinkage increased with an increase in the chamber pressure, because the increased pressure caused the banana surface to dry slowly. Regarding the texture of dried bananas, pressure and temperature did not significantly affect the hardness of bananas, but hardness increased with an increase in sample thickness.

Another version of FIR-VACUUM system was developed and tested by Mongpraneet et al. (2002a,b) using Welsh onion as a test material. Radiation

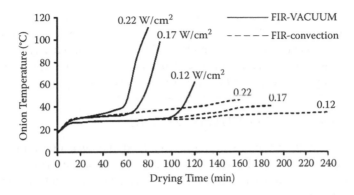

FIGURE 8.6 Temperature evolution of onions dried by FIR-VACUUM and FIR-convection at different radiation intensities. (From Mongpraneet et al. 2002b. *Transactions of the ASAE* 45:1529–2535. With permission.)

intensities in the range of 0.1 to 2.4 W/cm^2 were employed by regulating the power inputs to a far-IR radiator in the range of 40 to 100 W. The absolute chamber pressure of 2.5 kPa was set throughout the study. The results showed that the rate of moisture reduction of the sample decreased with an increase in the power input to the far-IR radiator. This is expected as higher power input implies higher radiation intensity. However, when a very low level of power input (40 W in this case) was applied, the required final moisture content of onion could not be achieved.

The temperature evolution of the onion undergoing FIR-VACUUM is depicted in Figure 8.6. It can be seen that the temperature of the sample rose rapidly during the early stage of the drying process. After this period, the temperature of the sample remained constant for a certain period of time. Longer periods of constant sample temperature and lower levels of the sample temperature were noted when drying was carried out at lower radiation intensities. After the period of constant sample temperature, the temperature of the sample rose again. It should be noted that when drying was performed at higher radiation intensity, the final temperature of the sample was higher. It is seen in Figure 8.6 that the temperature of the sample after the period of constant temperature in the case of FIR-VACUUM was much higher than that of the sample undergoing combined far-IR radiation and convection drying (FIR-convection drying). This may be the reason that Welsh onion dried by FIR-VACUUM required much less drying time than with FIR-convection drying.

Radiation intensity was found to be the main factor influencing the dried product quality. In the case of rehydration ability, it was found that rehydration ability of dried Welsh onion was lower when higher radiation intensity was used. This is because under this condition, dense layers on the surface were formed, leading to a decrease in rehydration capacity, as also noted in the case of FIR-VACUUM-dried carrot (Nimmol et al., 2005). This implies that lower radiation intensity should be adopted to obtain dried product with a higher degree of rehydration ability. However, the opposite phenomenon was observed in the case of FIR-convection drying.

In terms of shrinkage, it was observed that shrinkage of FIR-VACUUM-dried sample increased as the radiation intensity decreased. This is because at lower radiation intensity, moisture within the sample gradually and continuously moved to the surface, leading to a continuous collapse of the sample structure. The opposite phenomenon was again observed in the case of FIR-convection drying.

8.4   ADVANCES IN VACUUM IR DRYING

Although the use of vacuum IR drying to dry food and agricultural materials has several advantages, as noted earlier, it is possible to enhance the drying process further, both in terms of energy efficiency and dried product quality. Combination of vacuum IR drying with other advanced drying technologics, especially superheated stream drying (SSD), which involves the use of superheated steam as the drying medium, is among the most attractive alternatives. SSD has several advantages, especially in terms of the dried product quality, which is important to food industry (Devahastin and Suvarnakuta, 2004). However, SSD typically involves a high-temperature environment (higher than 100°C at atmospheric or higher pressures). Therefore, SSD is not suitable for heat-sensitive foods and agricultural materials (e.g., fruits and vegetables).

Recently, low-pressure (or subatmospheric pressure) superheated steam drying has been proposed and applied to many heat-sensitive foods and biomaterials (Devahastin and Suvarnakuta, 2008). Since superheated steam in the case of LPSSD can be produced at temperatures below 100°C due to the reduced pressure environment, quality degradation of products undergoing LPSSD could be alleviated. However, LPSSD is a rather slow drying process.

To help accelerate the LPSSD process, a combination of LPSSD and far-IR radiation was proposed by Nimmol et al. (2007a,b) as a novel drying technology for foods and biomaterials. It should be noted that this combined system is indeed a drying system combining the concept of SSD and vacuum IR drying.

A schematic diagram of the combined far-IR radiation and low-pressure superheated steam drying (FIR-LPSSD) system, which was developed from the LPSSD system of Devahastin et al. (2004), is shown in Figure 8.7. The operation of the far-IR radiator was controlled via the temperature of the drying medium measured at 30 mm above the sample surface. By using banana slices as a test material, drying experiments were performed at the temperatures of 70°C, 80°C, and 90°C and absolute chamber pressures of 7 and 10 kPa. Comparison was also made with similar sets of data obtained from LPSSD and vacuum-drying experiments conducted in the same drying chamber.

Figure 8.8 shows the drying curves of banana slices subjected to different drying methods under various conditions. In the case of FIR-LPSSD, it is seen from Figure 8.8a that, as expected, drying time decreased with an increase in drying temperature. In addition, drying time also decreased with a decrease in drying pressure, which is again as expected. It can also be seen that the rates of moisture reduction were more affected by the temperature than by the pressure when drying was performed at

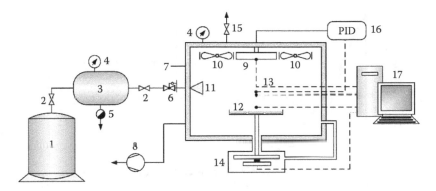

FIGURE 8.7 A schematic diagram of a combined far-IR radiation and low-pressure super-heated steam drying system. (1) Boiler, (2) steam valve, (3) steam reservoir, (4) pressure gauge, (5) steam trap, (6) steam regulator, (7) drying chamber, (8) vacuum pump, (9) far-infrared radiator, (10) electric fans, (11) steam inlet and distributor, (12) sample holder, (13) thermo-couples, (14) load cell, (15) vacuum break-up valve, (16) PID controller, (17) PC with data acquisition card. (From Nimmol et al. 2007b. *Journal of Food Engineering* 81:624–633. With permission.)

80°C and 90°C. This may be because temperature is the dominant factor influencing superheated steam's thermal properties, especially at higher drying temperatures.

It should be noted that, although the drying chamber was preheated via the use of the far-IR radiator during the first 5 min of the process, a small amount of steam condensation still occurred and could be observed over a short period; the results are not shown in Figure 8.8a, however. Moreover, when drying was performed at 70°C, the sample could not reach the required final moisture content during the first 250 min of drying even at the lowest drying pressure tested (7 kPa). This is because of an excessive amount of steam condensation in the drying chamber. This phenomenon was also observed in the case of drying at 10 kPa (data not shown in Figure 8.8a).

In the case of FIR-VACUUM (see Figure 8.8b), the phenomenon was similar to that of FIR-LPSSD; drying at higher temperatures and lower pressures required shorter drying time. Unlike FIR-LPSSD, however, the sample dried at the lowest drying temperature (70°C) could reach the required final moisture content. This is because there was no steam condensation. It was again found that the effect of chamber pressure was not obvious at all drying temperatures.

In the case of LPSSD, it was found from Figure 8.8c that the drying time was shorter when drying was carried out at higher temperatures and lower pressures. Similar to FIR-LPSSD, steam condensation occurred during the initial stage of the process under all drying conditions, and the sample could not reach the required final moisture content when drying was performed at 70°C.

As can be seen from Figures 8.8a and 8.8c, the differences in the rates of moisture reduction of the samples dried at 80°C and 90°C during FIR-LPSSD were more clearly observed compared with those of the samples dried by LPSSD at 80°C and 90°C. This is because extra heat generated by the use of far-IR during FIR-LPSSD

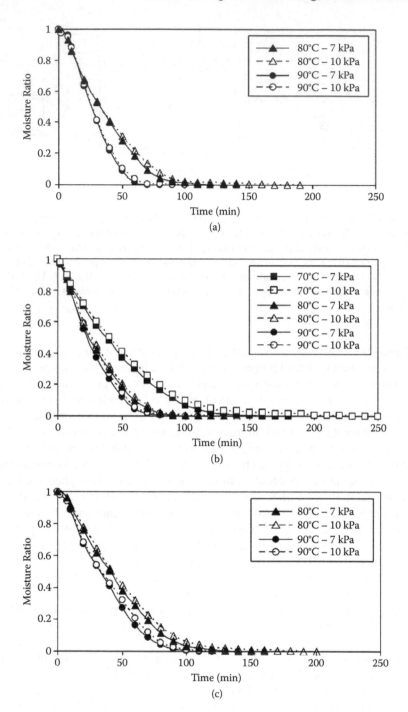

FIGURE 8.8 Drying curves of bananas undergoing (a) FIR-LPSSD, (b) FIR-VACUUM, and (c) LPSSD. (From Nimmol et al. 2007a. *Applied Thermal Engineering* 27:2483–2494. With permission.)

TABLE 8.1

Specific Energy Consumption and Total Steam Consumption of Different Drying Methods

Drying Method	Temp (°C)	Pressure (kPa)	Drying Time (min)	Energy Consumption (kWh)		Specific Energy Consumption (kWh/kg water)		Total Steam Consumption[a] (kg)
				E_{vacuum}	$E_{radiation/heater}$	SFC_{vacuum}	$SFC_{radiator/heater}$	
FIR-LPSSD	70	7	N/A	N/A	N/A	N/A	N/A	N/A
		10	N/A	N/A	N/A	N/A	N/A	N/A
	80	7	140	3.50	0.49	128.57	18.00	60.67
		10	190	4.75	0.53	174.49	19.47	82.33
	90	7	90	2.25	0.41	82.65	15.06	39.00
		10	100	2.50	0.42	91.84	15.43	43.33
FIR-VACUUM	70	7	185	4.63	0.20	169.90	7.35	—
		10	255	6.38	0.23	234.19	8.45	—
	80	7	130	3.25	0.25	119.39	9.18	—
		10	145	3.63	0.27	133.16	9.92	—
	90	7	110	2.75	0.33	101.02	12.12	—
		10	120	3.00	0.34	110.21	12.49	—
LPSSD	70	7	N/A	N/A	N/A	N/A	N/A	N/A
		10	N/A	N/A	N/A	N/A	N/A	N/A
	80	7	160	4.00	0.54	146.90	19.84	69.33
		10	200	5.00	0.71	183.68	26.08	86.67
	90	7	115	2.88	0.61	105.61	22.41	49.83
		10	135	3.38	0.68	123.98	24.98	58.50

Source: Data from Nimmol et al. 2007a. *Applied Thermal Engineering* 27:2483–2494.

Note: N/A implies that the required final moisture content was not obtainable.

[a] The flow rate of steam into the drying chamber was maintained at about 26 kg/h.

at 90°C caused much larger temperature gradients within the sample, resulting in a higher rate of moisture removal.

It is again illustrated in Figure 8.8 (and also Table 8.1) that the samples dried by FIR-LPSSD and FIR-VACUUM required less drying time than those dried by LPSSD under all drying conditions. This is because of the extra heating caused by the use of far-IR. In addition, it was observed that the samples dried by FIR-VACUUM required less drying time than those dried by FIR-LPSSD at lower drying temperatures (70°C and 80°C). However, FIR-LPSSD required a shorter drying time (higher drying rates) when drying was conducted at 90°C at all drying pressures tested. This is due to the sharp increase in the differences between the superheated steam temperature and the sample temperature in the case of FIR-LPSSD. On the other hand, the differences between the air temperature and the sample temperature in the case of FIR-VACUUM increased only slightly as the drying temperature increased. This suggests that the effective inversion temperature, as defined by Suvarnakuta et al.

(2005), calculated from the overall drying rates (the rates in both constant rate and falling rate periods) is somewhere between 80°C and 90°C.

Comparing with hot air drying (Maskan, 2000; Demirel and Turhan, 2003; Karim and Hawlader, 2005; Nguyen and Price, 2007), it is found that FIR-LPSSD and FIR-VACUUM require much less drying time (faster drying processes). This indicates that moisture transfer within the sample is more rapid during the process applying far-IR.

Figure 8.9 shows the plots of the drying rate versus moisture content of banana slices subjected to different drying methods under various conditions. It can be seen from this figure that drying rates increased with an increase in the drying temperature and a decrease in the drying pressure in all cases.

In the case of FIR-LPSSD and FIR-VACUUM (see Figures 8.9a and 8.9b), it was found that drying rates increased rapidly during an initial stage of drying (warming-up period). Since the radiation absorptivity of foods increases with an increase in moisture content (Sandu, 1986), thermal energy obtained from far-IR was more absorbed by banana slices during the initial stage of the process, when the sample moisture content was still high. However, Pathare and Sharma (2006), who investigated the use of IR-convective drying for onion, stated that this warming-up period was not observed. This is probably because forced convection in the drying chamber accelerated the cooling effect, which reduced the temperature of the IR radiator and of the sample.

After the warming-up period, the constant drying rate period was observed; the duration of this period varied with the drying techniques and conditions, however. In the case of FIR-LPSSD, the period of constant drying rate was very short when drying was performed at the highest temperature (90°C). For FIR-VACUUM, the period of constant drying rate was almost not observed under all drying conditions, even at the lowest drying temperature (70°C). The drying process after this period took place in the falling rate period, as indicated by a steady decrease in the drying rates.

The aforementioned results are contrary to those reported by many researchers such as Maskan (2000) and Nguyen and Price (2007), who reported that drying of banana slices using hot air was in the falling rate period without the period of constant drying rate. Doymaz (2007) also reported that no period of constant drying rate was observed when microwave was applied. Although many researchers have reported that the entire drying process takes place in the falling rate period during hot air drying of many food products, the period of constant drying rate was clearly observed by Demirel and Turhan (2003) during drying of banana slices using hot air at low drying temperature (40°C). This implies that drying conditions employed are an important factor influencing the drying rate.

The characteristics of FIR-LPSSD are indeed similar to those of LPSSD (Figure 8.9c). However, the periods of constant drying rate are more obviously observed in the case of LPSSD. It should be noted that under the same drying conditions, drying rates of FIR-LPSSD and FIR-VACUUM are higher than that of LPSSD, especially when compared with that of FIR-LPSSD at 90°C. This is again because of the influence of extra heating caused by the application of far-IR.

The changes of the moisture ratio and temperature of banana slices undergoing FIR-LPSSD and FIR-VACUUM under different conditions are shown in Figures 8.10

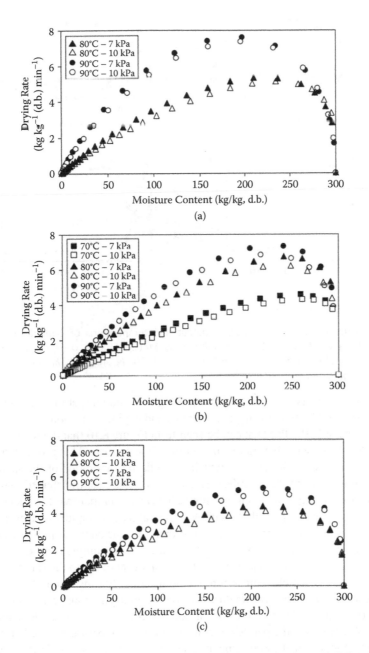

FIGURE 8.9 Drying rate curves of bananas undergoing (a) FIR-LPSSD, (b) FIR-VACUUM, and (c) LPSSD. (From Nimmol et al. 2007a. *Applied Thermal Engineering* 27:2483–2494. With permission.)

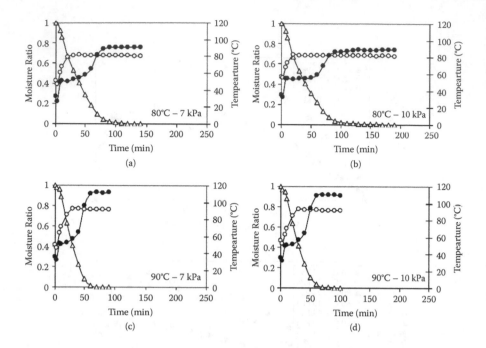

FIGURE 8.10 Changes in moisture content and temperature of bananas undergoing FIR-LPSSD under different drying conditions. △ Moisture ratio; ○ drying medium temperature; ● sample temperature. (From Nimmol et al. 2007a. *Applied Thermal Engineering* 27:2483–2494. With permission.)

and 8.11, respectively. As revealed by these figures, the temperature evolution patterns are affected by both the drying methods and drying conditions.

In the case of FIR-LPSSD, it can be seen from Figure 8.10 that the temperature of the sample fell suddenly within the first 3 min of the process. This is due to the rapid reduction of the chamber pressure, which led to flash evaporation of the sample surface moisture (Sun and Wang, 2006). This indicated that heat transferred to the samples by convection and radiation was not sufficient to compensate for latent heat of vaporization. After this period, the sample temperature rose rapidly to a level slightly higher than the boiling point of water corresponding to the chamber pressure and then remained unchanged at this level until the surface of the sample started to dry. The aforementioned phenomenon is different from that observed with the sample dried by LPSSD, where the sample temperature would be at the boiling point of water corresponding to the chamber pressure during the constant drying rate period (Devahastin et al., 2004). This is due to the effect of additional energy from far-IR radiation. It was also observed that the periods of constant sample temperature were longer when drying was conducted at lower temperatures and higher pressures. This observation was consistent with the period of constant drying rates (see Figure 8.10a). Once the sample surface began to dry at the end of the period of constant sample temperature, the sample temperature rose steadily until the level higher than the predetermined

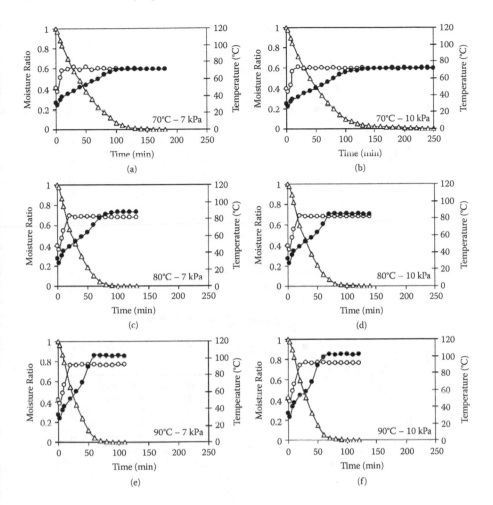

FIGURE 8.11 Changes in moisture content and temperature of bananas undergoing FIR-VACUUM under different drying conditions. Δ Moisture ratio; ○ drying medium temperature; ● sample temperature. (From Nimmol et al. 2007a. *Applied Thermal Engineering* 27:2483–2494. With permission.)

medium temperature was reached. It should be noted that an increase in the sample temperature beyond the predetermined medium temperature is due to the effect of radiation heat transfer obtained from the far-IR radiator. This phenomenon is different from that of the sample dried by LPSSD, where the temperature of the sample rose steadily and eventually approached the drying medium temperature.

In the case of FIR-VACUUM (see Figure 8.11), it was found that the sample temperature suddenly dropped during the initial period of the process. However, the sample temperature after this period steadily rose to a level higher than the predetermined medium temperature, as noted earlier in the case of FIR-LPSSD, without the period of constant sample temperature. This result was in agreement with that

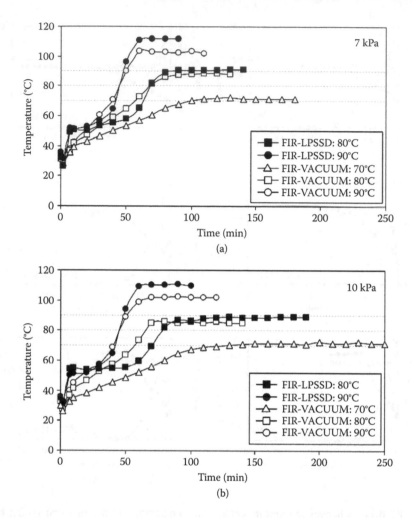

FIGURE 8.12 Comparison of changes in temperature of bananas undergoing FIR-LPSSD and FIR-VACUUM. (From Nimmol et al. 2007a. *Applied Thermal Engineering* 27:2483–2494. With permission.)

of drying rates; no period of constant drying rate was observed (see Figure 8.11b). After this period, the sample temperature remained almost unchanged as in the case of FIR-LPSSD.

To investigate the effect of radiation intensity, the temperature of the samples dried by FIR-LPSSD and FIR-VACUUM were again compared. It is seen in Figure 8.12 that, at the same predetermined medium temperature, the sample temperature during the later stage of FIR-LPSSD was higher than that in the case of FIR-VACUUM. This is because in the case of FIR-LPSSD the radiation intensity at the position (30 mm above the sample surface) of the thermocouple that was used to send the signal to a PID controller was less intense due to the higher absorptivity of superheated

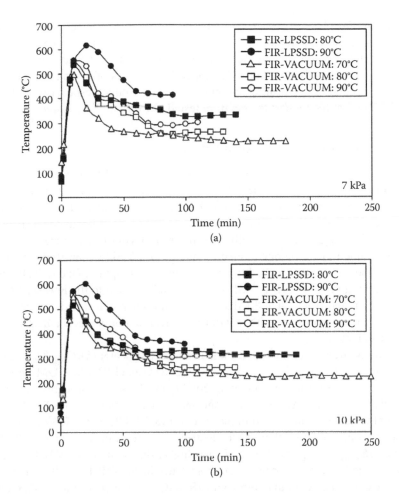

FIGURE 8.13 Evolution of surface temperature of far-IR radiator during FIR-LPSSD and FIR-VACUUM. (From Nimmol et al. 2007a. *Applied Thermal Engineering* 27:2483–2494. With permission.)

steam compared with that of air. The far-IR radiator was thus used more often during FIR-LPSSD to maintain the desired level of the drying medium temperature, leading to higher surface temperature of the far-IR radiator as shown in Figure 8.13. Consequently, the radiation intensity, which depends on the surface temperature of the far-IR radiator, experienced by the sample undergoing FIR-LPSSD was greater, and, hence, the sample temperature was higher.

It was also observed that in the case of FIR-VACUUM at the lowest temperature (70°C), the sample temperature during the later stage of the process was much closer to the drying medium temperature (not much higher than the drying medium temperature) compared with that of the samples dried at higher temperatures (see Figure 8.12). This is probably because when drying was performed at a lower temperature (70°C in this case), the radiation intensity was lower, as indicated by the lower surface temperature of the far-IR radiator. Since the sample temperature was not dictated by the drying medium temperature but by the radiation intensity, the very high sample temperatures developed in the cases of drying at higher temperatures resulted in overheating and burning of the product, especially in the case of FIR-LPSSD at the highest temperature tested (90°C).

8.4.1 ENERGY CONSUMPTION OF THE PROCESS

Table 8.1 lists the specific energy consumption (SEC) of different drying methods under different conditions; in the case where superheated steam was used, the total steam consumption is also given (Thomkapanish et al., 2007). In the case of the specific energy consumption of a vacuum pump (SEC_{vacuum}), it was found from this table that SEC_{vacuum} decreased with a decrease in drying time. Since the vacuum pump consumed the same rate of electric energy (1.5 kWh) under all drying conditions, the shorter drying time led to lower energy consumption. Due to its being the shortest drying process, FIR-LPSSD at 90°C and 7 kPa led to the lowest SEC_{vacuum}. It should be noted from Table 8.1 that the vacuum pump consumed a large amount of electric energy compared with that of the far-IR radiator or electric heater, as will be discussed later.

Regarding the SEC of the far-IR radiator or electric heater ($SEC_{radiator/heater}$), it was found that the values of $SEC_{radiator/heater}$ of FIR-LPSSD and FIR-VACUUM were lower than those of LPSSD under all drying conditions. This may probably be because FIR-LPSSD and FIR-VACUUM took shorter drying times than LPSSD; the power rating of the far-IR radiator used in the case of FIR-LPSSD and FIR-VACUUM was also lower compared with that of the electric heater used in the case of LPSSD. In addition, $SEC_{radiator/heater}$ of FIR-LPSSD was higher than that of FIR-VACUUM. This is because the far-IR radiator was used more often during FIR-LPSSD, as mentioned earlier. It should be noted that although FIR-VACUUM at 70°C and 7 kPa required a longer drying time (185 min) than almost all other drying methods and under all other conditions except FIR-VACUUM at 70°C and 10 kPa (255 min), $SEC_{radiator/heater}$ of FIR-VACUUM under this condition (70°C and 7 kPa) was the lowest. This is because the radiation intensity under this condition was much lower than that under the other conditions, as indicated by a lower surface temperature of the far-IR radiator (see Figure 8.13a), resulting in a lower electric energy consumption of the far-IR radiator. However, the electric energy consumption of the vacuum pump under this condition was still rather high.

In the case of drying with application of superheated steam to the drying chamber (FIR-LPSSD and LPSSD), it is seen from Table 8.1 that the total steam consumption increased with an increase in the drying time, as expected. Although superheated steam drying generally provides products with higher quality (Mujumdar, 2000; Devahastin and Suvarnakuta, 2004, 2008), the cost of superheated steam should also be taken into consideration.

8.4.2 Quality of Dried Products

Table 8.2 lists the color changes of banana slices dried by FIR-LPSSD and FIR-VACUUM in terms of the relative change in the lightness ($\Delta L/L_0$), relative change of the redness ($\Delta a/a_0$), and relative change of the yellowness ($\Delta b/b_0$). The measured initial values of the lightness (L_0), redness (a_0), and yellowness (b_0) of fresh banana, which was used as a test material, were in the ranges of 67.12 to 73.06, −3.75 to −4.49, and 18.42 to 22.20, respectively.

Since no oxygen was available in the case of FIR-LPSSD experiments, nonenzymatic browning reaction was considered to be the main cause of color changes of banana slices. Since in the case of FIR-VACUUM a very small amount of oxygen was also available in the drying chamber due to the very low chamber pressure, enzymatic browning was also considered to be negligible.

TABLE 8.2

Effects of Drying Methods, Drying Temperature, and Pressure on Color Changes of Dried Bananas

Drying Method	Temperature (°C)	Pressure (kPa)	$\Delta L/L_0$	$\Delta a/a_0$	$\Delta b/b_0$
FIR-LPSSD	70	7	N/A	N/A	N/A
		10	N/A	N/A	N/A
	80	7	−0.154 ± 0.026[e]	−0.804 ± 0.172[b]	0.255 ± 0.016[d]
		10	−0.220 ± 0.023[f]	−0.830 ± 0.144[b]	0.257 ± 0.017[d]
	90	7	−0.334 ± 0.022[h]	−2.424 ± 0.202[d]	0.335 ± 0.027[f]
		10	−0.318 ± 0.023[g]	−2.375 ± 0.217[d]	0.327 ± 0.020[ef]
FIR-VACUUM	70	7	−0.035 ± 0.005[ab]	−0.430 ± 0.036[a]	0.213 ± 0.026[ab]
		10	−0.034 ± 0.004[ab]	−0.429 ± 0.031[a]	0.210 ± 0.028[a]
	80	7	−0.048 ± 0.005[bc]	−0.513 ± 0.039[a]	0.253 ± 0.017[d]
		10	−0.046 ± 0.004[abc]	−0.524 ± 0.038[a]	0.251 ± 0.017[d]
	90	7	−0.144 ± 0.016[de]	−1.119 ± 0.195[c]	0.313 ± 0.020[e]
		10	−0.132 ± 0.015[d]	−1.168 ± 0.157[c]	0.309 ± 0.018[e]
LPSSD	70	7	N/A	N/A	N/A
		10	N/A	N/A	N/A
	80	7	−0.032 ± 0.005[a]	−0.416 ± 0.056[a]	0.212 ± 0.029[ab]
		10	−0.032 ± 0.002[a]	−0.422 ± 0.038[a]	0.210 ± 0.022[a]
	90	7	−0.053 ± 0.005[c]	−0.532 ± 0.052[b]	0.233 ± 0.034[bc]
		10	−0.049 ± 0.005[c]	−0.514 ± 0.051[b]	0.226 ± 0.020[ab]

Source: Data from Nimmol, C. 2007. Development of a combined low-pressure superheated steam and far-infrared radiation drying system. Ph.D. thesis, School of Energy, Environment and Materials, King Mongkut's University of Technology Thonburi, Bangkok, Thailand.

Note: N/A implies that the required final moisture content was not obtainable. Values in the same column with different superscripts mean that the values are significantly different ($p < 0.05$).

In the case of lightness, it was found that drying temperature and drying methods were the significant factors influencing the changes of lightness. Drying at higher temperatures yielded darker dried bananas. In addition, bananas dried by FIR-LPSSD were obviously darker than those dried by FIR-VACUUM under all drying conditions. This is because the temperature of bananas undergoing FIR-LPSSD increased more rapidly and stayed at higher level than in the case of FIR-VACUUM, as can be seen in Figure 8.12a. The high temperature led to a higher level of nonenzymatic browning reaction (Krokida et al., 2000; Chua et al., 2002). However, it was observed that the drying pressure had little effect on the lightness of dried banana samples.

The redness of dried banana slices was also significantly affected by both drying temperature and drying methods. It was observed that all dried bananas were redder than the fresh ones. Also, drying at higher temperatures yielded redder dried bananas than drying at lower temperatures. This is again because nonenzymatic browning reactions are accelerated by higher temperature. Considering the effects of the drying methods, it was found that FIR-LPSSD yielded dried bananas with larger increase in redness than the product obtained from FIR-VACUUM. This is because in the case of FIR-LPSSD, the sample temperature increased more rapidly and stayed at higher level than in the case of FIR-VACUUM, as described earlier. Consequently, bananas dried by FIR-LPSSD experienced more severe conditions for a longer period of time than the sample undergoing FIR-VACUUM.

In the case of yellowness, it was found that banana slices dried at higher temperatures tended to have higher values of yellowness. The effects of chamber pressure and drying methods on the changes of yellowness were not significant, however.

It is also seen in Table 8.2 that all dried bananas had higher values of color changes compared with those undergoing LPSSD alone, especially in the case of lightness and redness. This is because banana slices undergoing FIR-LPSSD and FIR-VACUUM were subjected to higher temperature for longer period of time, especially during the later stage of the process, as noted earlier.

Table 8.3 shows the results of shrinkage (in terms of the area shrinkage; volumetric shrinkage was not calculated as bananas were reported to shrink mainly only in the radial direction) and rehydration behavior of banana slices dried by different methods and under different conditions. In the case of FIR-LPSSD, it was found that the area shrinkage of banana slices dried at 80°C was significantly higher than those dried at 90°C. This is probably because case-hardening (rigid layers) on the sample perimeter, which retarded shrinkage (area change) of the sample, occurred least under this condition. This phenomenon is in agreement with that reported by Mongpraneet et al. (2002b), who observed that shrinkage of Welsh onion was higher at a lower level of radiation intensity and, hence, lower sample temperature. It was found, however, that the drying methods and chamber pressure did not significantly affect the degree of shrinkage at the same drying temperature (80°C and 90°C). Comparing these results with those from the product obtained by LPSSD, it was found that shrinkage of banana slices dried by far-IR-assisted processes was less. This is because case-hardening developed faster during FIR-LPSSD and FIR-VACUUM due to higher sample temperature. However, statistical analysis showed that the effects of drying temperature and drying techniques on shrinkage were not significant. It should be noted, however, that case-hardening, even on the slice perimeter, is sometimes not

TABLE 8.3

Effects of Drying Methods, Drying Temperature, and Pressure on Shrinkage and Rehydration Ratio of Dried Bananas

Drying Method	Temperature (°C)	Pressure (kPa)	Area Shrinkage (%)	Rehydration Ratio
FIR-LPSSD	70	7	N/A	N/A
		10	N/A	N/A
	80	7	18.38 ± 1.34[b]	1.55 ± 0.04[c]
		10	18.81 ± 1.61[b]	1.54 ± 0.04[c]
	90	7	15.20 ± 1.01[a]	1.54 ± 0.03[c]
		10	14.14 ± 1.20[a]	1.52 ± 0.04[c]
FIR-VACUUM	70	7	18.07 ± 1.40[b]	1.31 ± 0.06[a]
		10	18.17 ± 0.86[b]	1.31 ± 0.07[a]
	80	7	18.04 ± 1.74[b]	1.39 ± 0.07[b]
		10	18.80 ± 1.23[h]	1.40 ± 0.04[b]
	90	7	14.50 ± 1.52[a]	1.54 ± 0.06[c]
		10	14.70 ± 1.07[a]	1.54 ± 0.07[c]
LPSSD	70	7	N/A	N/A
		10	N/A	N/A
	80	7	18.74 ± 0.89[b]	1.36 ± 0.06[ab]
		10	19.14 ± 1.06[b]	1.35 ± 0.09[ab]
	90	7	14.90 ± 0.75[a]	1.33 + 0.05[ab]
		10	14.10 ± 1.16[a]	1.35 ± 0.07[ab]

Source: Data from Nimmol, C. 2007. Development of a combined low-pressure superheated steam and far-infrared radiation drying system. Ph.D. thesis, School of Energy, Environment and Materials, King Mongkut's University of Technology Thonburi, Bangkok, Thailand.

Note: N/A implies that the required final moisture content was not obtainable. Values in the same column with different superscripts mean that the values are significantly different ($p < 0.05$).

desired, since it would affect the overall texture (not hardness and crispness values as measured instrumentally but overall sensory characteristics of the whole slice) and appearance of the dried product.

Regarding rehydration behavior, it can be clearly seen from Table 8.3 that banana slices dried at higher temperatures had higher rehydration ability than those dried at lower temperatures; the effect was more pronounced in the case of FIR-VACUUM. This is because higher drying temperatures led to dried banana slices with more porous structure, thus facilitating rehydration ability. However, in general, banana slices dried by FIR-LPSSD had higher rehydration ability than those dried by FIR-VACUUM. This is because the temperature of bananas dried by FIR-LPSSD rose more rapidly to the level close to the boiling temperature (as can be seen in Figure 8.10). Consequently, moisture in bananas rapidly boiled, leading to rigorous

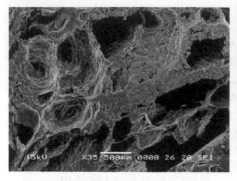

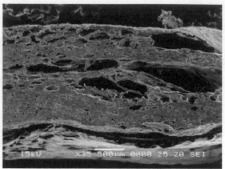

(a) FIR-LPSSD at 80°C and 7 kPa   (b) FIR-VACUUM at 80°C and 7 kPa

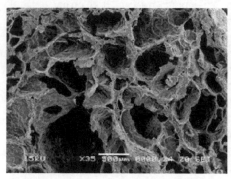

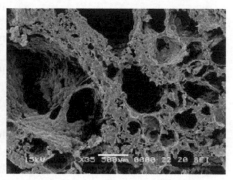

(c) FIR-LPSSD at 90°C and 7 kPa   (d) FIR-VACUUM at 90°C and 7 kPa

FIGURE 8.14 SEM photographs showing cross sections of dried bananas. (From Nimmol et al. 2007b. *Journal of Food Engineering* 81:624–633. With permission.)

evolution of steam within the samples. Therefore, larger and more pores were developed compared with the samples dried by FIR-VACUUM. However, the rehydration abilities of banana slices dried by FIR-LPSSD and FIR-VACUUM were not significantly different in the case of drying at 90°C. This is probably because moisture within the sample dried by FIR-VACUUM boiled as vigorously as in the case of FIR-LPSSD at this higher temperature. It was also observed that the rehydration abilities of FIR-LPSSD- and FIR-VACUUM-dried samples were significantly higher than that of the LPSSD dried. This is because of the development of larger and more pores within the samples during FIR-LPSSD and FIR-VACUUM.

The preceding analysis of results could be confirmed by scanning electron microscopy (SEM), selected results of which are shown in Figure 8.14, which illustrates cross sections of banana slices dried by FIR-LPSSD and FIR-VACUUM under some selected conditions. It is clearly seen from these figures that when drying was performed at lower temperature (80°C in this case), banana slices dried by FIR-LPSSD (see Figure 8.14a) had larger and more pores compared with those dried by FIR-VACUUM (see Figure 8.14b). This may be why banana slices dried

TABLE 8.4
Effects of Drying Methods, Drying Temperature, and Pressure on Maximum Force (Hardness) and Number of Peaks (Crispness) of Dried Bananas

Drying Method	Temperature (°C)	Pressure (kPa)	Maximum Force (N)	Number of Peaks
FIR-LPSSD	70	7	N/A	N/A
		10	N/A	N/A
	80	7	17.09 ± 3.15^a	37 ± 3^e
		10	17.30 ± 3.60^a	36 ± 4^{de}
	90	7	16.39 ± 3.57^a	38 ± 4^e
		10	16.89 ± 4.58^a	38 ± 5^e
FIR-VACUUM	70	7	18.44 ± 3.80^a	22 ± 4^a
		10	19.12 ± 4.07^a	21 ± 5^a
	80	7	19.95 ± 3.55^a	25 ± 5^b
		10	18.16 ± 4.51^a	26 ± 5^b
	90	7	16.72 ± 3.19^a	36 ± 3^e
		10	17.81 ± 3.63^a	36 ± 4^{de}
LPSSD	70	7	N/A	N/A
		10	N/A	N/A
	80	7	19.37 ± 3.57^a	28 ± 1^b
		10	19.57 ± 4.03^a	27 ± 1^b
	90	7	19.05 ± 3.86^a	31 ± 3^c
		10	19.27 ± 3.66^a	32 ± 3^{cd}

Source: Data from Nimmol, C. 2007. Development of a combined low-pressure superheated steam and far-infrared radiation drying system. Ph.D. thesis, School of Energy, Environment and Materials, King Mongkut's University of Technology Thonburi, Bangkok, Thailand.

Note: N/A implies that the required final moisture content was not obtainable. Values in the same column with different superscripts mean that the values are significantly different ($p < 0.05$).

by FIR-LPSSD had higher rehydration ability than those dried by FIR-VACUUM when drying was performed at 80°C. However, the structures of banana slices dried by both methods were similar in the case of drying at 90°C (see Figures 8.14c and 8.14d). Consequently, as expected, the rehydration abilities of bananas dried by FIR-LPSSD and FIR-VACUUM were not significantly different.

Table 8.4 shows the results of texture of dried banana slices in terms of the maximum force (hardness) and the number of peaks in the force–deformation curve (crispness). In the case of hardness, it was found that banana slices dried by FIR-VACUUM were harder than those dried by FIR-LPSSD. This is because FIR-VACUUM, especially at lower drying temperature (80°C in this case), yielded dried banana slices with more dense structure (smaller and less pores), as can be seen in Figures 8.14a and 8.14b. However, statistical analysis showed that the effects of drying temperature and chamber pressure as well as drying methods on the hardness were not significant. It should be noted that, at the same drying temperature, the hardness of dried banana slices was lower than that of the sample dried by LPSSD. This is probably because FIR-LPSSD and FIR-VACUUM yielded dried banana slices with more porous structure than that of sample dried by LPSSD.

In terms of crispness, it can be seen again from Table 8.4 that FIR-LPSSD yielded dried banana slices with a larger number of peaks (hence indicating that the product was crispier) compared with FIR-VACUUM, especially at 80°C. This is due to the larger and more pores that occurred during FIR-LPSSD. However, the effects of drying methods on the number of peaks were not significant when drying was performed at 90°C for the reasons mentioned earlier. It was found that FIR-LPSSD and FIR-VACUUM gave dried banana slices with slightly larger number of peaks compared with those dried only by LPSSD. This is again due to the larger and more pores developed during FIR-LPSSD and FIR-VACUUM.

The results of hardness and crispness of dried banana slices were also compared with those of commercially available banana chips (Fruit King™), which had the values of the maximum force and the number of peaks of 55.72 ± 5.48 N and 16 ± 3, respectively. It was found that banana slices dried by far-IR-assisted techniques had lower values of maximum force and larger number of peaks than those of the commercially available banana chips.

The total porosity of banana slices dried by different drying techniques (both with and without the use of far-IR) was also analyzed using x-ray microtomography coupled with image analysis (Léonard et al., 2008). The porosity values shown in Table 8.5 indicate that an increase in the drying temperature generally led to an increase in the porosity of the samples. On the other hand, at the same drying

TABLE 8.5

Total Porosity of Bananas Undergoing Various Drying Methods under Different Drying Conditions

Drying Method	Drying Temperature (°C)	Porosity
LPSSD	80	0.42 ± 0.05
	90	0.53 ± 0.06
FIR-LPSSD	80	0.55 ± 0.06
	90	0.70 ± 0.08
VACUUM	80	0.54 ± 0.05
	90	0.46 ± 0.05
FIR-VACUUM	80	0.57 ± 0.06
	90	0.63 ± 0.07

Source: Data from Léonard et al. 2008. *Journal of Food Engineering* 85:154–162.

temperature, the use of far-IR clearly resulted in an increase of the sample porosity. For example, at 90°C a relative augmentation of about 32% and 37% was observed in the case of LPSSD and VACUUM, respectively, when far-IR was applied.

It should be noted from Table 8.5 that when drying was performed at a higher temperature (90°C), the porosity value of the FIR-LPSSD sample was higher than that of the FIR-VACUUM sample. This is due to a rapid increase of the sample temperature during the initial stage of FIR-LPSSD, resulting in rigorous boiling of moisture within the sample. However, the results were opposite when drying was performed at a lower temperature (80°C). This may probably be due to the effect of an inversion phenomenon (Suvarnakuta et al., 2005), which happened somewhere between 80°C and 90°C in this case (Nimmol et al., 2007a,b; Thomkapanish et al., 2007). This effect could also be viewed from the evolution of the sample temperature. Although a rapid increase of the sample temperature during an initial stage was also observed in the cases of FIR-LPSSD at 80°C, the period of constant sample temperature occurring afterward was clearly found to be longer than that at 90°C, as noted earlier. Consequently, moisture within the sample had less chance to boil vigorously, leading to a lower degree of porosity.

8.5 CONCLUDING REMARKS

Vacuum drying along with a proper use of far-IR radiation could improve not only the energy efficiency of the process but also the dried product quality when compared with a conventional drying process. Besides the basic vacuum drying system, this chapter also reviews some advanced developments of vacuum IR drying, that is, a novel drying system combining the advantages of vacuum IR drying and superheated steam drying. For the development of a commercial-scale FIR-LPSSD where a large amount of dried product is to be produced, evaluation of an appropriate position, size, and power rating of a far-IR radiator must be carefully made. In addition, multiple-point steam injection system should be utilized to allow uniform steam distribution within the chamber.

REFERENCES

Abe, T., and Afzal, T.M. 1997. Thin-layer infrared radiation drying of rough rice. *Journal of Agricultural Engineering Research* 67:289–297.

Abe, T., and Afzal, T.M. 1998. Diffusion in potato during far infrared radiation drying. *Journal of Food Engineering* 37:353–365.

Afzal, T.M., Abe, T., and Hikida, Y. 1999. Energy and quality aspects during combined FIR-convection drying of barley. *Journal of Food Engineering* 42:117–182.

Audebert, P., and Temmar, A. 1997. Vacuum drying of Oakwood: Moisture strains and drying process. *Drying Technology* 15:2281–2302.

Chua, K.J., Hawlader, M.N.A., Chou, S.K., and Ho, J.C. 2002. On the study of time-varying temperature drying-effect on drying kinetics and product quality. *Drying Technology* 20:1559–1577.

Demirel, D., and Turhan, M. 2003. Air-drying behavior of dwarf Cavendish and Gros Michel banana slices. *Journal of Food Engineering* 59:1–11.

Devahastin, S., and Suvarnakuta, P. 2004. Superheated steam drying of food products. In *Dehydration of Products of Biological Origin*, Ed. Mujumdar, A.S., 493–512. Enfield: Science Publishers.

Devahastin, S., and Suvarnakuta, P. 2008. Low-pressure superheated steam drying of food products. In *Drying Technologies in Food Processing,* Eds. Chen, X.D. and Mujumdar, A.S., 160–189. West Sussex: Wiley-Blackwell.

Devahastin, S., Suvarnakuta, P., Soponronnarit, S., and Mujumdar, A.S. 2004. A comparative study of low-pressure superheated steam and vacuum drying of a heat-sensitive material. *Drying Technology* 22:1845–1867.

Doymaz, I. 2007. The kinetics of forced convective air-drying of pumpkin slices. *Journal of Food Engineering* 79:243–248.

Ginzburg, A.S., 1969. Application of Infrared Radiation in Food Processing, Chemical and Process Engineering Series. London: Leonard Hill.

Itoh, K. 1986. Drying of vegetable by far infrared radiation. *Shokuhin Kikai Souchi* 23:45–53.

Jomaa, W., and Baixeras, O. 1997. Discontinuous vacuum drying of Oakwood: Modelling and experimental investigations. *Drying Technology* 15: 2129–2144.

Karim, M.A., and Hawlader, M.N.A. 2005. Drying characteristics of banana: Theoretical modeling and experimental validation. *Journal of Food Engineering* 70:35–45.

Krokida, M.K., Karathanos, V.T., and Maroulis, Z.B. 2000. Effect of osmotic dehydration on color and sorption characteristics of apple and banana. *Drying Technology* 18:937–950.

Léonard, A., Blacher, S., Nimmol, C., and Devahastin, S. 2008. Effect of far infrared radiation assisted drying on microstructure of banana slices: An illustrative use of X-ray microtomography in microstructural evaluation of a food product. *Journal of Food Engineering* 85:154–162.

Maskan, M. 2000. Microwave/air and microwave finish drying of banana. *Journal of Food Engineering* 44:71–78.

Mogi, T., Ito, T., and Yamamoto, I. 1990. Research and development of rapid drying technique for vegetables by use of heating combination with far infrared rays and infrared rays. *Technical Report of Japan Food Industry Center* 16:17–32.

Mongpraneet, S., Abe, T., and Tsurusaki, T. 2002a. Accelerated drying of welsh onion by far infrared radiation under vacuum conditions. *Journal of Food Engineering* 55:147–156.

Mongpraneet, S., Abe, T., and Tsurusaki, T. 2002b. Far infrared-vacuum and convection drying of welsh onion. *Transactions of the ASAE* 45:1529–2535.

Mujumdar, A.S. 2000. Superheated steam drying—Technology for the future. In *Mujumdar's Practical Guide to Industrial Drying*, Ed. Devahastin, S., 115–138. Brossard: Exergex.

Nguyen, M.H., and Price W.E. 2007. Air-drying of banana: Influence of experimental parameters, slab thickness, banana maturity and harvesting season. *Journal of Food Engineering* 79:200–207.

Niibori, F., and Motoi, S. 1988. Evaporation of vegetables used by infrared rays. *Food Processing* 23:38–42.

Nimmol, C. 2007. Development of a combined low-pressure superheated steam and far-infrared radiation drying system. Ph.D. thesis, School of Energy, Environment and Materials, King Mongkut's University of Technology Thonburi, Bangkok, Thailand.

Nimmol, C., Devahastin, S., Swasdisevi, T., and Soponronnarit, S. 2005. Drying kinetics and quality of carrot dried by far-infrared radiation under vacuum condition. Paper presented at the International Agricultural Engineering Conference, December 6–9, in Bangkok, Thailand.

Nimmol, C., Devahastin, S., Swasdisevi, T., and Soponronnarit, S. 2007a. Drying and heat transfer behavior of banana undergoing combined low-pressure superheated steam and far-infrared radiation drying. *Applied Thermal Engineering* 27:2483–2494.

Nimmol, C., Devahastin, S., Swasdisevi, T., and Soponronnarit, S. 2007b. Drying of banana slices using combined low-pressure superheated steam and far-infrared radiation. *Journal of Food Engineering* 81:624–633.

Pathare, P.B., and Sharma, G.P. 2006. Effective moisture diffusivity of onion slices undergoing infrared convective drying. *Biosystems Engineering* 93:285–291.

Ratti, C., and Mujumdar, A.S. 1995. Infrared drying. In *Handbook of Industrial Drying: Volume 1* (2nd ed.), Ed. Mujumdar, A.S., 567–588. New York: Marcel Dekker.

Sandu, C. 1986. Infrared radiative drying in food engineering: A process analysis. *Biotechnology Progress* 2:109–119.

Sun, D.W., and Wang, L.J. 2006. Development of a mathematical model for vacuum cooling of cooked meats. *Journal of Food Engineering* 77:379–385.

Suvarnakuta, P., Devahastin, S., Soponronnarit, S., and Mujumdar, A.S. 2005. Drying kinetics and inversion temperature in a low-pressure superheated steam-drying system. *Industrial and Engineering Chemistry Research* 44:1934–1941.

Swasdisevi, T., Devahastin, S., Ngamchum, R., and Soponronnarit, S. 2007. Optimization of a drying process using infrared-vacuum drying of Cavendish banana slices. *Songklanakarin Journal of Science and Technology* 29:809–816.

Thomkapanish, O., Suvarnakuta, P., and Devahastin, S. 2007. Study of intermittent low-pressure superheated steam and vacuum drying of a heat-sensitive material. *Drying Technology* 25:205–223.

Wu, L., Orikasa, T., Ogawa, Y., and Tagawa, A. 2007. Vacuum drying characteristics of eggplants. *Journal of Food Engineering* 83:422–429.

9 Infrared Dry Blanching

Zhongli Pan and Griffiths Gregory Atungulu

CONTENTS

9.1 Introduction 169
9.2 Overview of Blanching Methods 170
9.3 Infrared Dry Blanching 171
    9.3.1 Absorption Band Characteristics of Chemical Groups of Foods 171
    9.3.2 Process Parameters 171
9.4 Characterization of Processing Conditions 172
    9.4.1 Case Study of Apple Slice Processing 172
        9.4.1.1 Continuous Heating Mode 172
        9.4.1.2 Intermittent Heating Mode 182
        9.4.1.3 Comparison between Continuous and Intermittent
                Heating Modes 191
9.5 Advances of Infrared Dry Blanching 192
    9.5.1 Equipment Description 192
    9.5.2 Performance of the Pilot-Scale Infrared Dry Blancher 193
9.6 Conclusion 198
Acknowledgment 198
Nomenclature 198
References 199

9.1 INTRODUCTION

For most processed fruits and vegetables, blanching is essential to inactivate the enzymes responsible for quality deterioration of fruits and vegetables in storage. Blanching also serves the purposes of microbial population reduction, color stabilization, and facilitation for further processing and handling. A newer method of simultaneous infrared (IR) dry blanching and dehydration (SIRDBD) utilizes efficient IR heating to combine blanching and dehydration into a one-step process that is simpler and more energy efficient than conventional methods. During IR heating, IR radiation energy with specific wavelengths penetrates into product and directly heats water or desired components to achieve the purposes of blanching and drying. Water absorbs heat energy very efficiently in the range of medium-infrared (MIR) and far-infrared (FIR) wavelengths with peak wavelengths at 3, 4.7, and 6 μm (Ginzberg, 1969). Since the MIR and FIR energy does not heat the air and medium, the energy transfer is highly efficient.

In this chapter, the applications of IR energy for blanching are reviewed and discussed in detail. The uses and advantages of IR dry blanching for fruits and vegetables are discussed with specific case study references. The development status and merit of SIRDBD as a dry blanching process that uses no water or steam (Pan and McHugh, 2004), unlike conventional methods, are clearly demonstrated for this novel and promising technology in the fruits and vegetable industry.

9.2   OVERVIEW OF BLANCHING METHODS

A variety of the blanching methods that have been developed and studied for fruits and vegetables include hot water blanching, steam blanching, electromagnetic heating, ohmic heating, surface methodology, and high hydrostatic pressure. Hot water and steam blanching are widely used. These conventional methods of blanching have many drawbacks such as low energy efficiency, long processing time, quality deterioration, and environmental problems. However, they are typically used in the food industry due to their high feasibility and low initial capital cost. However, the blanching process is an energy-intensive unit operation and can account for as high as one-third of the total energy required for processing fruits and vegetables (Ramaswamy, 2005). Water blanching is generally the immersion type or a spray on a conveyor, while steam blanching involves belt or chain conveyors that move the product through a tunnel containing live steam. Most of the water-soluble nutrients, such as ascorbic acid, can be lost in water blanching. The prevalence of water blanchers in the industry necessitates the comparison of different types of blanching for their energy utilization. On the basis of a theoretical requirement of 134 kg of steam per 10^3 kg of raw vegetables, the energy efficiency of a steam blancher was estimated at 5%, a hydrostatic steam blancher at 27%, and a water blancher at 60% (Bomben, 1977). One of the modifications of water blanching is low-temperature long-time (LTLT) blanching, often conducted at around 55°C to 70°C, to improve the firmness of the products (Alvarez and Canet, 1999; Garcia-Reverter et al., 1994). In contrast, high-temperature short-time (HTST) blanching also attracted research interest in the 1980s due to its reduced energy consumption and uniform final products (Drake and Carmichael, 1986; Sanchez-Pineda-Infantas et al., 1994). Stepwise blanching, a mixed two-step process, was developed to combine the advantages of both LTLT and HTST. The first step of LTLT blanching is followed by water cooling and conventional blanching of HTST (SanJuan et al., 2001).

Microwave blanching showed many advantages over conventional blanching, such as reduced processing time, energy and water usage, improved product quality, and retention of water-soluble nutrients. However, uneven heating is the major problem (Ponne et al., 1994). It is hard to control the heating uniformity of the process since the energy distribution within food varies significantly (Ramesh et al., 2002). This attribute limits the industrial application of microwave blanching. The solution is to combine microwave heating with conventional blanching methods to increase the energy efficiency of the process and avoid uneven temperature distribution in products (Devece et al., 1999). Other less frequently used blanching methods include ohmic heating, hot gas blanching, surface methodology, and high-pressure blanching. They are studied mostly in academia but show little promise in commercialization due to the complexity of the processes and/or high capital cost of the equipment at present.

Some studies have been done to investigate the efficacy of IR heating for blanching purposes (Gómez et al., 2004). IR has been successfully used to inactivate enzymes responsible for the development of off-flavors in peas prior to the freezing process (Van Zuilichen et al., 1986) as well as other enzymes (Sawai et al., 2003). Gómez et al. (2004) suggested that a heat treatment as short as 20 s at 100°C may be accompanied by cell damage in as much as 70% of the tissue. The literature indicated that treatment of carrot slices with FIR for a few seconds damaged cells only in the first half-millimeter from the surface and preserved most of the texture characteristic of the raw tissue. In addition to preserving the relative electrolyte leakage, IR treatment has the advantage of high retention of vitamins and minerals (Belyaev et al., 1985). Technological alternatives for inactivating enzymes near the surface of the product and at the same time to preserve tissue integrity are crucial.

9.3   INFRARED DRY BLANCHING

9.3.1   ABSORPTION BAND CHARACTERISTICS OF CHEMICAL GROUPS OF FOODS

When IR radiation energy impinges upon a food surface, IR is absorbed by organic materials at discrete frequencies corresponding to intramolecular transitions between energy levels. The mechanisms for energy absorption are associated with the wavelength range of the incident radiation energy (Rosenthal, 1992; Sakai and Hanazawa, 1994; Sandu, 1986). Table 9.1 shows the IR absorption band characteristics of chemical groups relevant to the heating of foods.

During IR heating, the IR absorption properties of foodstuffs depend mainly on three factors: water content, thickness, and physicochemical nature of the product (Sandu, 1986; Krust et al., 1962; Ginzburg, 1969).

9.3.2   PROCESS PARAMETERS

A simultaneous IR dry blanching and dehydration approach for blanching and partially dehydrating fruits and vegetables results in high product quality (Pan and McHugh, 2004). The technology utilizes catalytic IR radiation (CIR) energy to inactivate enzymes in fruits and vegetables as well as to remove a certain amount of

---

TABLE 9.1

The Infrared Absorption Band Characteristics of Chemical Groups Relevant to the Heating of Food

Chemical Group	Absorption Wavelength (µm)	Relevant Food Component
Hydroxyl group (O–H)	2.7–3.3	Water, carbohydrates
Aliphatic carbon-hydrogen bond	3.25–3.7	Fats, carbohydrates, proteins
Carbonyl group (C=O) (ester)	5.71–5.76	Fats
Carbonyl group (C=O) (amide)	5.92	Proteins
Nitrogen-hydrogen group (–NH–)	2.83–3.33	Proteins
Carbon-carbon double bond (C=C)	4.44–4.76	Unsaturated fats

moisture at the same time. The combined one-step process of SIRDBD leads to a simpler process and higher energy and process efficiency than the conventional two-step process. The mechanism of CIR has been illustrated in detail in recent publications (Zhu and Pan, 2009). Primarily, the platinum catalyst inside the CIR emitter accelerates the oxidation of natural or propane gas, resulting in medium- and far-IR energy with peak wavelengths between 3 and 6 μm, which matches reasonably well the three absorption peaks of water in that wavelength range.

SIRDBD can be operated in two heating modes, continuous or intermittent heating. During continuous heating, the radiation intensity is maintained constant by retaining a continuous supply of natural gas to the CIR emitter. For quick come-up and moisture removal or enzyme inactivation, continuous heating is advantageous since it delivers a constant high energy to the surface of the product. However, prolonged heating can cause severe surface discoloration (Zhu and Pan, 2009). The use of intermittent heating (Sandu, 1986; Zhu et al., 2010) has been shown to solve the problem of limited penetration of FIR and the application of FIR on thick materials. Intermittent heating is normally achieved by keeping product temperature constant through turning the natural gas or electricity supply on and off. The advantages of intermittent heating have been well recognized in terms of energy savings and improved product quality, since the desired processing temperature can be maintained (Chua and Chou, 2003).

Key processing parameters important to SIRDBD process performance include radiation intensity, product surface temperature, product thickness, and processing time. These parameters affect process characteristics such as heating rate, drying kinetics, and the ultimate final product quality. For most fruit and vegetable processing, the relevant processing quality aspects may include surface color, moisture reduction, texture, and residual polyphenol oxidase (PPO) and peroxidase (POD) activities.

9.4 CHARACTERIZATION OF PROCESSING CONDITIONS

9.4.1 CASE STUDY OF APPLE SLICE PROCESSING

Determining the appropriate processing conditions is very important for achieving desirable product quality, in terms of reasonable residual enzyme activity and moisture reduction while best preserving color. The processing parameters of SIRDBD with respect to characteristics and quality of apple slices have been studied extensively by Zhu and Pan (2009) and Zhu et al. (2010), who used a lab-scale IR dry blancher/dehydrator equipped with two catalytic IR emitters powered by natural gas to provide double-sided heating of apple slices during continuous and intermittent heating modes.

9.4.1.1 Continuous Heating Mode

Zhu and Pan (2009) studied the effects of processing parameters, including radiation intensities (3000–5000 W/m^2), heating times up to 20 min, and various sample thicknesses of 5–13 mm in order to determine their influence on process characteristics such as heating and drying rates, product temperature, moisture reduction, residual PPO and POD activities, and surface color change (ΔE) for the SIRDBD processing of apple slices with continuous heating mode. According to their findings, higher radiation intensity and/or thinner slices resulted in faster increase of surface and center temperatures than lower radiation intensity and/or thicker slices (Figure 9.1). In

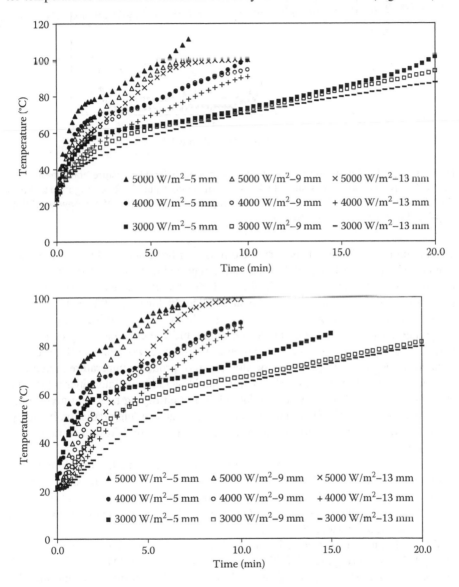

FIGURE 9.1 Surface temperature (top) and center temperature (bottom) of apple slices at different heating times, IR intensities, and slice thicknesses during continuous IR heating mode. (From Zhu, Y. and Z. Pan. 2009. *Journal of Food Engineering*, 90(4):441–452. Copyright 2009. Reprinted with permission of Elsevier.)

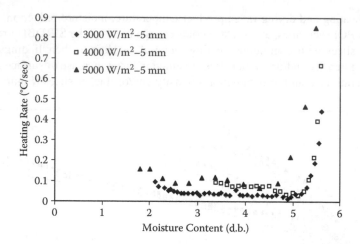

FIGURE 9.2 Surface heating rates of apple slices at different moisture contents at various IR intensities and various slice thicknesses during continuous IR heating mode. (From Zhu, Y. and Z. Pan. 2009. *Journal of Food Engineering*, 90(4):441–452. Copyright 2009. Reprinted with permission of Elsevier.)

general, higher radiation intensity resulted in higher heating rates than low radiation intensity (Figure 9.2). The high heating rates suggested that the absorbed energy was mostly used for heating up the apple slices. Higher radiation intensity reduces moisture faster than lower intensity for the same slice thickness. At the same radiation intensity, thinner slices lost moisture faster than thicker slices. Drying rates under all processing conditions increased in the beginning due to the increase of product temperature and stabilized after a certain amount of moisture was removed (Figure 9.3). This phenomenon could be due to the slowdown of product temperature increase in the middle stage of the process. The medium and low radiation intensities (4000 and 3000 W/m²) resulted in a period of relatively constant drying rates, but high radiation intensity (5000 W/m²) did not. This may be because at high radiation intensity, moisture evaporation was driven much faster than moisture diffusion from interior to the surface layer.

Many empirical and semi-empirical models have been used for modeling drying of thin layers of foods and biological materials. Of these studied models, Page model displays the greatest simplicity and good fit in describing the IR drying of thin layers of fruits and vegetables (Togrul, 2005; Gabel et al., 2006; Kumar et al., 2006; Sacilik and Elicin, 2006; Togrul, 2006a,b).

Zhu and Pan (2009) indicated that the Page model (Equation 9.1) described the drying behavior during the SIRDBD process ($r^2 = 0.9994$ and RMSE = 0.0047) well, except for some extreme processing conditions of radiation intensity and material thickness (5000 W/m²–5 mm and 3000 W/m²–13 mm):

$$MR = \exp(-kt^n) \tag{9.1}$$

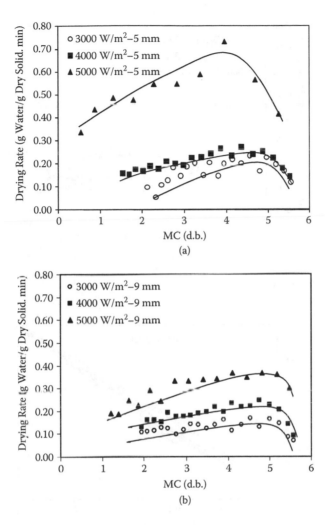

FIGURE 9.3 Drying rates of apple slices at different moisture contents at various IR intensities and various slice thicknesses during continuous IR heating mode: (a) 5 mm, (b) 9 mm, (c) 13 mm. (From Zhu, Y. and Z. Pan. 2009. *Journal of Food Engineering*, 90(4):441–452. Copyright 2009. Reprinted with permission of Elsevier.)

where *MR* is the moisture ratio, *t* is time, and *k* and *n* are empirical coefficients. Excluding extreme conditions, the correlation between experimental and predicted moisture contents resulted in an *r²* of 0.9821 (Figure 9.4).

A logarithmic type of equation (Equation 9.2) can be used for the regression of the empirical coefficients (*coc*) of the Page model (*k* and *n*) with processing variables such as slice thickness (*H*) and target surface temperature (*T_s*) (Togrul, 2006):

$$coc = a + b\ln(T_s) + c\ln(H) \tag{9.2}$$

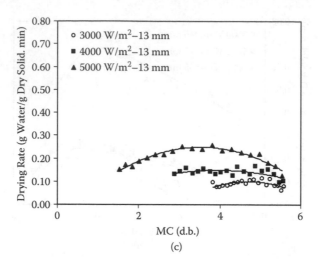

FIGURE 9.3   (continued).

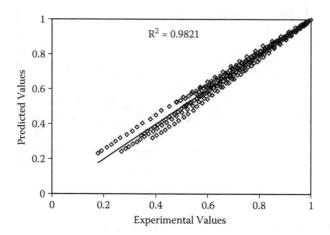

FIGURE 9.4   Correlation between the measured and predicted moisture contents (predicted values from the Page model with constants and coefficients derived from linear type model). (From Zhu, Y. and Z. Pan. 2009. *Journal of Food Engineering*, 90(4):441–452. Copyright 2009. Reprinted with permission of Elsevier.)

Following their experiments, Zhu and Pan (2009) expressed the Page model for predicting moisture ratio during the SIRDBD processing of apple slices as

$$MR = \exp\left[-(0.0143 + 5.80 \times 10^{-6}I - 0.002H)t^{(0.8645 + 0.00012I - 0.004H)}\right] \qquad (9.3)$$

Other factors of further interest in the drying kinetics during the SIRDBD process include the effective diffusivity and activation energy. According to Crank (1975):

$$\ln(MR) = \ln\left(\frac{M}{M_o}\right) = \ln\left(\frac{8}{\pi^2}\right) - \left[\pi^2 \frac{D_{eff}t}{H^2}\right] \tag{9.4}$$

where D_{eff} is the effective moisture diffusivity (m²s⁻¹). The activation energy (E_a) of the process following the Arrhenius relationship can be obtained according to Equation 9.5:

$$D_{eff} = D_o \exp\left(-\frac{E_a}{RT_s}\right) \tag{9.5}$$

where D_o is the diffusivity value for an infinite moisture content (m/s), T_s is the absolute surface temperature (K), and R represents the universal gas constant (kJ mol⁻¹ K⁻¹).

To apply the often-used enzyme inactivation kinetic models for describing the blanching process of SIRDBD, an overall rate constant ($k_{Ts,Tc,H}$) was reported as the enzyme inactivation rate constant at certain surface and center temperatures for apple slice with certain thickness (Zhu and Pan, 2009). A first-order kinetics model describes well the inactivation of PPO in apple slices according to Equation 9.6:

$$\log(A/A_o) = -\left(k_{Ts,Tc,H}/2.303\right)/t_{equiv} \tag{9.6}$$

where A_o is the original enzyme activity, A is the enzyme activity, and t_{equiv} is the equivalent processing time such that

$$t_{equiv} = t \times 10^{(T_c - T_{c,ref})/z} \tag{9.7}$$

where t is the processing time, T_c is the center temperature of the product, and $T_{c,ref}$ is the reference center temperature. A z value of 15°C according to Matsui et al. (2007) can be used as the initial value for calculating the equivalent processing time. The decimal reduction time $D_{Ts,Tc,H}$ is calculated as

$$k_{Ts,Tc,H} = 2.303/D_{Ts,Tc,H} \tag{9.8}$$

The subscripts T_s, T_c, and H denote surface temperature, center temperature, and slice thickness, respectively. The residual PPO activities under various processing conditions during the continuous SIRDBD processing are presented in Figure 9.5. Heating at high radiation intensity (5000 W/m²) generally inactivated PPO much faster than at low radiation intensity (3000 W/m²). At the same radiation intensity, thinner slices were blanched more quickly than thicker slices. Lee and Smith (1979) indicated that the dimension of the product determined the required blanching time. The authors showed that PPO activity in small beets (less than 3.7 cm in diameter) could be completely destroyed by 10 min in boiling water, while large beets (greater than 10 cm in diameter) needed more than 40 min of blanching for complete inactivation of PPO. Most of the enzyme activity was destroyed in the middle stage of processing when

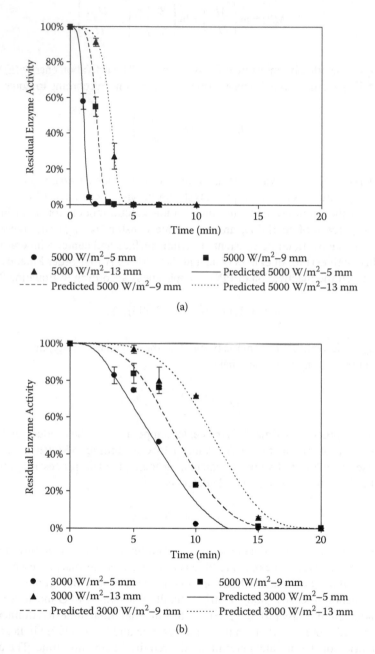

FIGURE 9.5 Residual PPO activities (solid symbols) and predicted enzyme activities (lines) of apple slices at different heating times with various IR intensities: (a) 5000 W/m² and (b) 3000 W/m². (From Zhu, Y. and Z. Pan. 2009. *Journal of Food Engineering*, 90(4):441–452. Copyright 2009. Reprinted with permission of Elsevier.)

TABLE 9.2

Summary of Model Parameters for PPO Inactivation during the Continuous Heating Mode of the SIRDBD Processing of Apple Slices

Slice Thickness (mm)	Radiation Intensity (W/m²)	Final Surface Temperature T_s (°C)	Reference Center Temperature $T_{c,ref}$ (°C)	$k_{Ts,Tc,H}$ (s⁻¹)	$D_{Ts,Tc,H}$ (min)
5	3000	84.7	74.2	0.0191	2.008
	4000	99.8	89.6	0.7674	0.050
	5000	111.4	97.5	3.4946	0.011
9	3000	93.6	81.4	0.1989	0.193
	4000	94.5	89.2	0.7611	0.050
	5000	100.5	97.1	2.7249	0.014
13	3000	87.5	79.6	0.2063	0.186
	4000	90.8	87.4	0.5361	0.072
	5000	100.1	99.3	4.1424	0.009

Source: Zhu, Y. and Z. Pan. 2009. *Journal of Food Engineering*, 90(4):441–452. Copyright 2009. Reprinted with permission of Elsevier.

apple slices reached the favorable temperature for PPO inactivation. It has been suggested that a significant thermal inactivation of PPO could only occur when the temperatures were higher than 40°C (Nicolas et al., 1994). The same author indicated that a partially purified extract of PPO from apple had a half-life of 12 min at 65°C and was destroyed at 80°C. It was also reported that the optimum temperature for Anamur banana PPO was 30°C, while the activity decreased to about 20% at 70°C (Unal, 2007). Zhu and Pan (2009) demonstrated that most of the PPO activity was destroyed when the center temperatures of apple slices were in the range of 60°C to 80°C and that when 2 log reduction of PPO was achieved, the slices were heated to the center temperatures over 75°C. Accordingly, to achieve more than 2 log reduction of PPO, the required center temperature of the apple slices is about 75°C.

The residual PPO activity showed strong correlation with equivalent processing time in a logarithmic scale diagram (Zhu and Pan, 2009). The rate constants ($k_{Ts,Tc,H}$) and decimal reduction times ($D_{Ts,Tc,H}$) are summarized in Table 9.2. The rate constants were highly dependent on the surface and center temperatures, and high product temperature could result in faster inactivation of enzymes and, thus, a high reaction rate constant. The good performance of first-order kinetics model in curve fitting of PPO inactivation also suggested that the thermal inactivation of PPO during the process followed a trend of exponential decay (Figure 9.5).

A number of studies have shown that the fractional conversion model (Equation 9.9) performed best for describing thermal inactivation of heat-resistant enzymes such as POD (Broeck et al., 1999; Soliva-Fortuny et al., 2002; Fachin et al., 2003; Zhu and Pan, 2009):

$$A_t = A_\infty + (A_o - A_\infty)\exp(-k_{Ts,Tc,H}t_{equiv}) \tag{9.9}$$

This model assumes that there is nonzero or residual enzyme activity (A_∞) upon prolonged heating (Weemaes et al., 1998). Zhu and Pan (2009) reported that the inactivation of POD in apple slices showed the same trend as that for PPO. Their results also showed that the required center temperature for achieving 1 log reduction of POD was about 75°C. Compared to PPO, POD was more heat resistant. For low radiation intensity, such as 3000 W/m^2, a heating time of 15 to 20 min was needed to achieve 1 log reduction of POD activities, compared to 2 log reduction of PPO. POD is one of the most heat-resistant enzymes in most fruits and vegetables, so that it often serves as an enzymatic indicator of blanching (Gunes and Bayindirli, 1993; Sheu and Chen, 1991). When POD is inactivated, most of the other enzymes in the fruits and vegetables may not be able to survive (Halpin and Lee, 1987). However, the direct relationship between residual POD activity and quality deterioration of fruits and vegetables is not clear (Lee et al., 1989). In addition, the processing time required to achieve more than 1 log reduction of POD resulted in severe darkening of apple slice surface for certain processing conditions. Therefore, whether POD is an appropriate blanching indicator for apple products is still questionable.

The fractional conversion model performed well for modeling the inactivation of POD (Figure 9.6). It has often been reported that the inactivation of POD deviated from simple first-order kinetics, whereas the fractional conversion model showed success for describing inactivation of POD, extracted or purified from many sources such as carrot and tomato, broccoli, green asparagus, and carrots (Morales-Blancas et al., 2002); carrot (Soysal and Soylemez, 2005); pumpkin (Goncalves et al., 2007); and watercress (Cruz et al., 2006). In the past, isozymes of POD with different heat tolerances have been isolated, purified, and identified (e.g., by electrophoresis or chromatography; Quarta and Arnone, 1987; Valderrama and Clemente, 2004). Table 9.3 summarizes the model parameters for POD inactivation and shows they were affected by both radiation intensity and apple slice thickness. High product temperatures, generated by high radiation intensity, led to high rate constants but low values of decimal reduction times. The thermal stability of POD significantly reduced when the radiation intensity (or product temperatures) was increased. In addition, based on the model parameters ($k_{Ts,Tc,H}$ and $D_{Ts,Tc,H}$), POD also showed greater thermal stability than PPO. The residual enzyme activity (A_∞) was dependent on the thickness of the apple slice and the product temperatures. Thick slices generally had more residual activities than thin slices. Meanwhile, high product temperatures result in less residual enzyme activities in apple slices. Since the residual enzyme activity is related to the heat-resistant isozyme of the enzyme, more thermal energy is needed to destroy the enzyme for thick slices than for thin slices. Both the slice thickness and product temperature affect the residual POD activity.

During SIRDBD, surface color change of product could be partially caused by dehydration of the surface layer. Because moisture diffusion from the interior of the product is usually much slower than surface moisture evaporation, the surface could be dried and even get burned quickly at high radiation intensity. Figure 9.7 shows the overall color change (ΔE) of the top surface of apple slices when heated with different IR intensities during the SIRDBD process.

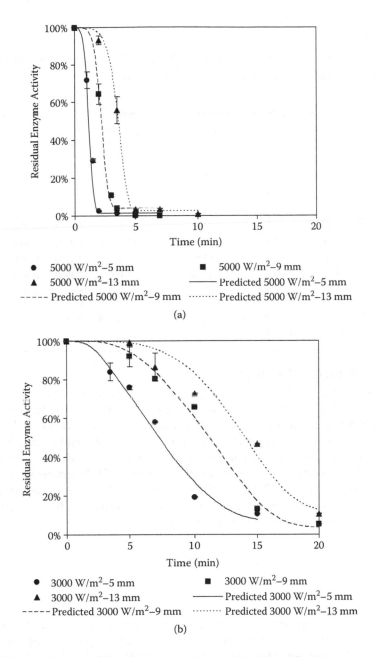

FIGURE 9.6 Residual POD activities (solid symbols) and predicted enzyme activities (lines) of apple slices at different heating times and at various IR intensities: (a) 5000 W/m² and (b) 3000 W/m² during continuous IR heating mode. (From Zhu, Y. and Z. Pan. 2009. *Journal of Food Engineering*, 90(4):441–452. Copyright 2009. Reprinted with permission of Elsevier.)

TABLE 9.3

Summary of Model Parameters for POD Inactivation during the Continuous Heating Mode of the SIRDBD Processing of Apple Slices

Slice Thickness (mm)	Radiation Intensity (W/m²)	Residual Activity (A_∞) (%)	Final Surface Temperature T_s (°C)	Reference Center Temperature $T_{c,ref}$ (°C)	$k_{Ts,Tc,H}$ (s⁻¹)	r^2	$D_{Ts,Tc,H}$ (min)
5	3000	7.01	84.7	74.2	0.0186	0.9822	2.064
	4000	2.49	99.8	89.6	0.3776	0.9982	0.102
	5000	1.37	111.4	97.5	2.1836	0.9944	0.018
9	3000	3.36	93.6	81.3	0.0245	0.9920	1.570
	4000	3.41	94.5	89.2	0.3112	0.9902	0.123
	5000	3.91	100.5	97.1	3.0541	0.9956	0.013
13	3000	11.09	87.5	79.6	0.0158	0.9635	1.906
	4000	10.66	90.8	87.4	0.3969	0.9797	0.097
	5000	2.56	100.1	99.3	5.1151	0.9983	0.008

Source: Zhu, Y. and Z. Pan. 2009. *Journal of Food Engineering*, 90(4):441–452. Copyright 2009. Reprinted with permission of Elsevier.

9.4.1.2 Intermittent Heating Mode

Zhu et al. (2010) studied the processing and quality characteristics of apple slices subjected to SIRDBD with intermittent heating. The moisture reductions of the slices followed a second-order polynomial relationship. The mathematical relationships between moisture reductions (M-R) and processing times during SIRDBD with the intermittent heating mode are presented in Table 9.4. The Page model best described the drying curves of the slices subjected to SIRDBD processing ($r^2 = 0.99$, root mean squared error (RMSE) = 0.013). The empirical drying equation (Equation 9.10) demonstrates good predictability with a strong correlation between predicted and experimental values ($r^2 = 0.99$, RMSE = 0.016).

$$MR = \exp\left\{\left[0.2982 - 0.0456 \times \ln(T_s) - 0.0320 \times \ln(H)\right] t^{\left[-3.6229 + 1.1457 \times \ln(T_s) - 0.1607 \times \ln(H)\right]}\right\}$$

(9.10)

The effective diffusivities for various processing conditions were in the range of 2.12 to 4.50×10^{-9} m² s⁻¹, which is at the high end of the general range of 10^{-9} to 10^{-11} m² s⁻¹ for air drying of food materials (Table 9.5). The effective moisture diffusivities increased as the surface temperature increased or as slice thickness increased (Afzal and Abe, 1998). As slice thickness increased, the activation energy decreased. The values of activation energy obtained for intermittent SIRDBD processing are slightly higher than the activation energy resulting from hot air drying of Red Delicious apples with thickness of 4 mm (19.96 to 22. 62 kJ/mol; Kaya et al., 2007) and FIR drying of 1–2 mm thick carrot slices (22.43 kJ/mol; Togrul, 2006).

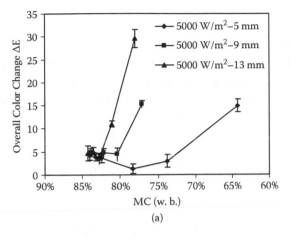

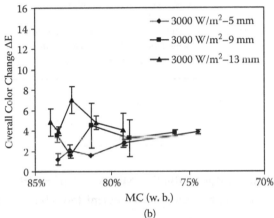

FIGURE 9.7 Overall color change (ΔE) of the top surface of apple slices at heating with different IR intensities: (a) 5000 W/m² and (b) 3000 W/m², during continuous IR heating mode. (From Zhu, Y. and Z. Pan. 2009. *Journal of Food Engineering*, 90(4):441–452. Copyright 2009. Reprinted with permission of Elsevier.)

The effective moisture diffusivity represents overall mass transport of moisture in the material, including liquid diffusion, vapor diffusion, or any other possible mass transfer mechanism (Afzal and Abe, 1998). The higher effective diffusivity for thicker slices may be due to the decrease in activation energy when slice thickness increases. Afzal and Abe (1998) suggested that decreased activation energy with increased potato slice thickness indicated that the penetration of IR radiation into biological materials causes the water molecule to vibrate. Therefore, the molecules require less energy to transfer from a porous material in the mobilized state.

During the intermittent SIRDBD process, the first-order kinetics model (Equation 9.6) describes well the inactivation of PPO in apple slices (Unal and Sener, 2006; Unal, 2007; Matsui et al, 2007). The residual PPO activities after application of various processing conditions are presented in Figure 9.8. Thin slices and/or high surface

TABLE 9.4

Mathematical Relationships between Moisture Reductions (M-R) and Processing Times during SIRDBD with the Intermittent Heating Mode

Surface Temp (°C)	Slice Thickness (mm)	Relationship between M-R and Time (*t*)	Moisture Reduction (%) for 1 log Reduction of POD
70	5	M-R = −0.0785*t^2 + 4.8995*t	59.48
75	5	M-R = −0.1096*t^2 + 5.6187*t	36.04
80	5	M-R = −0.1123*t^2 + 6.0583*t	34.23
70	9	M-R = −0.0310*t^2 + 2.6344*t	50.97
75	9	M-R = −0.0361*t^2 + 2.8068*t	26.60
80	9	M-R = −0.0362*t^2 + 3.0752*t	21.48
70	13	M-R = −0.0160*t^2 + 1.6015*t	33.95
75	13	M-R = −0.0181*t^2 + 1.6956*t	26.97
80	13	M-R = −0.0164*t^2 + 1.7300*t	19.78

Source: Pan 2007. Processing and Quality Characteristics of Apple Slices under Simultaneous Infrared Dry-blanching and Dehydration (SIRDBD). Dissertation (Zhu, Y.), Biological and Agricultural Engineering, University of California, Davis.

TABLE 9.5

Average Center Temperature, Effective Moisture Diffusivities, and Activation Energies of Apple Slices Processed with Intermittent Heating of IR

Slice Thickness (mm)	Target Surface Temperature (°C)	Average Center Temperature (°C)	Effective Diffusivity (D_{eff}) × 10^{-9} m² s⁻¹	Activation Energy (E_a) kJ/mol
5	70	67.4 ± 1.3	2.13	48.80
	75	70.1 ± 1.3	2.49	
	80	74.4 ± 1.7	3.02	
9	70	66.5 ± 0.8	2.99	39.22
	75	70.5 ± 1.3	3.43	
	80	76.3 ± 1.0	4.39	
13	70	65.0 ± 0.8	3.54	26.28
	75	67.9 ± 1.3	4.14	
	80	73.4 ± 0.8	4.50	

Source: Zhu et al. 2010. *Journal of Food Engineering*, 97 (1):8–16. Copyright 2010. Reprinted with permission of Elsevier.

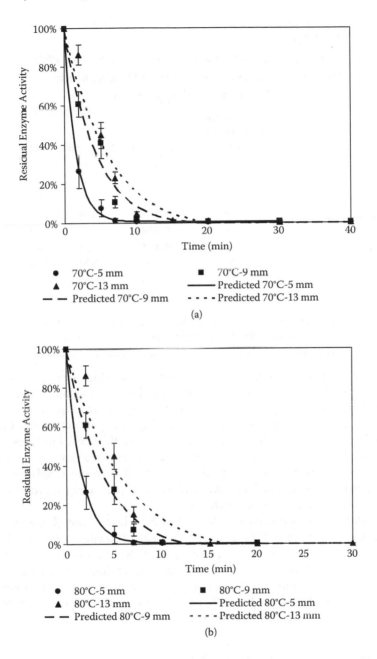

FIGURE 9.8 Residual PPO activities (solid symbols) and predicted enzyme activities (lines) of apple slices at different heating times with various target surface temperatures: (a) 70°C and (b) 80°C. (From Zhu et al. 2010. *Journal of Food Engineering*, 97 (1):8–16. Copyright 2010. Reprinted with permission of Elsevier.)

TABLE 9.6

Summary of Thermal Inactivation Parameters for PPO Inactivation during the Intermittent Heating Mode of the SIRDBD Processing of Apple Slices

Slice Thickness (mm)	Target Surface Temperature T_s (°C)	$k_{Ts,Tc,H}$ (s^{-1})	$D_{Ts,Tc,H}$ (min)	r^2
5	70	0.00921	4.17	0.97
	75	0.01036	3.70	0.99
	80	0.01152	3.33	0.98
9	70	0.0055	6.94	0.91
	75	0.0064	5.95	0.95
	80	0.0071	5.38	0.91
13	70	0.0046	8.33	0.94
	75	0.0055	6.94	0.93
	80	0.0067	5.75	0.90

Source: Zhu et al. 2010. *Journal of Food Engineering*, 97 (1):8–16. Copyright 2010. Reprinted with permission of Elsevier.

temperature result in faster inactivation of enzymes than thick slices and/or low surface temperature. However, the slice thickness had a stronger impact on PPO inactivation than surface temperature. This may be temperature range dependent. The rate constants and decimal reduction times were highly dependent on slice thickness and product temperatures (Table 9.6). When the slice thickness increases, the rate constants are reduced, while the decimal reduction times are elevated significantly. Thick slices tend to have lower center temperatures and, correspondingly, the enzyme inactivation rate is slower in thick slices and, therefore, the required processing time to achieve 90% PPO inactivation increases.

A similar trend was observed in the case of POD inactivation (Figure 9.9). POD is more heat tolerant than PPO. The biphasic model (Equation 9.11), which is also known as the 2-fraction model (Ling and Lund, 1978; Saraiva et al., 1996; Rodrigo et al., 1997), best fits the POD inactivation data. According to the biphasic model's assumption, there are only two types of isozymes involved, where one is heat resistant (E_R) and the other is heat labile (E_L). Each fraction of the enzyme is assumed to follow first-order kinetics and can be mathematically expressed as

$$\frac{A}{A_0}(\%) = \left[\frac{K_R E_{Ro} e^{-k_R t}}{K_R E_{Ro} + K_L E_{Lo}}\right] \times 100 \tag{9.11}$$

where E_{R0} and E_{L0} are the initial concentration of the heat-resistant and heat-labile isozyme fractions, respectively; K_R and K_L are the reaction rate constants for the respective isozyme fractions with the substrate; k_R and k_L (min^{-1}) are the first-order

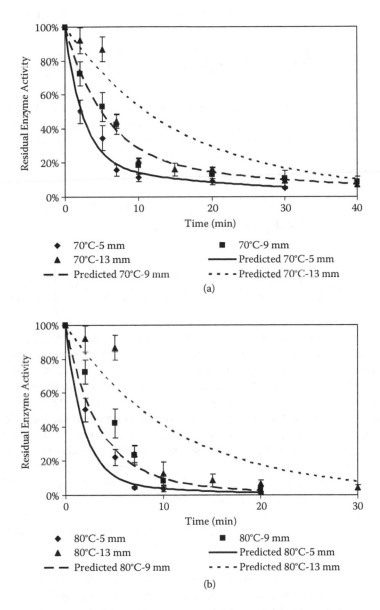

FIGURE 9.9 Residual POD activities (solid symbols) and predicted enzyme activities (lines) of apple slices at different heating times with various target surface temperature: (a) 70°C and (b) 80°C. (From Zhu et al. 2010. *Journal of Food Engineering*, 97 (1):8–16. Copyright 2010. Reprinted with permission of Elsevier.)

rate constants for thermal inactivation of the respective isozyme fractions. The limiting condition of long heating time is applied to obtain the special rate constant of heat-resistant isozyme fraction by rearranging Equation 9.11 to obtain Equation 9.12:

$$\log\left[\frac{A}{A_0}(\%)\right] = \log\left[\frac{K_R E_{Ro}}{K_R E_{Ro} + K_L E_{Lo}} \times 100\right] - \frac{k_R}{2.303}t \qquad (9.12)$$

The portion of the shallow slope straight line of the curve has a slope equal to $-k_R/2.303$ and, upon extrapolating the line to time zero, the intercept would be equal to $K_R E_{Ro}/(K_L E_{Lo} + K_R E_{Ro})$. Knowing the proportion of the heat-resistant isozyme fraction and plugging the value into Equation 9.13, we obtain the rate constant for the heat-labile isozyme fraction based on Equation 9.14:

$$A(\%) - \left[\frac{K_R E_{Ro}}{K_R E_{Ro} + K_L E_{Lo}}\right] \times 100 = \left[\frac{K_L E_{Lo} e^{-k_L t}}{K_R E_{Ro} + K_L E_{Lo}}\right] \times 100 \qquad (9.13)$$

$$\log\left[A(\%) - \frac{K_R E_{Ro}}{K_R E_{Ro} + K_L E_{Lo}} \times 100\right] = \log\left[\frac{K_L E_{Lo}}{K_R E_{Ro} + K_L E_{Lo}} \times 100\right] - \frac{k_L}{2.303}t \qquad (9.14)$$

When the proportion of the heat-resistant isozyme fraction was subtracted from the original activity values and the data replotted on a logarithmic scale against heating time, the slope of the resulting straight line was $-k_L/2.303$. In this way, the overall rate constants for both heat-labile and heat-resistant isozymes are estimated. These rate constants have the constraints of product slice surface, center temperatures, and product slice thickness. Fitting enzyme inactivation curves into the biphasic model resulted in rate constants ($k_{Ts,Tc,H}$) and decimal reduction times ($D_{Ts,Tc,H}$) for two isozymes of POD with different heat tolerances, namely, the heat-labile and heat-resistant isozymes (Table 9.7). The fraction of the heat-resistant isozyme of POD decreased as surface temperature increased, while it increased when slice thickness increased (Morales-Blancas et al., 2002). Ordinarily, the fraction of heat-resistant POD varied greatly in different materials and in different parts of the same material. The inactivation rate constants and decimal reduction times determined for both heat-labile and heat-resistant isozymes showed dependence on product temperature and slice thickness. The increase in product temperature or decrease in slice thickness resulted in elevation of rate constants and reduction of decimal reduction time. The effect of slice thickness on the rate constants was more significant for heat-labile isozymes than for heat-resistant isozymes. Although these model parameters may not be directly used without the constraints of product temperature and product thickness, they provided valuable information for the characterization of the blanching process of SIRDBD and recommendations for appropriate processing conditions. Accordingly, the required processing times to achieve a 1 log reduction of POD for various surface temperatures and slice thicknesses could be obtained (Table 9.8).

TABLE 9.7
Summary of Thermal Inactivation Parameters for POD Inactivation during the Intermittent Heating Mode of the SIRDBD Processing of Apple Slices

Slice Thickness (mm)	Target Surface Temperature (°C)	Heat-Resistant Isozyme Fraction (%)	Heat-Labile Isozyme			Heat-Resistant Isozyme		
			$k_{Ts,Tc,H}$ × 10^4 (s^{-1})	$D_{Ts,Tc,H}$ (min)	r^2	$k_{Ts,Tc,H}$ × 10^4 (s^{-1})	$D_{Ts,Tc,H}$ (min)	r^2
5	70	19.82	69.09	5.56	0.96	6.91	55.56	0.95
	75	15.15	80.61	4.76	0.82	9.21	41.67	1.00
	80	9.40	94.42	4.07	0.99	16.12	23.81	0.91
9	70	22.77	32.24	11.90	0.98	4.61	83.33	1.00
	75	21.45	43.76	8.77	0.99	11.52	33.33	0.99
	80	22.16	57.58	6.67	0.88	18.42	20.83	0.96
13	70	29.89	5.90	65.10	0.98	5.99	64.10	0.96
	75	26.03	6.45	59.52	0.82	7.85	48.88	0.96
	80	19.56	6.86	55.93	0.98	8.57	44.80	0.96

Source: Zhu et al. 2010. *Journal of Food Engineering*, 97 (1):8–16. Copyright 2010. Reprinted with permission of Elsevier.

TABLE 9.8
Mathematical Relationships between Residual PPO Activities and Processing Times during SIRDBD with Intermittent Heating Mode

Slice Thickness (mm)	Radiation Intensity (W/m²)	Residual PPO Activity (A_t)	Time for 1 log Reduction of POD (min)	Top Surface Color Change (ΔE)
5	70	Log (A_t) = −0.0003*t + 1.2971	16.5	4.344[a]
	75	Log (A_t) = −0.0004*t + 1.1804	7.5	2.274[a]
	80	Log (A_t) = −0.0007*t + 1.2501	6.4	3.269[a]
9	70	Log (A_t) = −0.0002*t + 1.3573	29.8	2.444[a]
	75	Log (A_t) = −0.0005*t + 1.3314	11.0	3.383[ab]
	80	Log (A_t) = −0.0008*t + 1.3455	7.7	5.880[b]
13	70	Log (A_t) = −0.00026*t + 1.4755	30.5	2.998[a]
	75	Log (A_t) = −0.00034*t + 1.4155	20.3	2.463[a]
	80	Log (A_t) = −0.00037*t + 1.2913	13.1	4.309[a]

Source: Pan 2007. Processing and Quality Characteristics of Apple Slices under Simultaneous Infrared Dry-blanching and Dehydration (SIRDBD). Dissertation (Zhu, Y.), Biological and Agricultural Engineering, University of California, Davis.

Note: Values within the same column for the same slice thickness followed by different letters are significantly different at $p < 0.05$.

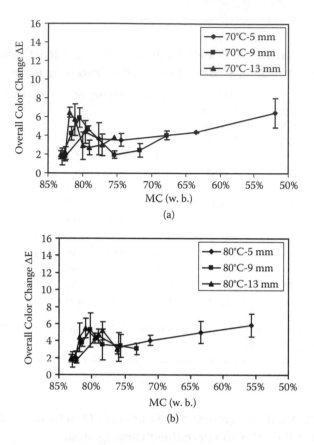

FIGURE 9.10 Overall color change (ΔE) of the top surface of apple slices at various target surface temperatures: (a) 70°C and (b) 80°C. (From Zhu et al. 2010. *Journal of Food Engineering*, 97 (1):8–16. Copyright 2010. Reprinted with permission of Elsevier.)

The overall color change (ΔE) of treated apple slices is plotted against moisture content on wet basis (w.b.) for various processing conditions in Figure 9.10. A similar trend in overall color change was reported in combined convective and microwave drying of potatoes (Chua and Chou, 2005). The first increase of ΔE values corresponded to enzymatic browning due to the high residual PPO activity in apple slices, and resulted in a dramatic increase of a value. When intermittent heating proceeded, the reduced moisture content on the surface of the apple slices induced enhanced reflectivity of the samples and, accordingly, an increase in L values. Similar increases in L values were also observed during processing of apple slices by SIRDBD in continuous heating mode (Zhu and Pan, 2009). Since the increase of a value became steady, while the L value increased again, the ΔE values dropped after they reached a peak. When surface moisture was further removed from the slices, the b value began to increase significantly, indicating yellowing, which contributed to the second increase of ΔE values (Krokida et al., 2000; Fernandez et al., 2005; Rababah et al., 2005). The increase in b value could be due to the concentration of certain yellowish phytochemicals (possibly phenolic compounds) when a large amount of moisture was removed after prolonged heating. Rababah et al. (2005) reported that total phenolics and anthocyanins increased 150% and 245%, respectively, due to concentration during drying of apple slices at 40°C for 24 h. Nonenzymatic browning of the product may also contribute to the increase in *b* value.

9.4.1.3 Comparison between Continuous and Intermittent Heating Modes

The effect of continuous and intermittent heating modes of SIRDBD on the center temperatures of samples, residual PPO and POD activities, moisture reductions, and overall surface color changes has been studied for different thicknesses of apple slices (Pan, 2007). Figure 9.11 shows the comparison of quality and processing characteristics between continuous heating (4000 W/m²) and intermittent heating (75°C) for 5-mm-thick apple slices. The center temperatures of the slice at the heat-up stage are very close for both heating modes. PPO is inactivated quickly in both heating modes, and the required processing times to achieve similar residual PPO do not deviate from each other much. It takes much longer with intermittent heating than with continuous heating to achieve the same level of POD inactivation. The prolonged heating for achieving more than a 1 log reduction of POD in the intermittent heating mode causes a large moisture reduction and significant increase of overall surface color change. In studies involving treatment of 9- and 13-mm-thick apple slices, it has been reported that continuous heating shows an obvious advantage over intermittent heating as it needs much shorter processing time to inactivate POD (Pan,

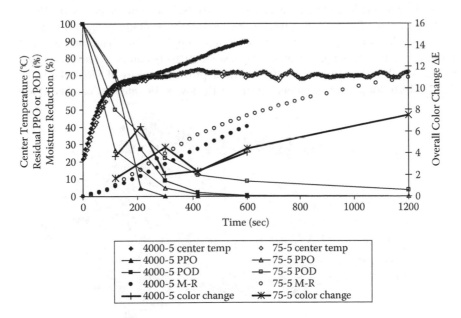

FIGURE 9.11 Comparison of quality and processing characteristics between continuous heating (4000 W/m²) and intermittent heating (75°C) for the 5 mm slice. (From Zhu, Y., 2007. Processing and quality characteristics of apple slices under simultaneous infrared dry-blanching and dehydration (SIRDBD). Dissertation, Biological and Agricultural Engineering, University of California, Davis.)

2007). Continuous heating at medium radiation intensity seems more advantageous than intermittent heating for blanching and partial dehydrating thin apple slices (5 mm thickness).

The SIRDBD process can serve the purposes of blanching alone and simultaneous blanching and dehydration. For certain fruits and vegetables, moisture removal may not be desired during blanching. In this case, quick blanching and limiting the moisture reduction as much as possible is necessary. Continuous heating may be beneficial for such an application. On the contrary, for certain scenarios, drying needs to be conducted after proper blanching, for instance, in the production of dehydrofrozen products. In such cases, intermittent heating is the best since it does not cause severe surface darkening. Because each heating mode has its own advantages and disadvantages, an appropriate heating mode and appropriate processing conditions need to be chosen based on the processing purpose and the property of the materials.

9.5   ADVANCES OF INFRARED DRY BLANCHING

9.5.1   EQUIPMENT DESCRIPTION

A mobile IR heating unit for SIRDBD processing equipped with an automatically controlled variable-speed conveyor belt and catalytic emitters powered with natural gas has been built by researchers at USDA ARS Western Research Center and University of California, Davis (Figure 9.12). The pilot-scale mobile IR heating equipment can be used for processing various vegetables and fruits. The equipment has an effective heating area of 1.5 × 4.6 m and weighs approximately 2000 kg. IR intensity incident on the products is adjusted by varying the gas supply or by changing the emitter positions. In this new equipment, the highest IR intensity can be achieved by setting the gas supply valves fully open (100%), which provides energy of 576,000 BTU/h ($\approx$ 607,714 kJ/h) for its eight emitters. Similarly, the lowest IR intensity can be achieved

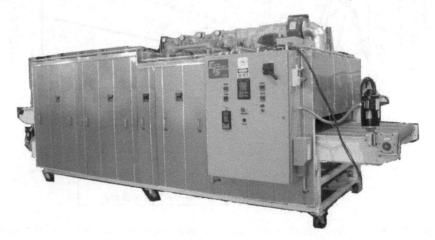

FIGURE 9.12 Mobile infrared heating equipment for processing of vegetables and fruits. (From Pan et al. 2009. Test results and performance of mobile infrared heating equipment for processing fruits and vegetables. USDA-ARS-WRRC Report.)

by setting the values at the lowest position (0%), which provides energy of 288,000 BTU/h ($\approx$ 303,857 kJ/h) for the eight emitters. Setting the controller to 0% at the panel will provide 50% actual gas flow to the emitters. Changing the setting from 0% to 100% at the panel will change the actual gas flow from 50% to 100%, respectively. A fan is installed on top of the unit and is used to remove air with high moisture from the heating chamber when it is necessary. During the blanching step, the fan is turned off to obtain high relative humidity in the chamber and thereby minimize moisture loss when excessive moisture loss is not desirable.

9.5.2 PERFORMANCE OF THE PILOT-SCALE INFRARED DRY BLANCHER

Several tests under different processing conditions (Table 9.9) of the pilot-scale IR dry blancher are reported for the blanching of vegetables such as potatoes, green bell peppers, onions, baby carrots, and carrot pomace (Pan et al., 2009). Russet potatoes slices having different thicknesses were exposed to IR heating under different operating conditions. The tests that best worked for slices having different thicknesses are tabulated in Table 9.10. Direct exposure to IR residence time is the time that the potato slices are directly passing through the top/bottom heating section, whereas the total residence time is the time that the potato slices remain in the unit. Figures 9.13–9.15 show potato slices of different thicknesses after blanching.

To regain the moisture loss caused by IR dry blanching, vegetables can be dipped into water after blanching to increase the moisture. Figures 9.16 and 9.17 show the effects of dipping treatment after dry blanching of russet potato and baby carrot slices. Unlike the treated samples, a slight browning reaction takes place in control samples. Dipping after blanching can improves the final appearance of diced potatoes. The sliced baby carrots had a nice appearance after exposure to IR heating. Dipping the samples into water after blanching eliminated the dry look of the slices' surfaces.

TABLE 9.9
Test Conditions for Potato Slices Using the New Commercial-Scale Infrared Dry Blancher

Test #	Gas Supply		Conveyor Speed (ft/min)	Fan	Residence Time (s)	
	Section 1	Section 2			Exposure to IR	Total
1	100%	OFF	1.654	ON	272	544
2	100%	OFF	3.175	OFF	141.5	283
3	100%	OFF	2.430	OFF	185	370
4	100%	50%	3.514	OFF	256	256
5	100%	OFF	2.739	OFF	164	328
6	100%	50%	4.017	OFF	224	224

Source: Pan et al. 2009. Test results and performance of mobile infrared heating equipment for processing fruits and vegetables. USDA-ARS-WRRC Report.

TABLE 9.10

Selected Best Test Conditions for Different Slice Thicknesses

Thickness (mm)	%Weight Reduction	Final MC (w.b.)	Belt Speed (ft/min)	Residence Time (s)		Emitters in Second Section
				IR Exposure	Total	
2.89 ± 0.34	29.63	73.29	3.175	141.5	283	OFF
6.42 ± 0.36	19.83	76.56	3.514	256	256	50%
	19.79	76.57	2.430	185	370	OFF
	15.93	77.65	2.739	164	328	OFF
9.03 ± 0.32	15.85	77.67	3.514	256	256	50%
	13.38	78.31	2.430	185	370	OFF
	10.29	79.05	2.739	164	328	OFF

Source: Pan et al. 2009. Test results and performance of mobile infrared heating equipment for processing fruits and vegetables. USDA-ARS-WRRC Report.

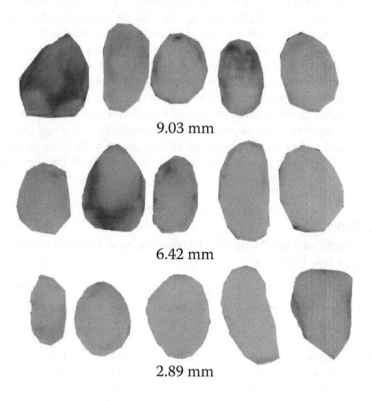

9.03 mm

6.42 mm

2.89 mm

FIGURE 9.13 Potato slices of different thicknesses 1 h after blanching (Test #2). (From Pan et al. 2009. Test results and performance of mobile infrared heating equipment for processing fruits and vegetables. USDA-ARS-WRRC Report.)

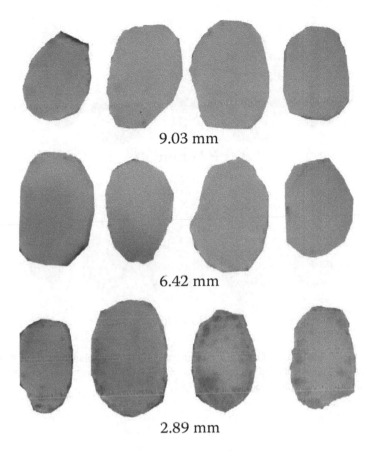

9.03 mm

6.42 mm

2.89 mm

FIGURE 9.14 Potato slices of different thicknesses 45 min after IR blanching (Test #3). (From Pan et al. 2009. Test results and performance of mobile infrared heating equipment for processing fruits and vegetables. USDA-ARS-WRRC Report.)

The blanched and dipped products visually are close to the control. Figures 9.18 and 9.19 show results of untreated (wet) and dry blanched and partially dehydrated green bell peppers and white onions, respectively, at different belt speeds.

Based on the test results from the pilot-scale SIRDBD equipment, IR heating can be used for both blanching of vegetables and for removal of a controlled amount of moisture through the use of appropriate equipment settings and operating conditions. During dry blanching of vegetables, moisture loss can be controlled by optimizing the processing conditions, such as emitter configuration, amount of natural gas to the emitters, belt speed, etc., and dipping in water after blanching to replenish the lost moisture. In order to avoid unnecessary energy use and to obtain high quality for either blanched or partially or fully dehydrated products, specific adjustment of the equipment settings based on each product should be made.

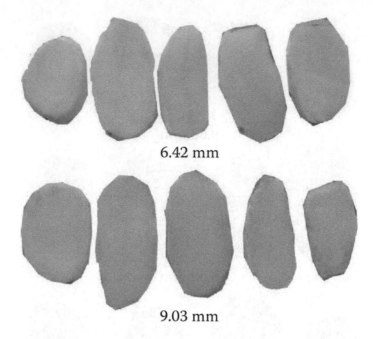

6.42 mm

9.03 mm

FIGURE 9.15 Potato slices of different thicknesses 30 min after blanching (Test #4). (From Pan et al. 2009. Test results and performance of mobile infrared heating equipment for processing fruits and vegetables. USDA-ARS-WRRC Report.)

FIGURE 9.16 Diced potatoes 30 min after dry blanching. (From Pan et al. 2009. Test results and performance of mobile infrared heating equipment for processing fruits and vegetables. USDA-ARS-WRRC Report.)

Blanched   Blanched + Dipped   Control

FIGURE 9.17 Blanched, blanched and dipped, and untreated (control) sliced carrots. (From Pan et al. 2009. Test results and performance of mobile infrared heating equipment for processing fruits and vegetables. USDA-ARS-WRRC Report.)

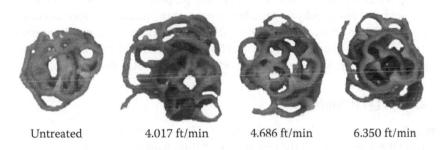

Untreated   4.017 ft/min   4.686 ft/min   6.350 ft/min

FIGURE 9.18 Untreated (wet) and dry blanched and partially dehydrated green bell peppers at different belt speeds. (From Pan et al. 2009. Test results and performance of mobile infrared heating equipment for processing fruits and vegetables. USDA-ARS-WRRC Report.)

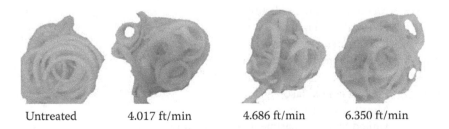

Untreated   4.017 ft/min   4.686 ft/min   6.350 ft/min

FIGURE 9.19 Untreated (wet) and dry blanched and partially dehydrated white onions at different belt speeds. (From Pan et al. 2009. Test results and performance of mobile infrared heating equipment for processing fruits and vegetables. USDA-ARS-WRRC Report.)

9.6   CONCLUSION

IR heating can be used for achieving simultaneous dry blanching and/or dehydration (SIRDBD) of fruit and vegetables through the use of appropriate equipment settings and operating conditions. The inactivation of PPO is achievable by SIRDBD and the moisture loss during the blanching and dehydration process can be controlled by selecting appropriate processing conditions or by replenishing the lost water of the blanched products through dipping in water or by spraying water onto the product. SIRDBD offers the food industry a novel waste-free, environmentally friendly technology to accomplish blanching and/or dehydration. Since this technology does not involve the addition of steam or water in the process of blanching, it has been named "IR dry blanching" (IRDB) technology. The industrialization and commercialization of IRDB is on and is intended to be a replacement for current steam, water, and/or microwave blanching methods to produce many kinds of value-added fruit- or vegetable-based dried, refrigerated, frozen, and dehydrofrozen food products.

Ordinarily, the application of any new technology presents a significant challenge to food technologist and food researchers. Although SIRDBD seems a simple operation, the heat and mass transfer processes are very complex and need more precise modeling for processing control. For blanching purposes, processing conditions are usually set up to inactivate enzymes, but other quality parameters, such as color and texture, are also to be monitored. Today, SIRDBD is still in its infant stage commercially. Nevertheless, there is huge potential for commercial applications. Opportunities clearly exist for innovative applications and new food product development.

ACKNOWLEDGMENT

The authors are extremely grateful to Dr. Yi Zhu, whose expertise was invaluable throughout all the stages of the preparation of this chapter.

NOMENCLATURE

SIRDBD	Simultaneous infrared radiation dry blanching and dehydration
LTLT	Low-temperature long-time
HTST	High-temperature short-time
MIR	Medium-infrared radiation
NIR	Near-infrared radiation
FIR	Far-infrared radiation
CIR	Catalytic infrared radiation
POD	Peroxidase
PPO	Polyphenol oxidase
M-R	Moisture reduction
USDA	United States Department of Agriculture
ARS	Agricultural Research Services
IR	Infrared radiation
IRDB	Infrared dry blanching

REFERENCES

Afzal, T. M., and T. Abe. 1998. Diffusion in potato during far infrared radiation drying. *Journal of Food Engineering* September 37(4):353–365.

Alvarez, M. D., and W. Canet. 1999. Optimization of stepwise blanching of frozen-thawed potato tissues (cv. Monalisa). *European Food Research and Technology—Zeitschrift fur Lebensmittel Untersuchung und Forschung A* 210(2):102–108.

Belyaev, M. I., A. I. Cherevko, V. S. Artemenko, and A. V. Paranich. 1985. Heat treatment of vegetables by IR radiation. *Pishchevaya Tekhnologiya* 6:49–52.

Bomben, J. L. 1977. Effluent generation, energy use, and cost of blanching. *Journal of Food Process and Engineering* 1(4):329–341.

Broeck, I. V., L. R. Ludikhuyze, A. M. Van Loey, C. A. Weemaes, and M. E. Hendrickx. 1999. Thermal and combined pressure-temperature inactivation of orange pectinesterase: influence of pH and additives. *Journal of Agricultural and Food Chemistry* 47:2950–2958.

Chua, K. J., and S. K. Chou. 2003. Low-cost drying methods for developing countries. *Trends in Food Science and Technology* 14(12):519–528.

Chua, K. J., and S. K. Chou. 2005. A comparative study between intermittent microwave and infrared drying of bioproducts. *International Journal of Food Science and Technology* 40(1):23–39.

Crank, J. 1975. *Mathematics of Diffusion* (2nd ed.), London: Oxford University Press. 513.

Cruz, R. M. S., M. C. Vieira, and C. L. M. Silva. 2006. Effect of heat and thermosonication treatments on peroxidase inactivation kinetics in watercress (*Nasturtium officinale*). *Journal of Food Engineering* 72(1):8–15.

Devece, C., J. N. Rodriguez-Lopez, L. G. Fenoll, J. Tudela, J. M. Catala, E. de-los Reyes, and F. Garcia-Canovas. 1999. Enzyme inactivation analysis for industrial blanching applications: Comparison of microwave, conventional, and combination heat treatments on mushroom polyphenoloxidase activity. *Journal of Agricultural and Food Chemistry* 47(11):4506–4551.

Drake, S. R., and D. M. Carmichael. 1986. Frozen vegetable quality as influenced by high temperature short time (HTST) steam blanching. *Journal of Food Science* 51(5):1378–1379.

Fachin, D., A. van Loey, B. L. Nguyen, I. Verlent, Indrawati, and M. E. Hendrickx. 2003. Inactivation kinetics of polygalacturonase in tomato juice. *Innovative Food Science and Emerging Technologies* 4:135–142.

Fernandez, L., C. Castillero, and J. M. Aguilera. 2005. An application of image analysis to dehydration of apple discs. *Journal of Food Engineering* 67(1–2):185–193.

Gabel, M., Z. Pan, K. S. P. Amaratunga, L. Harris, and J. F. Thompson. 2006. Catalytic infrared dehydration of onions. *Journal of Food Science* 71(9):351–358.

Garcia-Reverter, J., M. C. Bourne, and A. Mulet. 1994. Low temperature blanching affects firmness and rehydration of dried cauliflower florets. *Journal of Food Science* 59(6):1181–1183.

Ginzberg, A. S. 1969. *Application of Infra-Red Radiation in Food Processing*, London: Leonard Hill Book.

Gómez, F., Toledo, R. T., Wadso, L., Gekas, V., and I. Sjoholm. 2004. Isothermal calorimetry approach to evaluate tissue damage in carrot slices upon thermal processing. *Journal of Food Engineering* 65:165–173.

Goncalves, E. M., J. Pinheiro, M. Abreu, T. R. S. Brandao, and C. L. M. Silva. 2007. Modelling the kinetics of peroxidase inactivation, colour and textures of pumpkin (*Cucurbita maxima* L.) during blanching. *Journal of Food Engineering* 81:693–701.

Gunes, B., and A. Bayindirli. 1993. Peroxidase and lipoxygenase inactivation during blanching of green beans, green peas and carrots. *Lebensmittel Wissenschaft Technologie Food Science Technology* 26(5):406–410.

Halpin, B. E., and C. Y. Lee. 1987. Effect of blanching on enzyme activity and quality changes in green peas. *Journal of Food Science* 52(4):1002–1005.

Jayaraman, K. S., and D. K. Das Gupta. 1995. Drying of fruits and vegetables. In *Handbook of Industrial Drying*, Ed. A. S. Mujumdar. New York: Marcel Dekker.

Krokida, M. K., V. T. Karathanos, and Z. B. Maroulis. 2000. Effect of osmotic dehydration on viscoelastic properties of apple and banana. *Drying Technology* 18(4–5):951–966.

Krust, P. W., L. D. McGlauchlin, and R. B. McQuistan. 1962. *Elements of Infra-Red Technology*. New York: John Wiley & Sons.

Kumar, D. G. P., H. U. Hebbar, and M. N. Ramesh. 2006. Suitability of thin layer models for infrared-hot air-drying of onion slices. *Lebensmittel Wissenschaft und Technologie Food Science and Technology* 39(6):700–705.

Lee, C. Y., and N. L. Smith. 1979. Blanching effect on polyphenol oxidase activity in table beets. *Journal of Food Science* 44(1):82–83.

Lee, C. Y., N. L. Smith, and D. C. Hawbecker. 1989. Enzyme activity and quality of frozen green beans as affected by blanching and storage. *Journal of Food Quality* 11(4):279–287.

Ling, A. C., and D. B. Lund. 1978. Determining kinetic parameters for thermal inactivation of heat resistant and heat-labile isozymes from thermal destruction curves. *Journal of Food Science* 43:1307–1310.

Matsui, K. N., L. M. Granado, P. V. de Oliveira, and C. C Tadini. 2007. Peroxidase and polyphenol oxidase thermal inactivation by microwaves in green coconut water simulated solutions. *Lebensmittel Wissenschaft Technologie* 40:852–859.

Morales-Blancas, E. F., V. E. Chandia, and L. Cisneros-Zevallos. 2002. Thermal inactivation kinetics of peroxidase and lipoxygenase from broccoli, green asparagus and carrots. *Journal of Food Science* 67(1):146–154.

Nicolas, J. J., F. C. Richard-Forget, P. M. Goupy, M. J. Amiot, and S. Y. Aubert. 1994. Enzymatic browning reactions in apple and apple products. *Critical Reviews in Food Science and Nutrition* 34(2):109–157.

Pan, Z., Bingol, G., and T. McHugh. 2009. Test Results and Performance of Mobile Infrared Heating Equipment for Processing Fruits and Vegetables. USDA-ARS-WRRC Report.

Pan, Z., R. Khir, L. D. Godfrey, R. Lewis, J. F. Thompson, and A. Salim. 2008. Feasibility of simultaneous rough rice drying and disinfestations by infrared radiation heating and rice milling quality. *Journal of Food Engineering* 84(3):469–479.

Pan, Z., and T. H. McHugh. 2004. Novel infrared dry-blanching (IDB), infrared blanching, and infrared drying technologies for food processing. U.S. Patent Application 20060034981. Filed 8/13/2004, published 2/16/2006.

Ponne, C. T., T. Baysal, and D. Yuksel. 1994. Blanching leafy vegetables with electromagnetic energy. *Journal of Food Science* 59(5):1037–1041.

Quarta, R., and S. Arnone. 1987. Peroxidase polymorphism in apple cultivars. *Advances in Horticultural Science* 1(2):83–86.

Rababah, T. M., K. I. Ereifej, and L. Howard. 2005. Effect of ascorbic acid and dehydration on concentrations of total phenolics, antioxidant capacity, anthocyanins, and color in fruits. *Journal of Agricultural and Food Chemistry* 53(11):4444–4447.

Ramaswamy, H. 2005. Thermal processing of fruits. In *Processing Fruits: Science and Technology*, Eds. D. M. Barret, L. Somogyi, and H. Ramaswamy. CRC Press.

Ramesh, M. N., W. Wolf, D. Tevini, and A. Bognar. 2002. Microwave blanching of vegetables. *Journal of Food Science* 67(1):390–398.

Rodrigo, C., Rodrigo, M, Alvarruiz, A., and Frigola, A. 1997. Inactivation and regeneration kinetics of horseradish peroxidase heated at high temperatures. *Journal of Food Protection* 60(8):961–966.

Rosenthal, I. 1992. Electromagnetic radiations in food science. *Advanced Series in Agricultural Sciences* 19.

Sacilik, K., and A. K. Elicin. 2006. The thin layer drying characteristics of organic apple slices. *Journal of Food Engineering* 73:281–289.

Sakai, N., and T. Hanazawa. 1994. Applications and advances in far-infrared heating in Japan. *Trends in Food Science and Technology* 5:357–362.

Sanchez-Pineda-Infantas, M. T., G. Cano-Munoz, and J. R. Hermida-Bun. 1994. Various blanching treatments affect final texture and total fiber content of peeled processed asparagus. *Journal of Food Quality* 17(5):361–369.

Sandu, C. 1986. Infrared radiative drying in food engineering: A process analysis. *Biotechnology Progress* 2(3):109–119.

SanJuan, N., G. Clemente, J. Bon, and A. Mulet. 2001. The effect of blanching on the quality of dehydrated broccoli florets. *European Food Research and Technology—Zeitschrift fur Lebensmittel Untersuchung und Forschung A* 213(6):474–479.

Saraiva, J., J. C. Oliveira, A. Lemos, and M. Hendrickx. 1996. Analysis of the kinetic patterns of horseradish peroxidase thermal inactivation in sodium phosphate buffer solutions of different ionic strength. *Journal of Food Engineering* 31(3):223–231.

Sawai, J., K. Sagara, A. Hashimoto, H. Igarashi, and M. Shimizu. 2003. Inactivation characteristics shown by enzymes and bacteria treated with far-infrared radiative heating. *International Journal of Food Science and Technology* 38:661–667.

Sheu, S. C., and A. O. Chen. 1991. Lipoxygenase as blanching index for frozen vegetable soybeans. *Journal of Food Science: An Official Publication of the Institute of Food Technologists* 56(2):448–451.

Soliva-Fortuny, R. C., P. Elez-Martinez, M. Sebastian-Caldero, and O. Martin-Belloso. 2002. Kinetics of polyphenol oxidase activity inhibition and browning of avocado puree preserved by combined methods. *Journal of Food Engineering* 55(2):131–137.

Soysal, C., and Z. Soylemez. 2005. Kinetics and inactivation of carrot peroxidase by heat treatment. *Journal of Food Engineering* 68(3):349–356.

Togrul, H. 2005. Simple modeling of infrared drying of fresh apple slices. *Journal of Food Engineering* 71(3):311–323.

Togrul, H. 2006. Suitable drying model for infrared drying of carrot. *Journal of Food Engineering* 77(3):610–619.

Unal, M. U. 2007. Properties of polyphenol oxidase from Anamur banana (*Musa cavendishii*). *Food Chemistry* 100(3):909–913.

Unal, M. U., and A. Sener. 2006. Determination of some biochemical properties of polyphenol oxidase from Emir grape (*Vitis vinifera* L. cv. Emir). *Journal of the Science of Food and Agriculture* 86(14):2374–2379.

Valderrama, P., and E. Clemente. 2004. Isolation and thermostability of peroxidase isoenzymes from apple cultivars Gala and Fuji. *Food Chemistry* 87(4):601–606.

Van Zuilichem, D. J., K. Van't Riet, and W. Stolp. 1986. An overview of new infrared radiation processes for various agricultural products. In *Food Engineering and Process Applications, Transport Phenomena*; Eds. M. Le Maguer, and P. Jelen, Vol. 1. New York: Elsevier, pp. 595–610.

Weemaes, C. A., L. R. Ludikhuyze, I. van-den Broeck, and M. E. Hendrickx. 1998. Effect of pH on pressure and thermal inactivation of avocado polyphenol oxidase: A kinetic study. *Journal of Agricultural and Food Chemistry* 46(7):2785–2792.

Zhu, Y. (2007). Processing and Quality Characteristics of Apple Slices under Simultaneous Infrared Dry-blanching and Dehydration (SIRDBD). Dissertation, Biological and Agricultural Engineering, University of California, Davis.

Zhu, Y., and Z. Pan. 2009. Processing and quality characteristics of apple slices under simultaneous infrared dry-blanching and dehydration with continuous heating. *Journal of Food Engineering* 90(4):441–452.

Zhu, Y., Z. Pan., T. H. McHugh, and D. Barrett. 2010. Processing and quality characteristics of apple slices processed under simultaneous infrared dry-blanching and dehydration with intermittent heating. *Journal of Food Engineering* 97(1):8–16.

10 Infrared Baking and Roasting

Servet Gülüm Sumnu and Semin Ozge Ozkoc

CONTENTS

10.1 Introduction 203
10.2 Baking...204
    10.2.1 IR Baking...205
    10.2.2 IR-Assisted Baking...207
10.3 Roasting 215
    10.3.1 IR Roasting 216
    10.3.2 IR-Assisted Roasting 218
10.4 Conclusion 218
References..220

10.1 INTRODUCTION

Infrared (IR) radiation is the part of the electromagnetic spectrum that is predominantly responsible for the heating effect of the sun (Ranjan et al., 2002). IR radiation wavelengths (0.76–1000 µm) are found in the region of wavelength between visible light and radio waves and can be divided into three different categories: near-infrared radiation (NIR; wavelength, 0.76–2 µm), mid-infrared radiation (MIR; wavelength, 2–4 µm), and far-infrared radiation (FIR; wavelength, 4–1000 µm; Sepulveda and Barbosa-Canovas, 2003).

IR heating is transferred by radiation and often has a high temperature (500°C–3000°C). The penetration depth of IR radiation significantly affects the surface temperature and surface moisture of the products. Penetration depths of IR radiation can vary significantly for various food materials. Datta and Ni (2002) showed that as IR radiation penetration depth decreases, the surface temperature increases. The penetration depth depends on moisture content and changes with wavelength significantly (Almedia et al., 2006).

The use of IR radiation technology for processing of foods has numerous advantages, including reduced drying time, alternative energy source, increased energy efficiency, uniform temperature in the product during drying, high-quality finished products, reduced necessity for airflow across the product, high degree of process control parameters, space saving along with clean working environment, decreased chance of flavor loss, preservation of vitamins in food products, and absence of solute migration from inner to outer regions (Ranjan et al., 2002; Navari et al., 1992;

Mongpreneet et al., 2002). Two application areas of IR radiation are IR baking and roasting. In terms of baking, using only NIR radiation with radiation source of halogen lamp resulted in inferior quality. The high rate of heating resulted in sudden crust formation and the prevention of expansion of the cake batter (Sumnu et al., 2005). However, there are some studies in the literature in which breads baked by IR had comparable quality with conventionally baked ones (Wade, 1987; Skjölderbrand and Andersson, 1989; Martinez-Bustos et al., 1999; Shyu et al., 2008). IR heating can be combined with microwave heating or convective heating. The studies on IR-assisted baking are on quality, staling and physical, transport, and physicochemical properties of breads, cakes, or cookies (Demirekler et al., 2004; Keskin et al., 2004a,b, 2005, 2007; Sumnu et al., 2005, 2007; Sakiyan et al., 2007a,b, 2009; Datta et al., 2007; Ozkoc et al., 2009a,b). Products baked by IR–microwave combination heating were found to have a quality comparable with conventionally baked ones. There are studies in the literature in which IR or IR-assisted roasting are used for roasting of tea leaves, peanut, hazelnut, sesame, and fish (Mahajan and Pai, 1988; Sakai and Hanzawa, 1994; Lee et al., 2003, 2006; Kim et al., 2006; Kumar et al., 2009; Uysal et al., 2009).

This chapter will briefly summarize the baking and roasting applications of IR and IR-assisted heating.

10.2 BAKING

Baking is a complex process causing a series of physical, chemical, and biochemical changes in food, such as gelatinization of starch, denaturation of protein, liberation of carbon dioxide from leavening agents, volume expansion, evaporation of water, crust formation, and browning reactions, as a result of simultaneous heat and mass transfer within the product and with the environment inside the oven. During baking, heat is transferred to the dough/batter by radiation, convection, and conduction. Pei (1982) has classified conventional baking into four baking stages: formation of white crust, heat transmission from crust to interior, gelatinization or cooking process, and browning. The baking step is one of the most important stages in the production of bakery products, which determines the quality, palatibility, and consumability of final products. Changes in some bakery products during baking are now described.

In bread making, as a result of heat application during baking, the raw dough piece is transformed into a light, porous, readily digestible, and flavorful product. Bread production requires a carefully controlled baking process to achieve the quality attributes required. The most important parameters affecting final product quality can be summarized as the rate and amount of heat applied, the humidity level in the baking chamber, and baking time (Therdthai et al., 2002).

Similarly, during cake baking, the batter changes from an emulsion to a porous structure (Ngo and Taranto, 1986). During baking, as temperature increases, the vapor pressure of water and the rate of formation of carbon dioxide gas increases, which further diffuses into air bubbles, resulting in the expansion of the cake batter (Mizukoshi, 1983; Mizukoshi et al., 1980). Further increase in temperature causes starch gelatinization and protein denaturation, affecting the development of porous structure in the final product.

During cookie baking, changes in structure (formation of porous structure, crisp texture), taste, color, size, and moisture content of dough pieces are observed, resulting in a palatable end product (Broyart and Trystram, 2003).

Different heating technologies, such as IR, microwave, jet impingement, and combined heating technologies, have an important role in the baking industry. These technologies have short processing time, product quality enhancement, and energy-saving advantages when compared to conventional heating.

10.2.1 IR Baking

According to Skjöldebrand (2001), IR baking can be divided into three periods. In the first period, there is an increase in surface temperature to 100°C. Weight loss of the product in this period is very little. In the second period, mass transfer starts. An evaporation zone is formed, which moves toward the central parts. In the final period, the center temperature of the product reaches about 98°C–99°C at the end of baking. The duration of this period is about 25% of the total baking time.

IR heating is known to provide significant advantages over conventional heating. However, besides successful applications (Wade, 1987; Skjölderbrand and Andersson, 1989; Martinez-Bustos et al., 1999; Shyu et al., 2008), there are some applications in literature, in which inferior product quality was reported in the case of IR baking (Keskin et al., 2004a; Sumnu et al., 2005). Wade (1987) investigated the effectiveness of biscuit baking by NIR compared to biscuit baking in a pilot-scale forced-convection oven. As an NIR source, quartz–tungsten tubes were used. It was stated by the researcher that the products obtained from the two types of baking process resembled each other closely in appearance, physical dimensions, and eating properties, and baking time was considerably reduced (Wade, 1987). Skjölderbrand and Andersson (1989) compared crust characteristics of breads (crust color, crust thickness) baked in IR and conventional ovens. It was stated by the researchers that time for desirable crust color formation at the surface was shorter for breads baked in an IR oven. Additionally, thinner crust and softer crumb were observed for IR-baked breads (Skjölderbrand and Andersson, 1989). Martinez-Bustos et al. (1999) used an IR heat source (blackbody) for baking of wheat flour tortillas. They investigated the efficiency of the IR baking method on quality (rollability, puffing, layering, color, texture) of tortillas. They found that IR-baked tortillas showed good quality characteristics. Moreover, energy used by IR ovens was found to be less than that used in commercial, gas-fired ovens. Sakai and Mao (2006) compared the performance of FIR and conventional ovens for baking of rice cracker and stated that usage of FIR oven reduced baking time and energy cost by 66.7% and 54.5%, respectively. Sato et al. (1992) compared NIR and FIR heaters for baking of white bread and showed that the rate of color development was greater with FIR heater than that with NIR heaters, since FIR heating resulted in more rapid increase in surface temperature. The effect of FIR oven on qualities (texture, volume, staling rate, and sensory evaluation) of four bakery products (bun bread, toast, pound cake, and sponge cake) was investigated by Shyu et al. (2008), and the results were compared with the ones baked in an electrical oven. It was found that the hardness value for sponge cake baked in the FIR oven was lower than for cakes baked in the electric oven, after 7 days' storage. On the other hand, volume, water activity, staling rate, and sensory quality of bakery products were similar for both baking methods.

There are other studies in the literature where unacceptable quality was observed in products baked by IR heating. Heist and Cremer (1990) studied the effects of different heating modes (IR, forced air convection, and conventional deck oven) on sensory quality for baking of cookies. Cookies baked in a convection oven were more acceptable in terms of flavor characteristics, softness, uniformity in round shape, and intensity of surface browning. The effects of different baking methods (conventional, microwave, IR (halogen lamp as NIR source), IR–microwave combination) on quality of breads were studied by Keskin et al. (2004a). It was noted that using only the IR heating mode of the oven resulted in formation of very thick bread crust and production of samples with lower specific volume values and higher firmness values than conventionally baked ones. Figure 10.1 shows the lower volumes achieved in the presence of IR baking as compared to conventional baking. Sumnu et al. (2005) investigated the quality (weight loss, specific volume, color, firmness) of cakes baked in different ovens (microwave, IR, IR–microwave combination). It was stated that higher IR power provided browning at the surface of cakes in a shorter time. However, it was not recommended to use IR heating mode (halogen lamps were used as NIR radiation source) alone in cake baking because of the thick crust formation. Since crust thickness is a function of temperature, the source of IR and power of IR are critical in achieving good quality.

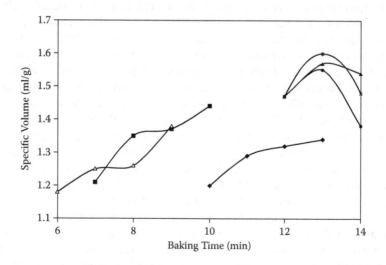

FIGURE 10.1 Effects of baking temperature (▲) 175°C; (✳) 200°C; (●) 225°C in conventional oven and halogen lamp power (◆) 50%; (■) 60%; (△) 70% in IR oven on variation of specific volume of breads during baking. (From Keskin et al. 2004a. *Food Research International* 37: 489–495. With permission from Elsevier.)

10.2.2 IR-Assisted Baking

In recent studies, combined heating technologies, such as IR–microwave heating technologies or IR–jet impingement heating technologies, have been used. The purpose of using combination technologies is to speed up processing time and improve product quality.

Several patents have been developed to provide surface browning and crispness by adding IR heat to microwaves (Eck and Buck, 1980; Fujii and Tsuda, 1987; Jung and Lee, 1992). A recent patent combines a microwave oven with halogen lamps (Lee, 2001). In this oven, the halogen lamp, 90% of the radiation energy of which has a wavelength of not longer than 1 µm, is used as the IR source. Both visible rays and IR rays are emitted from this lamp. Two upper halogen lamps are placed on the top wall of the cavity of the microwave oven. Two lower halogen lamps are installed on the bottom wall in such a way that they do not overlap the upper lamps. This oven is claimed to cook or heat the foods uniformly. An oven similar to that described in the patent mentioned earlier has been used in various baking studies in the literature. In these studies, in the design of the oven, there were only one halogen lamp located at the bottom and two halogen lamps placed at the top of the oven. These studies are about the investigation of the effect of IR–microwave combination baking on quality (texture, volume, porosity, color) of breads (Demirekler et al., 2004; Keskin et al., 2004a, 2007), cakes (Sumnu et al., 2005; Sakiyan et al., 2007b), rice cakes (Turabi et al., 2008a,b), and cookies (Keskin et al., 2005); on staling (Keskin et al., 2004b; Ozkoc et al., 2009a), pore characteristics (Datta et al., 2007; Ozkoc et al., 2009a) of breads, transport properties of breads (Keskin et al., 2007; Sumnu et al., 2007) and cakes (Sakiyan et al., 2007a), and on starch gelatinization of cookies (Keskin et al., 2005) and cakes (Sakiyan et al., 2009).

In IR–microwave combination ovens, two different heating mechanisms have been used together. Microwave heating speeds up the process, while IR heating promotes browning and crisping reactions. This technology has the potential to solve quality problems of microwave heating applications (when used alone). In microwave ovens, heat is absorbed by the food sample, and air surrounding the food is cold (Zuckerman and Miltz, 1997). The cool ambient temperature inside the microwave oven causes surface cooling of microwave-baked products, which prevents formation of Maillard reaction and caramelization products responsible for flavor and color (Decareau, 1992; Hegenbert, 1992). The disadvantages of microwave baking may be summarized as high moisture loss, soggy and gummy texture, lack of color, and crust formation and rapid staling (Sumnu, 2001). Combining microwaves with other modes of heating is a solution developed to overcome these problems and to provide more control of the moisture transport while simultaneously increasing the speed of heating.

In IR–microwave combination heating, IR heating can act at different times and at different spatial locations relative to microwave heating, which allows increasing the spatial uniformity and the overall rate of heating (Datta et al., 2005). Moreover, the penetration depth for NIR radiation is less than for microwave although they are of the same order of magnitude (Almedia et al., 2006). This shows that there will be a higher rate of heating near the surface that promotes color and crust formation. The selectivity of the combination heating can also be used to improve moisture distribution inside the food, by heating the surface of a food faster, which can help in removing moisture easily from the surface and keeping it crisp (Datta et al., 2005).

Keskin et al. (2004a) compared different heating modes in terms of their effect on bread quality. They found that breads baked in an IR–microwave combination oven had specific volume and color values comparable with the conventionally baked breads, but their weight loss and firmness values were still higher. Moreover, they stated that the IR–microwave combination oven reduced the conventional baking time of breads by about 75% under the studied baking conditions. In another study by Demirekler et al. (2004), it was found that breads baked in an IR–microwave combination oven for 5 min at 70% upper halogen lamp power, 50% lower halogen lamp power and 20% microwave power had comparable quality with conventionally baked ones in terms of color, textural characteristics, specific volume, and porosity. When an IR–microwave combination oven was used, 60% reduction in conventional baking time was obtained. In this study, with IR–microwave combination baking, lower and acceptable firmness and weight loss values were obtained, but less reduction in conventional baking time was achieved as compared to the study performed by Keskin et al. (2004a). The reason of this difference was the usage of water in the oven during baking to provide humidity in the study of Demirekler et al. (2004). Keskin et al. (2005) studied the effects of IR–microwave combination baking on the quality, gelatinization, and pasting properties of cookies. Microwave and IR power levels and baking time were found to be effective on moisture content, hardness, and spread ratio values of cookies. The best baking condition for cookies baked in an IR–microwave combination oven was determined as 5.5 min baking at 70% halogen lamp and 20% microwave power level. By using IR–microwave combination heating, a 50% reduction in conventional baking time was achieved. Sumnu et al. (2005) determined the quality of cakes baked in an IR–microwave combination oven. They stated that the best conditions for baking of cakes in an IR–microwave combination oven was 5 min at 70% halogen lamp and at 50% microwave power levels and, under these conditions, it was observed that the cake samples had similar color and firmness values as conventionally baked ones. Similar to the results of the study by Keskin et al. (2004a), baking in an IR–microwave combination oven under the given conditions reduced conventional baking time of cakes by about 75%. Turabi et al. (2008a) studied the effects of emulsifier content, halogen lamp power (as NIR source), and baking time on quality of gluten-free cakes containing 1% xanthan gum by using RSM. The increase in halogen lamp power decreased the volume of cakes since the thicker crust was formed on the cakes compressed the structure more, resulting in lower volumes (Figure 10.2a). Coded values of baking times and halogen lamp power of −1 to +1 means 7 to 8 min and 50% to 70%, respectively. The increase in both baking time and halogen lamp power increased total color change (Figure 10.2b). High halogen lamp power means higher temperatures, which enhance Maillard browning reactions.

Various studies discussed earlier have shown that conventional baking time was significantly reduced in IR–microwave combination heating (Keskin et al., 2004a, 2005; Demirekler et al., 2004; Sevimli et al., 2005; Turabi et al., 2008a). Table 10.1 shows a comparison of baking times in different ovens for different products.

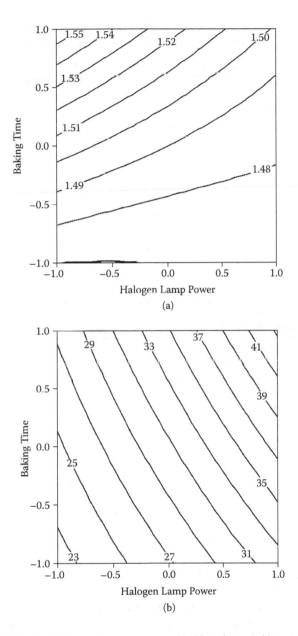

FIGURE 10.2 Effects of halogen lamp power and baking time on (a) specific volume and
(b) total color change of rice cakes formulated with xanthan gum. (From Turabi et al. 2008a.
Food and Bioprocess Technology 1: 64–73. With permission from Springer Science and
Business Media.)

TABLE 10.1

Comparison of Baking Conditions of Different Products and Percentage Reduction in Conventional Baking Time for Different Ovens

Product	Oven Type and Baking Conditions		Percentage Reduction in Conventional Baking Time	Reference
	IR Oven	Conventional Oven		
Baguettes	NIR, 50%, t: 2.25 min	T: 180°C t: 5 min	55%	Olsson et al., 2005
Tortilla	MIR, T: 549°C, t: 17 s	Hot griddle, T: 220°C, t: 3 min	~90%	Martinez-Bustos et al., 1999
	MIR, T: 549°C, t: 17 s	Three-tier, gas-fired oven $T_{1,2,3}$: 232°C, 273°C, 232°C (top, middle, bottom tier temperatures), t: 40 s	~58%	

	IR–Microwave Combination Oven			
Hamburger bread	UH: 70%, LH: 50% MW: 20%, t: 5 min	T: 200°C t: 13 min	~60%	Demirekler et al., 2004
Cake	UH: 60%, LH: 70% MW: 30%, t: 5 min	T: 175°C t: 24 min	~79%	Sevimli et al., 2005
Rice cake	UH: 60%, LH: 70% MW: 40 %, t: 7 min	T: 175°C t: 30 min	~75%	Turabi et al., 2008a
Cookie	UH and LH: 70% MW: 20%, t: 5.5 min	T: 205°C t: 11 min	~50%	Keskin et al., 2005

Note: UH: Upper halogen lamp power, LH: lower halogen lamp power, MW: microwave power, t: baking time, T: temperature.

In order to produce products baked in an IR–microwave combination oven with comparable volume, texture, and eating quality with conventionally prepared ones, new product development is required. In the literature, conventional formulations were improved or new formulations were designed by using some additives such as gums, emulsifiers, and enzymes to solve the problem of toughness or firmness in IR–microwave combination baked products. Keskin et al. (2004b) aimed to reduce the firmness of bread samples related to microwave and IR–microwave combination heating by using enzymes (fungal α-amylase, xylanase, lipase, protease). In microwave and IR–microwave combination baking, all of the enzymes were effective in reducing initial crumb firmness and firmness during storage and in increasing the specific volume of breads. They recommended the use of enzymes to reduce the firmness of bread samples baked in an IR–microwave combination oven. Keskin et al. (2007) used gums in bread formulations to overcome quality problems (high weight loss and high crumb firmness) of breads baked in an IR–microwave combination oven. Since the addition of xanthan–guar gum blend to the formulation resulted in an increase in specific volume and porosity and a decrease in firmness of breads, they recommended that a xanthan–guar blend be used in bread formulations for baking in IR–microwave combination ovens. Sakiyan et al. (2007b) investigated the effect of emulsifier, fat, and fat-replacer addition on physical properties (color and textural characteristics) of cakes during baking in an IR–microwave combination oven. Hardness and color values of samples baked in an IR–microwave combination oven were found to be dependent on formulation. The results showed that use of protein-based fat replacer (Simplesse™) in the formulation resulted in a very hard texture. Moreover, samples formulated with 25% Simplesse™ were found to have higher total color change (ΔE) values than those with 25% fat and 3% Purawave™-(emulsifier) added. Turabi et al. (2008b) compared the quality characteristics of rice cakes baked in an IR–microwave combination oven prepared using different gums with or without an emulsifier blend. They stated that the highest specific volume values were obtained for cakes containing only xanthan gum in the absence of emulsifier blend. It was found that emulsifier blend addition to the formulation improved the quality of rice cakes, providing higher specific volume and porosity values and lower crumb firmness values.

The physicochemical (rheology, optical, stability), sensory (texture, appearance, flavor), nutritional (bioavailability), and transport properties of foods are largely dependent on the type of components present, the interactions among them, and their structural organization (McClements, 2007). When viewed from the structural organization standpoint, crumb structure is one of the major quality attributes of bakery products. The relationship between crumb structure and crumb appearance may be self-evident, but crumb structure is also a determinant of loaf volume (Zghal et al., 1999) and texture (Pyler, 1988). Therefore, it may be concluded that knowing the structure of breads may be helpful to predict many of the quality properties of bread (Scanlon and Zghal, 2001). The baking process creates a hierarchical structure of gas cells, resulting in a wide spectrum of cell sizes, from macro- to microscale within the crumb (Liu and Scanlon, 2003). The studies related to crumb micro- and macrostructure can be summarized as follows. Datta et al. (2007) characterized pores in breads baked using various heating modes (IR–microwave combination, jet impingement–microwave combination, and jet impingement). They stated that pore-size distribution is bimodal, and many techniques in combination should be used to obtain comprehensive information. Total porosity values of breads baked in an IR–microwave combination oven were found to be the lowest, followed by jet impingement–microwave combination and jet impingement ovens, in increasing order. The highest fraction of pores in the closed-cell structure was observed for samples baked in an IR–microwave combination oven, followed by jet impingement–microwave combination and jet impingement ovens in decreasing order. According to scanning electron microscopy (SEM) analysis, it was stated that breads baked in a jet impingement oven looked quite different than the ones baked in IR–microwave and jet impingement–microwave combination ovens. Ozkoc et al. (2009a) compared the micro- and macrostructure of breads baked in an IR–microwave combination oven and conventional ovens. According to SEM analysis, pores in control breads baked in a conventional oven were found to be smaller, and had a spherical, oval-like shape with a homogeneous closed-cell structure. On the other hand, pores of breads baked in an IR–microwave combination oven were found to be close to each other, resulting in coalescence of the gas cells and channel formation (Figure 10.3).

The degree of starch gelatinization is also an important parameter affecting bakery product quality. Heating mechanisms and baking conditions are known to affect the degree of gelatinization of bakery products. The degree of gelatinization of some bakery products (bread, cake, and cookie) baked in an IR–microwave combination oven was investigated in terms of peak viscosities (one of the parameters of the rapid viscoanalyser ([RVA]; Keskin et al., 2005; Ozkoc et al., 2009b; Sakiyan et al., 2009; Table 10.2). In general, the highest peak viscosity was obtained for products baked in the microwave oven, meaning that less gelatinization was achieved. Peak viscosity values of IR–microwave combination and conventional ovens were comparable. Keskin et al. (2005) investigated the gelatinization and pasting properties of cookies baked in microwave and IR–microwave combination ovens by RVA. In general, cookies baked in an IR–microwave combination oven was found to have gelatinization levels comparable with those of conventionally baked ones, which might result in similar sensorial impact. Similarly, in a study by Ozkoc et al. (2009b), no difference was found between peak viscosity values of breads baked in conventional and IR–microwave combination ovens. Sakiyan et al. (2009) investigated the degree of starch gelatinization of cakes during baking in different ovens (microwave, IR–microwave combination, and conventional). It was found that as baking time increased, the gelatinization level increased significantly for all baking types through DSC (differential scanning calorimetry) and RVA results. They stated that the combination of IR heating with microwaves solved the insufficient starch gelatinization problem of microwave baking.

Using different heating modes and changing baking conditions are two of the known approaches to solving the staling problem in bakery products. Ozkoc et al. (2009b) studied the staling mechanism of bread samples baked in an IR–microwave combination oven by various physicochemical, mechanical, and rheological methods. The results demonstrated that the staling degrees of bread samples baked in an IR–microwave combination oven were found to be similar to the ones baked in a conventional oven, in terms of the results of FTIR (Fourier transform infrared spectroscopy) and retrogradation enthalpy values. Moreover, it was found that IR–microwave combination heating partially solved the rapid staling problem of microwave heating in terms of hardness, setback viscosity, and total mass crystallinity values. Figure 10.4 shows the total mass crystallinity values of breads baked in different ovens. Crystallinity is related to starch retrogradation. Samples baked in a microwave oven had significantly higher crystallinity values, meaning that these products staled more. The crystallinity values of breads baked in an IR–microwave combination oven were comparable with those from a conventional oven.

Olsson et al. (2005) investigated the effect of air jet impingement and NIR (alone or in combination) on color and crust development during postbaking of par-baked

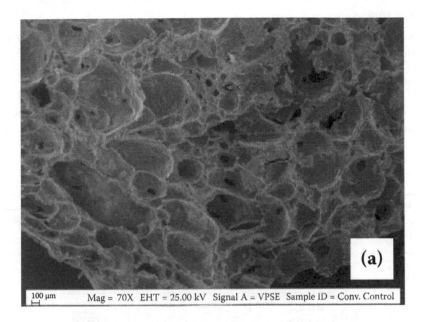

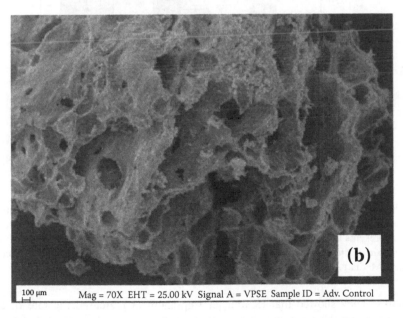

FIGURE 10.3 Microstructure of control breads at 70× magnification baked in (a) conventional and (b) IR–microwave combination ovens. (From Ozkoc et al. 2009a. *Food Hydrocolloids*. With permission from Elsevier.)

TABLE 10.2

Peak Viscosity of Some Bakery Products (Bread, Cake, and Cookie) Baked in Microwave, IR–Microwave Combination, and Conventional Ovens

	Oven Type			
Product	Microwave Oven	IR–Microwave Combination Oven	Conventional Oven	Reference
Bread (no-gum-added samples)	663 cP	290 cP	286 cP	Ozkoc et al., 2009b
Cake (nonfat samples, for the highest baking times)	127.50 cP	97.33 cP	74.00 cP	Sakiyan et al., 2009
Cake (25% fat-added samples, for the highest baking times)	147.33 cP	96.50 cP	80.00 cP	Sakiyan et al., 2009
Cookie (under the best baking conditions)	1969 cP	1834 cP	1655 cP	Keskin et al., 2005

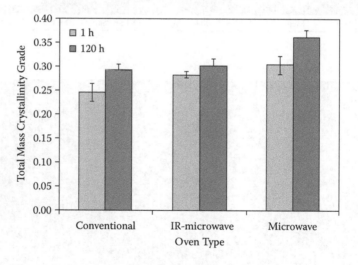

FIGURE 10.4 Variation in total mass crystallinity of breads baked in conventional, IR–microwave combination, and microwave ovens during storage. (Data from Figure 4, which represents total mass crystallinity of bread samples, in Ozkoc et al. 2009. *European Food Research and Technology* 228: 883–893. With permission from Springer Science and Business Media.)

baguettes. They showed that IR and jet impingement increased the rate of color development on the crust and shortened the baking time when compared to conventional heating. They concluded that IR–impingement combination heating provided control of heat and mass transfer during processing and achieved desired bread crust characteristics.

10.3   ROASTING

Roasting is a time–temperature dependent process that causes several changes such as heat exchange, chemical reactions, and drying. These changes lead to an improvement in flavor, color, texture, and appearance. In addition, free amino acids, peptides (Montavon et al., 2003), fatty acids, vitamin E, phytosterols, and lignans (Murkovic et al., 2004) are found to be changed during roasting process. By roasting, enzymes that cause nutrient loss can be inactivated, and undesirable microorganisms, toxins or allergens, and food contaminants can be destroyed (Ozdemir and Devres, 1999).

Since the enzymes responsible for browning are denatured due to the high temperatures employed during industrial roasting (>100°C), the possibility of enzymatic browning was considered to be negligible (Driscoll and Madamba, 1994). Therefore, the chemical reaction responsible for the enhancement of color, texture, flavor, and appearance is mainly nonenzymatic browning. Nonenzymatic browning is not a single reaction but a complex set of reactions. These reactions can be divided into three primary flavor-generating reactions: sugar caramelization, the Maillard reaction, and oxidation of ascorbic acid. The Maillard reaction involves a reaction between the carbonyl group of a reducing sugar with a free, uncharged amine group of an amino acid or protein with the loss of 1 mole of water (Buckholz et al., 1980; Mayer, 1985; Lopez et al., 1997). Thus, nonenzymatic browning may cause a decrease in the nutritive properties of the food associated with a certain decrease in protein digestibility and solubility of the product, loss of essential amino acids and vitamins, development of off-flavor, undesirable color, and textural changes, and increase in acidity (Villota and Hawkes, 1992; Labuza and Braisier, 1992; Jinap et al., 1998). Nonenzymatic browning products also have antioxidant properties. Antioxidant properties are related to the formation of phenolic-type structures and/or the metal chelating properties of melanoidins (O'Brien and Morrissey, 1989; Nicoli et al., 1991).

Since roasting is a time–temperature-dependent process, another similar process, drying, also occurs during roasting. Drying is considered to be mainly responsible for textural changes (Mayer, 1985; Saklar et al., 1999, 2001). Saklar et al. (1999) pointed out that moisture loss can be influenced by roasting temperature, roasting time, air velocity, product, and roaster characteristics, all of which are related to heating performance.

The main reason for applying heat treatment on certain foods is to promote flavor changes that fundamentally increase the overall palatability of the product. In this manner, peanuts, almonds, hazelnuts and other nuts, coffee, cocoa, and other similar products are subjected to heat treatments such as roasting. After a certain amount of heating, the volatile components are released, resulting in flavor enhancement. Free amino acids and monosaccharides are essential flavor precursors for the development of unique flavors during roasting (Mason et al., 1969). Among the amino acids, aspartic acid, glutamic acid, glutamine, histidine, asparagine, and phenylalanine are precursors of flavor together with threonine, tyrosine, and lysine. It was stated previously that sucrose participated in flavor development through its inversion to glucose and fructose during the browning process (Reyes et al., 1982).

Schenker et al. (2002) studied the effect of roasting conditions on the formation of aroma compounds in coffee beans. They confirmed that roasting was the main flavor-determining process. The greatest aroma formation rates during roasting were found during the medium stages of product dehydration. At least one roasting phase at a medium temperature level was found to be essential to generate sufficient aroma intensity, while high-temperature conditions could alter the aroma profile and should be avoided.

During roasting, foods can undergo rancidity reactions that cause spoilage, because of the odd colors and flavors formed. The major oxidative reactions in foods are due to peroxidation of lipids. Lipid oxidation in foods is associated almost exclusively with unsaturated fatty acids, and it is often autocatalytic, with oxidation products themselves catalyzing the reaction so that the rate increases with time (Karel, 1985). Lipid hydrolysis is an enzymatic reaction catalyzed by lipase. The hydrolysis of the lipids results in a progressive increase of the food acidity, caused by the formation of free fatty acids. Therefore, lipid hydrolysis favors lipid oxidation because the fatty acids formed can be substrates of the oxidation reaction (Richardson, 1984).

10.3.1 IR ROASTING

IR roasting has the advantages over conventional electrically heated drum roasting, such as higher production speeds, more compact installations, and lower investment costs (Kumar et al., 2009).

Pan et al. (2008, 2009) recently reported their notable industrial intervention that used IR and sequential IR and hot air (SIRHA) for dry roasting of almonds. Typically, industrial dry roasting process of almonds with hot air uses common temperature ranging from 265°F (130°C) to 310°F (154°C), and it may take 40–45 and 10–15 min to obtain a light- to medium-roasted product at the lower temperature and higher temperature, respectively. The almond industry has been concerned about the current dry roasting processes because of (1) the failure to ensure pasteurization of the product with a minimum 5 log reduction of *Salmonella enteriditis* PT 30 (SE PT 30) as required by the FDA standards, and (2) hot air roasting requires a relatively long processing time, thereby increasing processing costs. Pan et al. (2008, 2009) addressed these key issues, and in their milestone research, reported key deliverables in their case study that included (1) the appropriate IR heating conditions to achieve the desired product temperatures with minimum heating/roasting time, (2) the pasteurization efficacy of IR alone compared to SIRHA roasting and traditional hot air roasting, (3) the quality of almond kernels produced, and (4) recommendations about scaling up of the technology for commercial application. Their study found that SIRHA roasting was a much superior roasting method for producing pasteurized roasted almonds, with the potential to reduce costs due to reduced roasting time compared to current hot air roasting. They recommended that roasting using IR alone could be suitable only for pasteurization that targets 4 log bacterial reduction.

IR heating has been successful in the fields of coffee and green tea roasting. Patents related to IR roasting date back to 1968 (Arnold, 1968). In this patent, whole pecan nut halves were coated with gum arabic, salt, and spices. Coated nuts were moved through an IR tunnel in which IR was applied from above and below the belt to roast the pecans and to dry the coating film simultaneously. There are various patents where IR heating was combined with heated air for roasting of coffee (Tamaki et al., 1989; Nakamura and Ito, 1991). In most of the conventional roasters for coffee beans, hot air is used (Sakai and Mao, 2006). Air is heated to 350°C–450°C in a combustion unit and sent to the roasting unit. IR roasting is developed as a substitute for hot air. In the roasting device of Tamaki et al. (1989), in the roasting body, there was a rotary drum to accommodate and stir the coffee beans, and there were IR heaters to heat the coffee beans by radiant heat. In addition, hot air was circulated in the rotary drum into the roasting body in order to utilize the thermal energy effectively. Roasted coffee prepared in that roaster was expected to have high quality with respect to odor and taste. The roaster of Nakamura and Ito (1991) was designed to increase uniformity of roasting. A coffee bean roasting device included a heating oven and a roasting drum. IR heaters were attached to inner walls of the heating oven. There was also a cooling drum located under the roasting drum. In this device, coffee beans were uniformly roasted by the IR heaters while being agitated by the blades in the form of thin plates located on the inner walls of the drum.

There are various studies in the literature in which IR radiation was applied. Mahajan and Pai (1988) roasted whole peanuts by IR lamp at temperatures between 150°C and 230°C for 1 to 15 min and compared the quality results with those from conventional roasting. They concluded that peanuts roasted by IR took less roasting time and lost more oil even though the moisture content was fairly higher after roasting. Sakai and Hanzawa (1994) reported on the performance of IR systems with conventional ovens for roasting fish and found that more than 25% energy was saved with IR heating.

In the study of Brown et al. (2001), a computer control system was developed to regulate a continuous IR roaster used to process cereal grains. Bulk average temperature was used to adjust IR panel temperature and to achieve acceptable degree of roast.

Lee et al. (2006) compared the effects of FIR and conventional roasting on total phenolic contents and antioxidant activities of water extracts of peanut hulls. FIR emitted radiation at a wavelength of 2 to 14 μm. FIR and conventional heating were conducted at 150°C for 5–60 min. Total phenolic content and antioxidant activities increased as time of conventional or IR heating increased. FIR radiation was more effective than conventional heat treatment to increase phenolic content. This could be because FIR may have ability to cleave covalent bonds and liberate antioxidants such as flavonoids, carotene, tannin, ascorbate, flavoprotein, or polyphenols from repeating polymers (Niwa et al., 1988). It has been shown that conventional heat treatment cannot cleave covalently bound phenolic compounds from rice hulls, but FIR application can (Lee et al., 2003).

Kim et al. (2006) investigated the effects of FIR roasting and drying on the physicochemical properties of green tea. FIR roasting was achieved in an FIR heater at 200°C–300°C for 3–4 min. For comparison, green tea leaves were roasted in a heating pan at 200°C–300°C for 3–4 min. FIR irradiation increased total phenolic content significantly. FIR radiation in roasting may denature enzymes such as polyphenol oxidases more effectively than general heating, resulting in prevention of oxidation or polymerization of polyphenol compounds. Epigallocatechin, epigallo-catechin gallate, and caffeine contents increased in IR-roasted samples.

Kumar et al. (2009) studied the effects of IR roasting on the formation of sesamol and quality of defatted flours from sesame seed. Roasting of sesame seeds degraded the lignan sesamolin to sesamol, which increased the oxidative stability of sesame oil synergistically with tocopherols. IR roasting was achieved with a quartz NIR heating source. The IR roasting of sesame seeds at 200°C for 30 min increased the conversion of sesamolin to sesamol from 51% to 82% as compared to conventional roasting. There were no significant differences in functional properties of flours obtained from IR and conventionally roasted sesame seeds.

10.3.2 IR-ASSISTED ROASTING

IR–microwave combination heating has the advantage of reduction of conventional processing time. This method can be an alternative to conventional roasting. There are limited studies in the literature about the usage of IR–microwave combination roasting. In the study of Uysal et al. (2009), roasting conditions in IR–microwave combination oven were optimized. An IR source halogen lamp that gives NIR radiation was used.

To find the effect of microwave power, upper halogen lamp power, lower halogen lamp power, and roasting time on quality parameters, second-order polynomial equations were developed. The model equations are shown in Table 10.3.

As upper and lower halogen lamp powers increased, a^* value increased but L^* value decreased, since more NIR radiation was applied to hazelnut from the top and the bottom, leading to more browning reactions. However, upper halogen lamp power was found to be more significant than lower halogen lamp power (Table 10.3). Microwave power and upper halogen lamp power were found to be more significant than roasting time and lower halogen lamp power (Table 10.3) in their impact on the texture of hazelnut. As microwave power and upper halogen lamp powers increased, the force required to break a hazelnut kernel decreased.

For hazelnuts, 2.5 min roasting time at 90% microwave power, 60% upper halogen lamp power, and 20% lower halogen lamp power were found to be optimum roasting conditions. Hazelnuts roasted at this optimum point had quality comparable with conventionally roasted ones (at 150°C for 20 min) with respect to color, texture, moisture content, and fatty acid composition (Table 10.4).

10.4   CONCLUSION

IR baking may have advantages or disadvantages depending on IR source and processing time. In recent years, IR has been combined with microwaves or hot air to improve the product quality. IR-assisted baking has promising applications in production of cakes, breads, and cookies. These products have quality and physicochemical properties comparable with conventionally baked products. IR heating has been applied successfully to coffee and tea roasting. Although there are various studies on IR roasting in the literature, more research is needed in the area of IR-assisted roasting.

TABLE 10.3

Regression Equations for Roasted Hazelnuts in IR–Microwave Combination Oven

Quality Parameter	Equation	R^2_{adj}
$L*$ value	$Y_1 = 65.6332 - 3.27629X_1{}^{****a} - 1.57039X_2{}^{***b} - 1.20355X_3{}^{**c} - 2.77034X_4{}^{****} - 2.60224X_1{}^{2****} - 0.977237X_4{}^{2**} - 2.2205X_1X_2{}^{***} - 2.34283X_2X_4{}^{***}$	0.88
$a*$ Value	$Y_2 = 14.2337 + 2.19797X_1{}^{****} + 1.58857X_2{}^{****} + 0.176584X_3{}^{*d} + 1.41392X_4{}^{****} + 0.271334X_1{}^{2*} + 0.467093X_4{}^{2*}$	0.87
Fracture force (N)	$Y_3 = 42.1593 - 6.29307X_1{}^{****} - 4.99526X_2{}^{****} - 1.5716X_3 + 2.5842X_4{}^{*} + 1.77275X_2{}^{2*} + 1.76414X_3{}^{2*} - 3.07594X_1X_2{}^{*} + 3.10211X_2X_3{}^{*} - 3.74642X_3X_4{}^{**}$	0.77
Moisture content (%)	$Y_4 = 1.44565 - 0.46378X_1{}^{****} - 0.300785X_2{}^{****} - 0.199925X_3{}^{***} - 0.41899X_4{}^{****} - 0.150939X_3{}^{2***} + 0.182540X_1X_2{}^{**} - 0.242470X_1X_3{}^{***} + 0.151965X_1X_4{}^{*}$	0.93

Source: Uysal et al. 2009. *Journal of Food Engineering* 90: 255–261. With permission from Elsevier.
Note: X_1: Microwave power, X_2: upper halogen lamp power, X_3: lower halogen lamp power and X_4: processing time.
[a] Significant at 0.01% level, [b]Significant at 0.1% level, [c]Significant at 1% level, [d]Significant at 5% level.

TABLE 10.4

Response Values and Fatty Acid Composition for Hazelnuts Roasted in IR–Microwave Combination Oven and in Conventional Oven

	IR–Microwave Combination Oven Roasting at the Optimum Point	Conventional Oven Roasting
$L*$ value	50.26	52.17
$a*$ Value	17.15	18.03
Fracture force (N)	22.16	21.53
Moisture content (%)	1.39	1.22
Oleic acid (C18:1) (%)	76.44	78.80
Linoleic acid (C18:2) (%)	13.00	10.50
Palmitic acid (C16:0) (%)	5.88	5.87
Stearic acid (C18:0) (%)	3.11	3.27

Source: Uysal et al. 2009. *Journal of Food Engineering* 90: 255–261. With permission from Elsevier.

REFERENCES

Almedia, M., K. E. Torrance, and A. K. Datta. 2006. Measurement of optical properties of foods in near and midinfrared radiation. *International Journal of Food Properties* 9: 651–654.

Arnold, F. W. 1968. Infrared roasting of coated nutmeats. U.S. Patent 3,383,220.

Brown, R. B., T. M. Rothwell, and V. J. Davidson. 2001. A fuzzy controller for infrared roasting of cereal grain. *Canadian Biosystems Engineering* 43: 3.9–3.15.

Broyart, B., and G. Trystram. 2003. Modeling of heat and mass transfer phenomena and quality changes during continuous biscuit baking using both deductive and inductive (neural network) modeling principles. *Food and Bioproducts Processing* 81: 316–326.

Buckholz, L. L., H. Daun, and E. Stier. 1980. Influence of roasting time on sensory attributes of fresh roasted peanuts. *Journal of Food Science* 45: 547–554.

Datta, A. K., S. S. R. Greedipalli, and M. F. Almeida. 2005. Microwave combination heating. *Food Technology* 59: 36–40.

Datta, A. K., and H. Ni. 2002. Infrared and hot air-assisted microwave heating of foods for control of surface moisture. *Journal of Food Engineering* 51: 355–364.

Datta, A. K., S. Sahin, G. Sumnu, and S. O. Keskin. 2007. Porous media characterization of breads baked using novel heating modes. *Journal of Food Engineering* 79: 106–116.

Decareau, R. V. 1992. *Microwave Foods: New Product Development.* CT: Food Nutrition Press Inc.

Demirekler, P., G. Sumnu, and S. Sahin. 2004. Optimization of bread baking in halogen lamp–microwave combination oven by response surface methodology. *European Food Research and Technology* 219: 341–7.

Driscoll, R. H., and P. S. Madamba. 1994. Modeling the browning kinetics of garlic. *Food Australia* 46: 66–71.

Eck, P., and R. G. Buck. 1980. Method of browning food in a microwave oven. U.S. Patent 4396817.

Fujii, M., and T. Tsuda. 1987. Microwave heating and infrared ray heating appliance. U.S. Patent 4803324.

Hegenbert, S. 1992. Microwave quality: Coming of age. *Food Product Design* 17: 29–52.

Heist, J., and L. M. Cremer. 1990. Sensory quality and energy use for baking of molasses cookies prepared with bleached and unbleached flour and baked in infrared, forced air convection and conventional deck ovens. *Journal of Food Science* 55: 1095–1101.

Jinap, S., W. I. Wan-Rosli, A. R. Russly, and L. M. Nordin. 1998. Effect of roasting time and temperature on volatile component profile during nib roasting of cocoa beans (*Theobroma cacao*). *Journal of the Science of Food and Agriculture* 77: 441–48.

Jung, K. H., and S. Y. Lee. 1992. Microwave ovens with infrared rays heating units. U.S. Patent 5310979.

Karel, M. 1985. Control of lipid oxidation in dried foods. In *Concentration and Drying of Foods*, Ed. D. MacCarthy, 37–51. New York: Elsevier Applied Science Publishers.

Keskin, S. O., S. Ozturk, S. Sahin, H. Koksel, and G. Sumnu. 2005. Halogen lamp-microwave combination baking of cookies. *European Food Research and Technology* 220: 546–551.

Keskin, S.O., G. Sumnu, and S. Sahin. 2004a. Bread baking in halogen lamp-microwave combination oven. *Food Research International* 37: 489–495.

Keskin, S.O., G. Sumnu, and S. Sahin. 2004b. Usage of enzymes in novel baking process. *Nahrung* 48: 156–160.

Keskin, S. O., G. Sumnu, and S. Sahin. 2007. A study on the effects of different gums on dielectric properties and quality of breads baked in infrared-microwave combination oven. *European Food Research and Technology* 224: 329–334.

Kim, S. Y., S. M. Jeong, and S. C. Jo. 2006. Application of far-infrared radiation in manufacturing of process green tea. *Journal of Agricultural and Food Chemistry* 54: 9943–9947.

Kumar, C. M., A. G. Rao, and S. A. Singh. 2009. Effect of infrared heating on the formation of sesamol and quality of defatted flours from *Sesamum indicum* L. *Journal of Food Science* 74: 105–111.

Labuza, T. P., and V. M. Braisier. 1992. The kinetics of nonenzymatic browning. In *Physical Chemistry of Foods*, Eds. H.G. Schwartzberg and R.W. Hartel, 595–649. New York: Dekker.

Lee, K. H. 2001. Microwave oven with halogen lamps. U.S. Patent 6,172,347.

Lee, S. C., S. M. Jeong, S. Y. Kim, H. R. Park, K. C. Nam, and D. U. Ahn. 2006. Effect of far infrared radiation and heat treatment on the antioxidant activity of water extracts from peanut hulls. *Food Chemistry* 94: 489–493.

Lee, S. C., J. H. Kim, S. M. Jeong, D. R. Kim, J. U. Ha, K. C. Nam, and D. U. Ahn. 2003. Effect of far infrared radiation on the antioxidant activity of rice hulls. *Journal of Agricultural and Food Chemistry* 51: 4400–4403.

Liu, Z., and M. G. Scanlon. 2003. Predicting mechanical properties of bread crumb. *Food and Bioproducts Processing: Transactions of Institute of Chemical Engineers Part C* 81: 224–238.

Lopez, A., M. T. Pique, J. Boatella, A. Romero, and J. Garcia. 1997. Influence of drying conditions on the hazelnut quality: III. Browning. *Drying Technology* 15: 989–1002.

Mahajan, G. N., and J. S. Pai 1988. Defatting of whole peanut kernels after infrared heating. *Food Chemistry* 27: 237–240.

Martinez-Bustos, F., S. E. Morales, Y. K. Chang, A. Herrera-Gomez, M. J. L. Martinez, L. Banos, M. E. Rodriguez, and M. H. E. Flores. 1999. Effect of infrared baking on wheat flour tortilla characteristics. *Cereal Chemistry* 76: 491–495.

Mason, M. E., J. A. Newell, B. R. Johnson, and P. E. Koehler. 1969. Nonvolatile flavor components of peanuts. *Journal of Agricultural and Food Chemistry* 17: 728–732.

Mayer, K. 1985. Infra-red roasting of nuts, particularly hazelnuts. *Confectionery Production* 51: 313.

McClements, D. J. 2007. *Understanding and Controlling the Microstructure of Complex Foods*, Ed. D. J. McClements. Cambridge: CRC Press.

Mizukoshi, M. 1983. Model studies of cake baking IV. Foam drainage in cake batter. *Cereal Chemistry* 60: 399–402.

Mizukoshi, M., H. Maeda, and H. Amano. 1980. Model studies of cake baking II. Expansion and heat set of cake batter during baking. *Cereal Chemistry* 57: 352–355.

Mongpreneet, S, T. Abe, and T. Tsurusaki. 2002. Accelerated drying of welsh onion by far infrared radiation under vacuum conditions. *Journal of Food Engineering* 55: 147–156.

Montavon, P., A. F. Mauron, and E. Duruz. 2003. Changes in green coffee protein profiles during roasting. *Journal of Agriculture and Food Chemistry* 51: 2335–2343.

Murkovic, M., V. Piironen, A. M. Lampi, T. Kraushofer, and G. Sontag. 2004. Changes in chemical composition of pumpkin seeds during the roasting process for production of pumpkin seed oil (Part 1, Non-volatile compounds). *Food Chemistry* 84: 359–365.

Nakamura, A., and T. Ito. 1991. Coffee bean roasting device. U.S. Patent 5,016,362.

Navari, P., J. Andrieu, and A. Gevaudan. 1992. Studies on infrared and convective drying of nonhygroscopic solids. In *Drying 92*, Ed. A. S. Mujumdar, 685–694. Amsterdam: Elsevier Science.

Ngo, W. H., and M. V. Taranto. 1986. Effect of sucrose level on the rheological properties of cake batters. *Cereal Foods World* 31: 317–322.

Nicoli, M. C., B. E. Elizalde, A. Pitotti, and C. R. Lerici. 1991. Effect of sugars and maillard reaction products on polyphenol oxidase and peroxidase activity in food. *Journal of Food Biochemistry* 15: 169–184.

Niwa, Y., T. Kanoh, T. Kasama, and M. Neigishi. 1988. Activation of antioxidant activity in natural medicinal products by heating, brewing and lipophillization. A new drug delivery system. *Drugs under Experimental and Clinical Research* 14: 361–372.

O'Brien, J., and P. A. Morrissey. 1989. Nutritional and toxicological aspects of the Maillard browning reaction in foods. *Critical Reviews in Food Science and Nutrition* 28: 211–248.

Olsson, E. E. M., A. C. Tragardh, and L. M. Ahrne. 2005. Effect of near-infrared radiation and jet impingement heat transfer on crust formation of bread. *Journal of Food Science* 70: E484–E491.

Ozdemir, M., and O. Y. Devres. 1999. The thin layer drying characteristics of hazelnuts during roasting. *Journal of Food Engineering* 42: 225–233.

Ozkoc, S. O., G. Sumnu, and S. Sahin. 2009a. The effects of gums on macro and micro-structure of breads baked in different ovens, *Food Hydrocolloids* 23: 2182–2189.

Ozkoc, S. O., G. Sumnu, S. Sahin, and E. Turabi. 2009b. Investigation of physicochemical properties of breads baked in microwave and infrared-microwave combination ovens during storage. *European Food Research and Technology* 228: 883–893.

Pan, Z., J. Yang, G. Bingol, and T. McHugh, 2009. Infrared Heating for Raw Almond Pasteurization. Report for Almond Board of California, 1–23.

Pan, Z., J. Yang, M. Brandl, H. T. McHugh, and G. Bingol. 2008. Infrared Heating for Improved Safety and Processing Efficiency of Dry-Roasted Almonds. Report for Almond Board of California, 1–20.

Pei, D. C. 1982. Microwave baking: New developments. *Bakers Digest* 56: 8–12.

Pyler, E. J. 1988. *Baking Science and Technology*. Kansas State, Sosland.

Ranjan, R., J. Irudayaraj, and S. Jun. 2002. Simulation of three-dimensional infrared drying using a set of three-coupled equations by the control volume method. *Transactions of ASAE* 45: 1661–1668.

Reyes, F. G., R. B. Poocharoen, and R. E. Wrolstad. 1982. Maillard browning reaction of sugar–glycine model systems: Changes in sugar concentration, color and appearance. *Journal of Food Science* 47: 1376–1377.

Richardson, T. 1984. Controlling acyl transfer reactions of hydrolases to alter food constituents. In *Chemical Changes in Food during Processing*, Eds. T. Richardson and J. W. Finley, 219–254. Westport: Avi Publishing Company, Inc.

Sakai, N., and T. Hanzawa. 1994. Applications and advances in far-infrared heating in Japan. *Trends in Food Science and Technology* 5: 357–362.

Sakai, N., and W. Mao. 2006. Infrared Heating. In *Thermal Food Processing: New Technologies and Quality Issues*, Ed. D.W. Sun, 493–525. Boca Raton, FL: CRC Press/Taylor and Francis.

Sakiyan, O., G. Sumnu, S. Sahin, and V. Meda. 2007a. Investigation of dielectric properties of different cake formulations during microwave and infrared-microwave combination baking. *Journal of Food Science* 72: E205–E213.

Sakiyan, O., G. Sumnu, S. Sahin, and V. Meda. 2007b. The effect of different formulations on physical properties of cakes baked with microwave and near infrared-microwave combinations. *Journal of Microwave Power and Electromagnetic Energy* 41: 17–23.

Sakiyan, O., G. Sumnu, S. Sahin, V. Meda, H. Koksel, and P. Chang. 2009. A study on degree of starch gelatinization in cakes baked in three different ovens. *Food and Bioprocess Technology*.

Saklar, S., S. Katnas, and S. Ungan, S. 2001. Determination of optimum hazelnut roasting conditions. *International Journal of Food Science and Technology* 36: 271–281.

Saklar, S., S. Ungan, and S. Katnas. 1999. Instrumental crispness and crunchiness of roasted hazelnuts and correlations with sensory assessment. *Journal of Food Science* 64: 1015–1019.

Sato, H., K. Hatae, and A. Shimada. 1992. Effects of radiant characteristics of crust formation and coloring process of food surface. Studies of radiative heating condition of food: Part 1. *Nippon Shokuhin Kogyo Gakkaishi* 39: 784–789.

Scanlon, M. G., and M. C. Zghal. 2001. Bread properties and crumb structure. *Food Research International* 34: 841–864.

Schenker, S., C. Heinemann, M. Huber, R. Pompizzi, R. Perren, and F. Escher. 2002. Impact of roasting conditions on the formation of aroma compounds in coffee beans. *Journal of Food Science* 67: 60–66.

Sepulveda, D. R., and G. V. Barbosa-Canovas. 2003. Heat transfer in food products. In *Transport Phenomena in Food Processing*, Eds. J. Welti-Chanes, J. F. Velez-Ruiz, and G. V. Barbosa-Canovas, 42. Boca Raton, FL: CRC Press.

Sevimli, K. M., G. Sumnu, and S. Sahin. 2005. Optimization of halogen lamp–microwave combination baking of cakes: A response surface methodology study. *European Food Research and Technology* 221: 61–68.

Shyu, Y. S., W. C. Sung, M. H. Chang, and J. Y. Hwang. 2008. Effect of far-infrared oven on the qualities of bakery products. *Journal of Culinary Science and Technology* 6: 105–118.

Skjöldebrand, C. 2001. Infrared heating. In *Thermal Technologies in Food Processing*, Ed. P. Richardson, 493–525. Cambridge, England: Woodhead Publishing Limited.

Skjöldebrand, C., and C. Andersson. 1989. A comparison of infrared bread baking and conventional baking. *Journal of Microwave Power and Electromagnetic Energy* 24: 91–101.

Sumnu, G. 2001. A review on microwave baking of foods. *International Journal of Food Science and Technology* 36: 117–127.

Sumnu, G., A. K. Datta, S. Sahin, S. O. Keskin, and V. Rakesh. 2007. Transport and related properties of breads baked using various heating modes. *Journal of Food Engineering* 78: 1382–1387.

Sumnu, G., S. Sahin, and K. M. Sevimli. 2005. Microwave, infrared and infrared-microwave combination baking of cakes. *Journal of Food Engineering* 71: 150–155.

Tamaki, Y., T. Kino, T. Ito, and A. Nakamura. 1989. Coffee beans roasting device. U.S. Patent 4,860,461.

Therdthai, N., W. Zhou, and T. Adamczak. 2002. Optimization of the temperature profile in bread baking. *Journal of Food Engineering* 55: 41–48.

Turabi, E., G. Sumnu, and S. Sahin. 2008a. Optimization of baking of rice cakes in infrared microwave combination oven by response surface methodology. *Food and Bioprocess Technology* 1: 64–73.

Turabi, E., G. Sumnu, and S. Sahin. 2008b. Rheological properties and quality of rice cakes formulated with different gums and an emulsifier blend. *Food Hydrocolloids* 22: 305–312.

Uysal, N., G. Sumnu, and S. Sahin. 2009. Optimization of microwave-infrared roasting of hazelnut. *Journal of Food Engineering* 90: 255–261.

Villota, R., and J. G. Hawkes. 1992. Reaction kinetics in food systems. In *Handbook of Food Engineering*, Eds. D. R. Heldman and D. B. Lund, 39–144. New York: Marcel Dekker.

Wade, P. 1987. Biscuit baking by near-infrared radiation. *Journal of Food Engineering* 6: 165–175.

Zghal, M. C., M. G. Scanlon, and H. D. Sapirstein. 1999. Prediction of bread crumb density by digital image analysis. *Cereal Chemistry* 76: 734–742.

Zuckerman, H., and J. Miltz. 1997. Prediction of dough browning in the microwave oven from temperatures at the susceptor/product interface. *Lebensmittel-Wissenschaft und Technologie* 30: 519–524.

11 Infrared Radiation Heating for Food Safety Improvement

Kathiravan Krishnamurthy, Soojin Jun,
Joseph Irudayaraj, and Ali Demirci

CONTENTS

11.1 Introduction 225
11.2 Pathogen Inactivation 226
    11.2.1 Effect of Power and Sample Temperature 226
    11.2.2 Effect of Peak Wavelength and Bandwidth 227
    11.2.3 Effect of Sample Depth 227
    11.2.4 Types of Microorganisms 228
    11.2.5 Types of Food Materials 228
11.3 Inactivation Mechanism 228
11.4 Selective Heating by IR Radiation 231
11.5 Thermal Death Kinetics Model 234
11.6 Conclusion and Future Research Potential 234
References 235

11.1 INTRODUCTION

Consumption of contaminated foods can cause food-borne illnesses. In the United States alone, foodborne diseases are estimated to cause approximately 76 million illnesses, 325,000 hospitalizations, and 5,000 deaths every year. There are several causes of food-borne illnesses, including bacteria, viruses, prions, parasites, toxins, and metals (Mead et al., 1999). A major public health concern of today is food contaminated with pathogenic microorganisms such as *Salmonella* spp., *Clostridium perfringens*, *Staphylococcus aureus*, *Campylobacter jejuni*, *Escherichia coli* O157:H7, and *Listeria monocytogenes* (Krishnamurthy et al., 2004). Buzby and Roberts (1997) and Mead et al. (1999) estimated that the annual dollar costs of foodborne illnesses, including medical expenses and lost wages and productivity, ranges from $6.5 to $34.9 billion. Therefore, it is crucial to inactivate the pathogenic microorganisms present in the food to ensure the safety of the products. There are several preservation methods available to effectively inactivate the pathogens. However, though effective in pathogen reduction, most of the preservation techniques result in quality deterioration. Therefore, there is always a need for investigating the efficacy of novel food processing technologies.

Traditionally, food materials undergo a heat treatment step to achieve pasteurization or sterilization condition. However, due to exposure to high temperature, there is a significant loss in the sensory and nutritional quality of the food. By reducing the total thermal energy supplied, quality changes due to temperature can be minimized. Infrared (IR) radiation heating is an efficient form of applying heat compared to conventional heating as it results in energy reduction of up to 245% (Afzal et al., 1999). IR heating also gained popularity due to its higher thermal efficiency and fast heating rate/response time in comparison to conventional heating (Krishnamurthy et al., 2008).

Although IR heating has been used in the food industry for over two centuries, its applications to food safety have not been widely investigated. IR heating has been widely applied to various thermal unit operations such as dehydration, frying, and pasteurization (Sakai and Hanzawa 1994; Krishnamurthy et al., 2010). However, in recent years, application of IR heating for pathogen inactivation has gained much attention.

11.2   PATHOGEN INACTIVATION

IR heating can be used to inactivate bacteria, spores, yeast, and mold in both liquid and solid foods. The efficacy of microbial inactivation by IR heating depends on the following parameters: IR power level, temperature of food sample, peak wavelength, and bandwidth of IR heating source, sample depth, types of microorganisms, physiological phase of microorganisms (exponential or stationary phase), and types of food materials. Therefore, several researchers have investigated the effects of these parameters on inactivation of pathogenic microorganisms as follows.

11.2.1 EFFECT OF POWER AND SAMPLE TEMPERATURE

Increase in the power of IR heating source produces more energy and, thus, total energy absorbed by microorganisms increases, leading to microbial inactivation. Sterilization of wheat surface was investigated by Hamanaka et al. (2000). The surface temperature increased rapidly as IR rays directly heated the surface without any need for conductors. Therefore, irradiating powers of 0.5, 1.0, 1.5, and 2.0 kW resulted in temperatures of 60°C, 80°C, 125°C, and 195°C inside the experimental device, and 45°C, 65°C, 95°C, and 120°C on the surface of the wheat stack, obtaining 0.83, 1.14, 1.18, and 1.90 $\log_{10}$ colony-forming units (CFU)/g total bacteria after a 60 s treatment, respectively. Dry heat inactivation of B. subtilis spores by IR radiation was investigated by Molin and Ostlund (1975). D values of B. subtilis at 120°C, 140°C, 160°C, and 180°C were 26 min, 66 s, 9.3 s, and 3.2 s, respectively. Shorter treatment time was enough to inactivate pathogens at higher temperatures, and the estimated Z value was 23°C. E. coli population was reduced by 0.76, 0.90, and 0.98 $\log_{10}$ after 2-min exposure to IR radiation when the temperature of the bacterial suspension was maintained at 56°C, 58°C, and 61°C, correspondingly (Sawai et al., 2003).

11.2.2 Effect of Peak Wavelength and Bandwidth

As indicated earlier, food and microbial components absorb certain wavelengths of IR radiation. Therefore, it is beneficial to investigate the absorption pattern of key components in order to ensure pathogen inactivation and minimize changes in food quality. It would be feasible to selectively heat the microorganisms present in food products without adversely increasing the temperature of sensitive food components. Jun and Irudayaraj (2003) utilized selective infrared heating in the wavelength range of 5.88 to 6.66 μm using optical bandpass filters for inactivation of *Aspergillus niger* and *Fusarium proliferatum* in corn meal. The selected wavelength denatures the protein in microorganisms, leading to a 40% increase in inactivation of *A. niger* and *F. proliferatum*, compared to normal IR heating. For instance, a 5 min treatment with nonselective and selective heating resulted in approximately 1.8 and 2.3 $\log_{10}$ CFU/g reduction of *A. niger*. Similarly, reductions of 1.4 and 1.95 $\log_{10}$ CFU/g of *F. proliferatum* were obtained with 5 min of nonselective and selective heating, respectively. Although the sample temperatures after selective or nonselective IR heating were identical, absorption of energy by fungal spores increased in selective heating, leading to a higher lethal rate (Jun and Irudayaraj, 2003).

Total energy decreases as the peak wavelength increases. Therefore, near-IR (NIR) radiation with short wavelength has a relatively higher energy level than far-IR (FIR) radiation with longer wavelength. Hamanaka et al. (2006) studied the inactivation efficacy of *Bacillus subtilis* treated with three IR heaters (A, B, and C) having different peak wavelengths (950, 1100, and 1150 nm) and radiant energies of 4.2, 3.7, and 3.2 $\mu W/cm^2/nm$, respectively. Air-dried *Bacillus subtilis* solution placed on a stainless steel petri dish was treated with IR heating after water activity adjustment using a desiccator. The surface temperature of the petri dish was 100°C after a 2 min exposure for all the heaters. Pathogen inactivation was higher with heater A than with heaters B and C, although the temperature was the same for all the heaters. For example, at a water activity of 0.7, decimal reduction times of heaters A, B, and C were approximately 4, 12, and 22 min, respectively. Therefore, it is obvious that inactivation efficiency is associated with the radiation spectrum (Hamanaka et al., 2006).

11.2.3 Effect of Sample Depth

The penetration depth of IR radiation is very low. An increase in sample depth slows down the bulk temperature increase of the food sample (Hashimoto et al., 1991). A 90% reduction in IR power was observed within a thin layer of 40 μm in bacterial suspension (Hashimoto et al., 1997). Therefore, the effect of IR radiation on microbial inactivation diminishes as sample thickness increases. Decreasing the sample depth also accelerates the inactivation of spores (Sawai et al., 1997) and *E. coli* and *S. aureus* (Hashimoto et al., 1992a). The ratio of number of injured cells to the number of survivors increased as the depth decreased. For example, the *S. aureus* population was reduced by approximately 2 and 5 $\log_{10}$ CFU/mL at 321 K, when the sample depths were 2.9 and 0.9 mm, respectively. Similarly, the *E. coli* population in the samples with 1.3 and 2.2 mm in depth showed approximately 1.33 and 1.66 log_{10} CFU/mL at 321 K.

11.2.4 TYPES OF MICROORGANISMS

Resistance of bacteria, yeasts, and molds to IR heating may be different due to their structural and compositional differences. In general, spores are more resistant than vegetative cells. When *Bacillus subtilis* spores in physiological saline were exposed to IR heating, the spore population increased up to five times in the first 2 min, followed by subsequent exponential reduction, resulting in shoulder and tailing effects. Upon IR heat treatment, vegetative cells were inactivated, followed by activation of spores. Then, vegetative cells formed from spores will be activated and, thus, spores will be inactivated. If inactivation occurs in sequence, there will be tailing and shoulder effects. An initial increase in *B. subtilis* population was caused by heat shock germination of spores. A 10-min treatment with IR heating resulted in a more than 90% reduction in the *B. subtilis* population (Daisuke et al., 2001). Hamanaka et al. (2006) also reported a shoulder effect where *B. subtilis* spores were germinated.

Cereal surface is often contaminated with spore formers such as *Bacillus*, *Aspergillus*, and *Penicillium*. Wheat was treated with IR heating at 2.0 kW for 30 s, followed by cooling for 4 h and again treated for 30 s with IR heating to obtain a 1.56 log_{10} CFU/g reduction. The irradiation helped in activation of spores into vegetative cells, and the second irradiation effectively inactivated spore formers. Furthermore, intermittent treatment can minimize the quality changes, as continuous treatment longer than 50 s resulted in discoloration of wheat surface (Hamanaka et al., 2000).

Naturally occurring yeasts in honey were completely inactivated with an 8 min IR heat treatment (Hebbar et al., 2003). The temperature of the honey was raised to 110°C after the treatment, resulting in microbial reduction of 3.85 log_{10} CFU/mL.

11.2.5 TYPES OF FOOD MATERIALS

As described earlier, IR radiation has poor penetration capacity. However, the surface temperature of food materials increases rapidly, and heat is transferred inside food materials by thermal conduction. Typical thermal conductivities of solid foods are much lower than those of liquid foods. Convective heat transfer, which occurs inside liquid foods under IR heating, can contribute to an increase in the lethality of microbes. A summary of the study pertinent to pathogen inactivation in different types of food materials such as solid, liquid, and nonfood materials is given in Table 11.1.

11.3   INACTIVATION MECHANISM

Inactivation of microorganisms by IR heating may include an inactivation mechanism similar to that of ultraviolet light (DNA damage) and microwave heating (induction heating) in addition to a thermal effect, as IR lies between ultraviolet and microwave in the electromagnetic spectrum (Hamanaka et al., 2000). Thermal

TABLE 11.1
Inactivation of Pathogenic Microorganisms by Infrared Heating

Pathogen	Food/Nonfood Material	Temperature/Energy	Time	Log Reduction[a]	Reference
		Solid Foods			
Monilia fructigena	Strawberry	~50°C[b]	10 s	2.5 to 5.2 log (estimated)	60
Total bacterial count	Wheat or soybean surface	1.5 kW	10 s	~3.0	47
Total aerobic plate count	Onion	80°C (average 2226 W/m²)	~24 min	$1.72 \pm 0.45 \log_{10}$ CFU/10g	35
Coliform counts	Onion	80°C (average 2226 W/m²)	~24 min	$4.04 \pm 0.47 \log_{10}$ CFU/10g	
Yeast and mold	Onion	80°C (average 2226 W/m²)	~24 min	$1.26 \pm 0.14 \log_{10}$ CFU/10g	
Natural bacterial microflora	Wheat surface	2.0 kW	60 s	$\sim 1.9 \log_{10}$ CFU/g	38
Listeria monocytogenes	Turkey frankfurters	70°C[b]	82.1 s	$3.5 \pm 0.4 \log_{10}$ CFU/cm²	65
		75°C[b]	92.1 s	$4.3 \pm 0.4 \log_{10}$ CFU/cm²	
		80°C[b]	103.2 s	$4.5 \pm 0.2 \log_{10}$ CFU/cm²	
Aspergillus niger spores	Corn meal	72°C	6 min	1.8	41
Aspergillus niger spores	Corn meal	68°C (with an optical filter: 5.45 to 12.23 μm)	6 min	2.3	
Fusarium proliferatum spores	Corn meal	72°C	6 min	1.4	
Fusarium proliferatum spores	Corn meal	68°C (with an optical filter: 5.45 to 12.23 μm)	6 min	1.95	
Listeria monocytogenes	Oil-browned deli turkey	399°C around product surface	75 s	$3.7 \log_{10}$ CFU/mL	69
		Liquid Foods			
Yeast	Honey	0.2 W/cm²	8 min	$\sim 3.85 \log_{10}$ CFU/mL[d]	48

(continued on next page)

TABLE 11.1 (continued)
Inactivation of Pathogenic Microorganisms by Infrared Heating

Pathogen	Food/Nonfood Material	Temperature/Energy	Time	Log Reduction[a]	Reference
		Nonfood Materials			
Bacillus subtilis	Stainless steel plate at water activity of 0.7	4.2 µW/cm^2/nm (peak wavelength: 950 nm)	4 min[e]		42
		3.7 µW/cm^2/nm (peak wavelength: 1100 nm)	22 min[e]		
		3.2 µW/cm^2/nm (peak wavelength: 1150 nm)	12 min[e]		
E. coli	Nutrient agar (depth = 0)	4.36 × 10^3	6 min	~ 2.30 to 2.48 log$_{10}$ CFU/plate$^+$	87
	(depth = 1 mm from surface)	4.36 × 10^3	6 min	~0.70 log$_{10}$ CFU/plate	
	(depth = 2 mm from surface)	4.36 × 10^3	6 min	~0.66 log$_{10}$ CFU/plate	
Bacillus subtilis spores	Steel plate	180°C	3.2 s[e]		39
E. coli	Phosphate buffer saline	3.22 kW/m^2	8 min	1.8 log$_{10}$ CFU/mL	43
E. coli	Phosphate buffer	61°C	2 min	0.98 log$_{10}$ CFU/mL	40
Aspergillus niger spores	Physiological suspension	1.0 kW	40 s	4.0 to 5.0	47
Bacillus subtilis spores	Physiological suspension	1.0 kW	10 s	~1.0	

a In (log$_{10}$ CFU/mL), unless specified.
b Surface temperature.
c Temperature of corn meal.
d No growth observed after treatment.
e D value.

inactivation can damage DNA, RNA, ribosome, cell envelope, and proteins in the microbial cell. Sawai et al. (1995) investigated the inactivation mechanism of *E. coli* treated with IR radiation in phosphate buffer saline. They proposed that sublethally injured cells will become more sensitive to an inhibitory agent that has an inhibitory action on the damaged portion of the cell. Four inhibitory agents, namely, penicillin (PCG; inhibits cell wall synthesis), chloramphenicol (CP; inhibits protein synthesis), rifampicin (RFP; inhibits RNA synthesis), and nalidixic acid (NA; inhibits DNA synthesis) were used for the enumeration of pathogens. An 8 min IR radiation at a wattage of 3.22 kW/m^2 resulted in approximately 1.8, 1.9, 2.7, and 3.2 log$_{10}$ reduction of *E. coli,* when NA-, PCG-, RFP-, and CP-enriched agars were used for enumeration, respectively. When no inhibitory agents were present, a 1.8 log reduction was obtained. This observation implies that approximately 0.1, 0.9, and 1.4 log reductions were caused by inhibitory actions of PCG, RFP, and CP, respectively. With conductive heating, similar damage was observed; however, RNA, protein, and cell wall showed more vulnerability to IR heating than conductive heating. The order of magnitude of infrared damage was as follows: protein > RNA > cell wall > DNA. RFP inhibits RNA polymerase in *E. coli*, and CP binds ribosomal subunits and inhibits peptidyltransferase reactions (Sawai et al., 1995).

Sawai et al. (1997) reported that for both stationary and exponential phase cells, sensitivity to NA increased as the sample temperature increased. However, there was only a small increase, indicating that minimal damage occurred in the DNA. In particular, exponential phase cells had more cell wall and membrane damage than stationary phase cells. However, more serious injuries to RNA polymerase occurred for stationary phase cells, compared to exponential phase cells (Sawai et al., 1997). Recently, Krishnamurthy et al. (2010) investigated the inactivation mechanism of IR heat treatment with the aid of transmission electron microscopy and Fourier transform infrared spectroscopy (FTIR). The authors suggested that IR heat treatment results in lack of cell wall, cell wall breakage, cytoplasmic content leakage, cytoplasm shrinkage and breakage in mesosome, and damage to the cytoplasm as evidenced by the transmission electron microscopy (Figures 11.1 and 11.2).

11.4   SELECTIVE HEATING BY IR RADIATION

Jun (2002) developed a novel selective FIR heating system, demonstrating the importance of optical properties besides thermal properties when electromagnetic radiation is used for processing. The system had the capability to selectively heat higher absorbing components to a greater extent using optical bandpass filters that can emit radiation in the spectral ranges as needed. In selective infrared heating, only specific wavelength regions are used and, thus, only a particular component of interest is heated. While the temperature of the component of interest increases, the temperature of the rest of the food material does not. In other words, selective heating results in localized heating of the component of interest. Similarly, there is potential to selectively heat-treat the microorganisms present in the food material rather than heat-treating the entire food material. This results in microbial inactivation while preserving the quality of the food material.

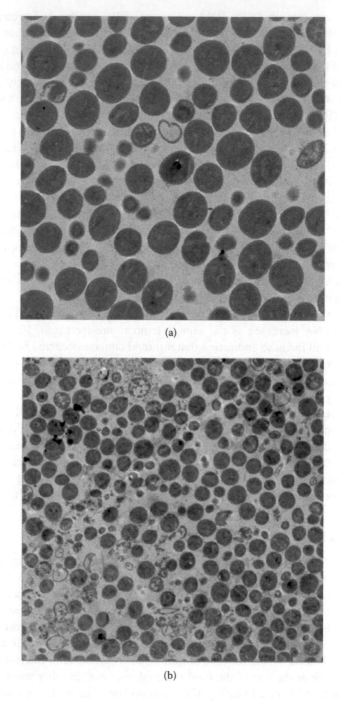

(a)

(b)

FIGURE 11.1 Effect of infrared heat treatment on bacterial cells. (a) Control sample and (b) infrared heat-treated sample (700°C lamp temperature, 20 min, 5 mL volume in buffer). (From Krishnamurthy et al., 2008.)

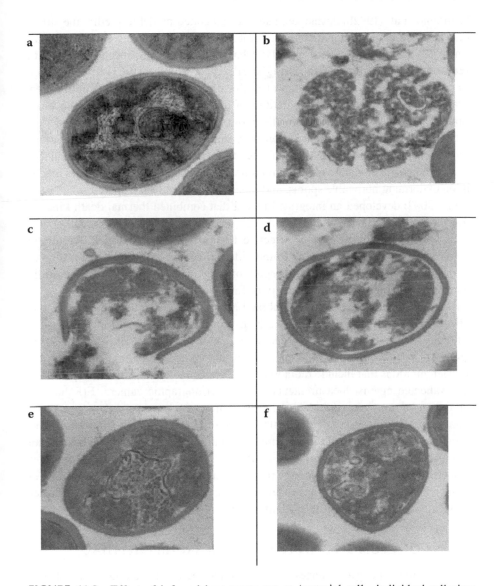

FIGURE 11.2 Effect of infrared heat treatment on bacterial cell—individual cell view. (a) Control sample (intact cell wall, cytoplasmic membrane, and mesosome); (b) lack of cell wall; (c) cell wall breakage and cytoplasm content leakage; (d) cytoplasm shrinkage; (e) breakage in mesosome; and (f) cytoplasm damage. (From Krishnamurthy et al., 2010. Microscopic and spectroscopic evaluation of inactivation of *Staphylococcus aureus* by pulsed UV light and infrared heating. *Food and Bioprocess Technology*. Available online: DOI: 10.1007/s11947-008-0084-8. With permission.)

11.5   THERMAL DEATH KINETICS MODEL

Hashimoto et al. (1992b) developed a simple integrated model to predict the survivors of *E. coli* under the predicted temperature distribution during FIR pasteurization. Analytical and numerical models of bacterial spores have been developed to predict microbial spore growth during sterilization. Stumbo (1965) first validated a model with first-order inactivation of uniformly activated spores during the sterilization process. To overcome limitations of traditional models in predicting spore populations during treatment, especially under ultra-high temperatures, new models that include spore activation have been proposed (Stumbo, 1965). The populations in a suspension of bacterial spores subjected to lethal heat treatment were simulated using a composite model involving simultaneous, independent activation and inactivation of dormant but viable spores, and inactivation of activated spores.

Jun (2002) developed an integrated model that combined thermal death kinetics with the IR heat transfer model, and could predict the survivors of fungal spores based on temperature prediction. Selective IR heating was found to differentially deliver a higher degree of lethality to individual fungal spores. The denaturation of the protein band as a target spectral region of selective heating might also partially contribute to an increase in the lethality of fungal spores.

Recently, Tanaka et al. (2007) combined Monte Carlo FIR radiation simulations with convection–diffusion airflow and heat transfer simulations to investigate the suitability of the method for surface decontamination in strawberries. The model was a powerful tool for fast and comprehensive evaluation and to address complex heating configurations that include radiation, convection, and conduction. Computations were validated against measurements with a thermographic camera. FIR heating obtained more uniform surface heating than air convection heating, with a maximum temperature well below the critical limit of about 50°C. To improve system functionality in terms of heating rate and temperature uniformity, several factors can be considered, that is, system rotation, optimized heating cycles, and different heater geometries. The projected modeling approach can be used to achieve such a goal in a comprehensive manner and the model should be extended to consider mass transfer and volumetric dissipation of the radiation power.

11.6   CONCLUSION AND FUTURE RESEARCH POTENTIAL

IR heating offers many advantages over convection heating, including greater energy efficiency, heat transfer rate, and heat flux, that save time and increase production line speed. IR heating is attractive primarily for surface pasteurization applications. IR heating can be effectively used for pathogen inactivation, as is evident from the previously mentioned research. Selective infrared heating is an area of high interest as only the pathogen of interest can be potentially heated rather than heating the entire food material. Selective heating can result in higher-quality food products that are safe to be consumed.

REFERENCES

Daisuke, H., Toshitaka, U., Wenzhong, H., and Yaunaga, E. 2001. The short-time infrared ray sterilization of the cereal surface. *Proceedings of IFAC Control Applications in Post-Harvest and Processing Technology*, Tokyo, Japan, pp. 195–201.

Gabel, M.M., Pan, Z., Amaratunga, K.S.P., Harris, L.J., and Thompson, J.F. 2006. Catalytic infrared dehydration of onions. *J. Food Sci.* 71(9):E351–E357.

Hamanaka, D., Dokan, S., Yasunaga, E., Kuroki, S., Uchino, T., and Akimoto, K. 2000. The sterilization effects on infrared ray of the agricultural products spoilage micro-organisms (part 1) An ASAE Meeting Presentation, Milwaukee, WI, July 9–12, No. 00 6090.

Hamanaka, D., Uchino, T., Furuse, N., Han, W., and Tanaka, S. 2006. Effect of the wavelength of infrared heaters on the inactivation of bacterial spores at various water activities. *Int J. Food Microbiol.* 108:281–285.

Hashimoto, A., Sawai, J., Igarashi, H., and Shimizu, M. 1991. Effect of far-infrared radiation on pasteurization of bacteria suspended in phosphate-buffered saline. *Kagaku Kogaku Ronbunshu* 17:627–633.

Hashimoto, A., Sawai, J., Igarashi, H., and Shimizu, M. 1992a. Effect of far-infrared irradiation on pasteurization of bacteria suspended in liquid medium below lethal temperature. *J. Chem. Eng. Japan* 25(3):275–281.

Hashimoto, A., Sawai, J., Igarashi, H., and Shimizu, M. 1992b. Far-infrared irradiation effect on pasteurization of bacteria on or within wet-solid medium. *J. Chem. Eng. Japan* 25(6):666–671.

Hebbar, H.U., Nandini, K.E., Lakshmi, M.C., and Subramanian, R. 2003. Microwave and infrared heat processing of honey and its quality. *Food Sci. Technol. Res.* 9:49–53.

Huang, L. 2004. Infrared surface pasteurization of turkey frankfurters *Innovative Food Sci. Emerging Technol* 5:345–351.

Jun, S. 2002. Selective far infrared heating of food systems. Ph.D. dissertation, the Pennsylvania State University.

Jun, S. and Irudayaraj, J. 2003. A dynamic fungal inactivation approach using selective infrared heating. *Trans. ASAE* 46(5):1407–1412.

Krishnamurthy, K. 2006. Decontamination of milk and water by pulsed UV light and infrared heating. Ph.D. Dissertation. Department of Agricultural and Biological Engineering, the Pennsylvania State University.

Krishnamurthy, K., Tewari, J.C., Irudyaraj, J., and Demirci, A. 2010. Microscopic and spectroscopic evaluation of inactivation of *Staphylococcus aureus* by pulsed UV light and infrared heating. *Food and Bioprocess Technology*.

Lentz, R.R., Pesheck, P.S., Anderson, G.R., DeMars, J., and Peck, T.R. 1995. Method of processing food utilizing infra-red radiation. U.S. Patent 5382441.

Mead, P.S., Slutsker, L., and Dietz, V. 1999. Food related illness and death in the United States. *Emerg. Infect. Dis.* 5:607–625.

Molin, G. and Ostlund, K. 1975. Dry heat inactivation of *Bacillus subtilis* spores by means of IR heating. *Antonie van Leeuwenhoek* 41(3):329–335.

Muriana, P., Gande, N., Robertson, W., Jordan, B., and Mitra, S. 2004. Effect of prepackage and postpackage pasteurization on postprocess elimination of *Listeria monocytogenes* on deli turkey products. *J. Food Prot.* 67 (11):2472–2479.

Sawai, J., Kojima, H., Igarashi, H., Hashimoto, A., Fujisawa, M., Kokugan, T., and Shimizu, M. 1997. Pasteurization of bacterial spores in liquid medium by far-infrared irradiation. *J. Chem. Eng. Japan* 30:170–172.

Sawai, J., Sagara, K., Hashimoto, A., Igarashi, H., and Shimizu, M. 2003. Inactivation characteristics shown by enzymes and bacteria treated with far-infrared radiative heating. *Int. J. Food Sci. Tech.* 38:661–667.

Sawai, J., Sagara, K., Igarashi, H., Hashimoto, A., Kokugan, T., and Shimizu, M. 1995. Injury of *Escherichia coli* in physiological phosphate buffered saline induced by far-infrared irradiation. *J. Chem. Eng. Jap* 28(3):294–299.

Stumbo, C.R. 1965. *Thermobacteriology in Food Processing*. New York: Academic Press.

12 Industrial Applications of Infrared Radiation Heating and Economic Benefits in Food and Agricultural Processing

Belgin S. Erdoğdu, İbrahim H. Ekiz, Ferruh Erdoğdu, Griffiths Gregory Atungulu, and Zhongli Pan

CONTENTS

12.1 Introduction 238
12.2 Development of IR Food and Agricultural Processing Technologies 238
    12.2.1 Equipment 238
    12.2.2 Drying 239
    12.2.3 Pasteurization 240
    12.2.4 Baking, Cooking, Blanching, and Roasting 241
    12.2.5 Peeling 243
    12.2.6 Thawing 243
12.3 Emerging Novel Industrial Applications in Food and Agricultural
    Processing 244
    12.3.1 Simultaneous Infrared Dry Blanching and Dehydration 244
    12.3.2 Combined Infrared and Freeze-Drying 244
    12.3.3 IR Dry Roasting and Pasteurization 244
    12.3.4 Simultaneous IR Drying and Disinfestations 249
    12.3.5 IR Radiation Dry Peeling 259
12.4 Economic Benefits 259
12.5 Industrialization Needs 263
12.6 Conclusion 263
References 264
Appendix 12.1: Energy Efficiency and Cost Analysis of a Small Commercial
IR Unit for Dry Blanching 267
    Energy Loss Due to Conversion of Natural Gas to Infrared Energy 267
    Assumptions 267

Nomenclature..267
Greek Symbols..268
Subscripts...268
Heat Losses by Natural Convection and Radiation from Faces of the
    Equipment..268
    A1. Natural Convection Heat Losses from Sides 268
    A2. Radiation Heat Loss from Sides 269
    B1. Heat Losses from Front Side...269
    B2. Radiation Heat Loss from Front Side 270
    C1. Heat Losses from Rear Side...270
    C2. Radiation Heat Loss from Rear Side..271
    D1. Heat Losses from Top and Bottom Sides..................................271
    D2. Radiation Heat Loss from Top and Bottom Sides......................273
In Summary 273

12.1   INTRODUCTION

Infrared (IR) radiation is the part of the electromagnetic spectrum lying between ultra-violet (UV) and microwave (MW) energy. It is normally classified into three regions: near-infrared (NIR), mid-infrared (MIR), and far-infrared (FIR), corresponding to the spectral ranges of 0.78 to 1.4 μm, 1.4 to 3 μm, and 3 to 1000 μm, respectively. The application of IR heating for food processing has the following advantageous features: (1) efficient heat transfer to the food, which reduces the processing time and energy costs; (2) air in the equipment is not heated by IR. Consequently, the air temperature may be kept at normal levels; (3) the possibility of designing compact and easily automated equipment with high controllability and safety; (4) uniformity of heating more than in conventional heating because of IR penetration; and (5) danger of product overheating because of rapid heating rates, which requires exact condition control (Sakai and Hanzawa, 1994). Because of these features, IR heating has been accepted as an important means of cooking, drying, roasting, baking, blanching, and pasteurization of food and agricultural products (Lloyd et al., 2003; Ranjan et al., 2002; Staack et al., 2008a).

This chapter will review the scientific developments in IR applications, demonstrate the status of selected industrial and pilot-scale IR applications, discuss the economic benefits of IR application in food and agricultural processes and, finally, provide conclusive remarks on the present needs for successful industrial implementation of IR technologies.

12.2   DEVELOPMENT OF IR FOOD AND AGRICULTURAL PROCESSING TECHNOLOGIES

12.2.1   EQUIPMENT

Industrial food applications of IR heating mainly use MIR and FIR for baking (roasting), drying, thawing, pasteurization, and blanching. For each of these operations, commercial equipment either exists or is currently at the pilot-scale development stage. Table 12.1 shows some commercial food applications of MIR and FIR heating equipments.

TABLE 12.1
Commercial Food Applications of MIR and FIR Heating Equipment

Equipment	Type of Application	Foods Processed
	Baking /Heating	
Gas oven	Batch	Bread
Tunnel oven	Continuous	Fish paste, oyster, egg, liver
Drum oven	Continuous	Coffee
Cooker	Batch	Sweet potato, chestnut
	Drying	
Tunnel dryer	Continuous	Vegetables, flaked dried bonito, semidried noodle
Drum dryer	Batch	Green tea
Freeze dryer	Continuous	Liquid foods
Pneumatic dryer	Continuous	Particulate foods
	Thawing	
Tunnel thawing equipment	Continuous	Frozen fish paste
Thawing device	Batch	Frozen sushi and other frozen foods, tuna
	Pasteurization	
Conveyor pasteurizer	Continuous	Packed foods
	Peeling	
Infrared dry peeling	Continuous and batch	Tomatoes, potatoes
	Blanching	
Infrared dry blancher	Continuous and batch	Strawberry, carrots, etc.

12.2.2 DRYING

IR energy is the oldest energy source for drying; it employs radiant energy from the sun, which is the oldest drying method (Fasina and Tyler, 2001; Ranjan et al., 2002). IR processes can be applied to drying of food products since the main components of food products have principal absorption bands at wavelengths greater than 2.5 μm (Sandu, 1986; Sakai and Hanzawa, 1994).

An increasing number of studies in the literature on IR drying applications of food products have been noted (Wang and Sheng, 2004). Sandu (1986) and Ratti and Mujumdar (1995) reported that IR drying processes have the advantages of requiring simple equipment, fast transient response, and significant energy savings. Higher heat transfer and heat flux rates leading to shorter drying times and higher drying rates are additional advantages of IR drying processes (Sandu, 1986; Afzal and Abe, 1997). Fu and Lien (1998) studied the effects of operational variables on the drying rate and quality of dehydrated shrimp using an FIR drying system. Afzal and Abe (1998) investigated the effect of IR heating on drying of potato. Das et al. (2003) applied IR heating to the drying and parboiling of paddy rice and recommended optimum process parameters for a vibration-aided process. Hidaka et al. (2004)

developed FIR radiation drying for grains. They determined the energy efficiency and conservation compared to an air drying system. Hebbar et al. (2004) developed a combined IR and hot air dryer for vegetables in which MIR heaters for radiative heating were used. The combined mode reduced frying time by 48% while consuming less energy (63%) compared to hot air drying. Gabel et al. (2006) used catalytic IR drying for dehydration of onions, indicating the greater drying rates and shorter drying times. Meeso et al. (2008) applied different strategies of FIR radiation to paddy rice drying. Shi et al. (2008) determined the effect of berry size and sodium hydroxide pretreatment on drying characteristics of blueberries during an IR radiation process. The combined use of IR heating with freeze-drying processes was also noted to reduce the processing times (Krishnamurthy et al., 2008). All these studies reported in the literature suggest that IR drying processes improve the final quality of the dried products with increased rehydration rates.

12.2.3 PASTEURIZATION

Applying IR heating for surface pasteurization purposes has the potential to become a common industrial practice. Exposing a food product to an IR heating source results in an increase in the surface temperature, and the heat is conducted to the interior by conduction. Because food products have lower thermal conductivity (<0.6 W/m²-K in most cases), the rate of heat transfer through the food products is rather slow. Hence, an intense heat might accumulate on the surface, causing the surface temperature to increase rapidly. If the IR exposure time is properly controlled, the surface temperature can be preferentially raised to a degree sufficient to inactivate target pathogenic microorganism without substantially increasing the interior temperature. IR heating inactivates the pathogen microorganisms by causing damage in their intracellular components such as DNA, RNA, ribosomes, cell envelope, and/or cell proteins (Krishnamurthy et al., 2008). Absorption of IR radiation by water molecules in the microorganism is a significant factor in the inactivation since the water absorbs the IR radiation, resulting in a rapid increase in temperature (Hamanaka et al., 2006a). The state, location, and amount of water inside the microorganism and bonding conditions are significant factors affecting the response of microorganism to IR heating (Hamanaka et al., 2006a). Even though IR energy is known to inactivate microorganisms by conventional thermal mechanisms (denaturation of proteins and nucleic acids), nonthermal effects on, for example, enzymes were also reported (Sakai and Mao, 2005). Sawai et al. (2000) determined that FIR heating was more effective in pasteurizing vegetative cells compared to thermal conductive heating.

Rosenthal et al. (1996) used IR heating for surface pasteurization of cottage cheese. Hamanaka et al. (2000) investigated the sterilization of wheat surface by using IR heating. Jun and Irudayaraj (2004) applied IR heating for disinfection of fungal spores in agricultural materials. Huang (2004) used IR heating as a possible intervention technology to pasteurize the surface of turkey frankfurters contaminated with *Listeria monocytogenes*. Use of IR heating for pasteurization of oysters, Japanese noodles, and secondary pasteurization of boiled fish paste was reported by Sakai and Mao (2005). Hamanaka et al. (2006b) developed an IR radiation heating grain sterilizer for a continuous process. They determined that excessive IR treatment to obtain certain disinfectant effects might lead to degradation of internal quality of wheat grains with higher initial moisture content. For the case of agricultural products with lower water activity, transmission becomes lower with lower surface reflection of IR energy, leading to higher absorption rates.

Tanaka et al. (2007) used FIR radiation as an alternative technique for surface decontamination of strawberry. FIR heating was determined to achieve more uniform surface heating compared to air convection heating, but the method was found to be inferior to heating in water due to efficacy and uniformity. Krishnamurthy et al. (2008) investigated the potential of IR heating for inactivating *Staphylococcus aureus* in milk processing. Huang and Sites (2008) developed an IR pasteurization process for inactivation of *Listeria monocytogenes* on hot dogs. Staack et al. (2008a; 2008b) investigated the effect of IR heating on quality and microbial decontamination in paprika powder. Maria et al. (2008) used IR heating to determine its effects on *Salmonella enteritidis* on almonds. It was determined that skin morphology, meat texture, and kernel colors of almonds were not distinguishable compared to the untreated samples, indicating the positive feasibility of IR heating technology for dry pasteurization of almonds. IR heating might also be used in process lines to pasteurize food-contact surfaces for elimination of fungal growth and microorganisms to improve the shelf life of food products; for example, pasteurization of baking trays.

12.2.4 BAKING, COOKING, BLANCHING, AND ROASTING

As an industrial process, IR heating, using IR ovens, was reported by Dagerskog (1979), and many examples of baking, cooking, blanching, and roasting using IR energy can be found in the literature. Wade (1987) used quartz-tungsten tubes emitting radiation of peak wavelength of 1.2 μm to bake biscuits, and it was found that a wide range of biscuit products (crackers, semisweet, and short dough types) could be baked using IR heating in approximately half the time required in a conventional oven. Skjoldebrand and Anderson (1989) used IR heating as a bread-baking method and reported the advantages of higher heat transfer efficiency, reduction in baking time to attain a desired quality, and rapid control of oven parameters.

Khan and Vandermey (1985) investigated the quality assessment of ground beef patties after IR processing in a tube broiler used for food service. Sheridan and Shilton (1999) evaluated the efficacy of cooking hamburger patties by IR heating at 2.7 and 4.0 μm. With a higher-temperature (lower-wavelength) energy source, a change in core temperature of patties was closer to the change in surface temperature, and cooking time was found to be shorter. In addition, roasting and browning of meat products' surfaces are significant processes in the meat process industry, and IR heating might be an important application for these purposes.

Sakai and Hanzawa (1994) reported the use of FIR heating to make boiling eggs without requiring the use of hot water. Cenkowski and Sosulski (1997) investigated the effects of IR heating on physical and cooking properties of lentils. It was determined that the IR heating was effective in gelatinization and solubilization of starch, leading to reduced cooking times. Cenkowski and Sosulski (1998) determined the effects of high-intensity IR heating on water hydration rate and cooking time of split peas and the changes in functional properties of the protein and starch. While the water hydration rate increased by 7%, cooking time decreased by one third, and proteins were denatured and starch granules pregelatinized. IR heating was found to be an effective technique for making instant split peas and expanding dry pea-based food. Fasina et al. (2001) determined the effects of IR heating on the properties of legume seeds, and functional characteristics of flours of the IR-processed seeds were determined to be superior to those obtained from untreated seeds.

Pan et al. (2005) evaluated the feasibility of MIR and FIR for blanching and dehydration of various fruits and vegetables. They determined IR radiation to be an effective method for dry-blanching of fruits and vegetables with high processing efficiency and significant processing time saving, leading to high-quality products without using any steam or water. Khir et al. (2006, 2007) studied the moisture removal characteristics of thin-layer rough rice heated by IR heating and cooled by natural and forced air. They determined that IR thin-layer drying of rough rice followed by cooling could be an effective approach to design of an IR rice dryer for improving the drying rates and reducing energy consumption. It was concluded that higher heating rates, faster drying, and better rice quality could be achieved by heating rough rice to about 60°C followed by tempering and slow cooling. Zhu and Pan (2009) applied IR heating in a continuous heating mode to achieve IR dry blanching and dehydration of apple slices, and concluded that continuous IR heating could be used as an alternative to the current processing methods to produce high-quality blanched and dried fruits and vegetables. The authors also established that during the IR dry blanching, it took a much longer time for intermittent IR heating than continuous IR heating to achieve the same level of peroxidase (POD) inactivation. The prolonged heating to achieve more than a 1 log reduction of POD under intermittent heating mode may cause a large moisture reduction and increase of overall surface color change in products. The IR dry blanching process can serve the purposes of blanching alone and simultaneous blanching and dehydration. For certain fruits and vegetables, moisture removal may not be desired during blanching. In this case, quick blanching and limiting the moisture reduction as much as possible is necessary. Therefore, continuous heating is beneficial for such an application. On the contrary, for certain scenarios, drying needs to be conducted after proper blanching, for instance, to produce dehydrofrozen products. In such a case, intermittent heating can be suggested since it does not cause severe surface darkening. Because both heating modes have their own advantages and disadvantages, an appropriate heating mode and appropriate processing conditions need to be chosen based on the processing purpose and the characteristics of the materials.

Conventional roasters use hot air for roasting, and air may be heated to 350°C–450°C to roast; for example, coffee beans. For the case of coffee beans, the roasting process might take up to 20 min (Sakai and Mao, 2005) for bean temperatures of around 180°C to 230°C. Due to the requirement of higher temperatures, IR energy can be used as an alternative to roasting. Sakai and Mao (2005) reported that formation of an aroma component in the beans was more pronounced in an FIR roaster due to uniform heating. Using FIR energy was also found to be more effective in roasting green tea leaves to produce tea with a better flavor (Sakai and Mao, 2005). Many other different applications of IR heating have been reported in the literature for heating (Sakai and Hanzawa, 1994; Fasina and Tyler, 2001; Metussin et al., 1992).

12.2.5 PEELING

The development of nonchemical peeling technology has recently been identified as a priority for economical and environmentally friendly processing of fruits and vegetables, especially for peeling purposes. The presently and widely industrialized lye peeling methods have posed threats to long term water supply, caused a salinity management crisis, and escalated the cost of water supply and wastewater management. The fruit and vegetable processors are challenged to look for technologies to reduce water and lye usage. Consequently, the application of IR heating for peeling has attracted strong interest from the food processing industry. Due to the higher efficiency of IR heating compared to conventional lye peeling and its inherent lower penetration depth, IR heating has shown suitability for peeling fruits and vegetables (Pan et al., 2009b; Li et al., 2009). IR dry peeling has emerged as a novel method for peeling tomatoes.

12.2.6 THAWING

The quality of frozen products has been improved significantly with advancements in freezing technologies. With notable exceptions such as ice cream, most frozen foods must be thawed before further use or consumption, and the quality might deteriorate during thawing. In a number of food-manufacturing operations, it is a common practice to begin with frozen foods as a raw material. For example, in manufacturing sausages, frozen meat is used as the raw material. Similarly, large blocks of frozen fish are processed into fillets for further processing. Different thawing and tempering methods are used for preparing frozen foods for further processing, and each has its own advantages and disadvantages. The main goal of the thawing processes is to keep thawing time to a minimum so that the least damage is caused to the quality of the food product. However, during thawing, a number of quality attributes may be adversely affected by moisture (drip) loss, change in structure of proteins, microbial growth, and textural changes.

Microwave thawing is one possible method to reduce thawing time, but, due to the significant differences in the dielectric properties of frozen and unfrozen food products, runaway heating or overheating near the surface might occur.

In the IR region, the absorption coefficients of ice and water are approximately same (Sakai and Mao, 2005), and this prevents runaway heating, making IR heating a possible thawing method. Sakai and Mao (2005) refer to the studies found in the literature for thawing tuna using FIR heating without drip losses and discoloration, where commercial thawing equipments and a refrigerator with a partial defrosting system using IR energy were developed. Evolution of surface temperature due to absorption of IR energy is a significant parameter to control in thawing using IR energy. Liu et al. (1999) investigated the FIR thawing conditions on the temperature distribution of frozen tuna during thawing.

12.3 EMERGING NOVEL INDUSTRIAL APPLICATIONS IN FOOD AND AGRICULTURAL PROCESSING

12.3.1 SIMULTANEOUS INFRARED DRY BLANCHING AND DEHYDRATION

In the food industry, blanching has become a very important unit operation prior to freezing, canning, and drying of fruits or vegetables in order to inactivate enzymes; modify food texture; preserve food color, flavor, and nutritional value; and remove trapped air. More recently, energy conservation and waste reduction requirements have called for the need to improve the design of blanching equipment. The latest development to address the urgent need is reported by Pan and McHugh (2004), who have developed a new method that uses IR radiation for dry blanching. Since this technology does not involve the addition of steam or water in the process of blanching, it has been named *IR dry-blanching* (IDB) technology and is intended to be a replacement for current steam, water, and/or microwave blanching methods to produce many kinds of value-added fruit- or vegetable-based dried, refrigerated, frozen, and dehydrofrozen food products. The details and merits of IDB technology are discussed in Chapter 9.

12.3.2 COMBINED INFRARED AND FREEZE-DRYING

Combining IR radiation and freeze-drying either simultaneously or sequentially are relatively new methods that are gaining great potential for industrial application. Especially, sequential IR radiation and freeze-drying (SIRFD) is one of the emerging novel combinations. The employment of catalytic IR (CIR) emitters for the SIRFD has been used as a new alternative to high-energy-efficient drying in which IR heating is used during the predrying stage to remove a significant amount of moisture in fruits and vegetables before freeze-drying is introduced in the final stage of the drying process. Because IR heating is a very energy-efficient drying method with significantly short drying time, the energy savings of SIRFD is significant compared to the current methods of drying. Using IR emitters, drying can be used to target various weight reductions during the predehydration step before introducing the samples into the freeze dryer. The details and merits of SIRFD technology are discussed in Chapter 7.

12.3.3 IR DRY ROASTING AND PASTEURIZATION

Dry roasting is a thermal process used popularly by the almond industry. At present, the typical dry-roasting process uses hot air, which is achieved via a continuous conveyor roaster or rotary roaster. The continuous conveyor roaster can be single stage or multiple stage, with a variety of temperatures. Common temperatures used for hot air roasting range from 130°C (265°F) to 154°C (310°F). At the lower temperature, it may take 40–45 min to obtain a light-to-medium-roasted almond, while at the higher temperature, it may take 10 to 15 min to obtain a light-to-medium-roasted product. At the same time, due to the outbreaks of salmonellosis that have been associated with whole raw almonds, the almond industry is pursuing a mandatory pasteurization plan and has determined that aggressive measures are necessary to prevent any other outbreaks of salmonellosis (Pan et al., 2009a). Several different technologies have been used or are under considerations for raw almond pasteurization, including propylene oxide (PPO) fumigation, FMC JSP-I almond surface pasteurization technology, Ventilex steam pasteurizer, vacuum steam, radio frequency, IR, and cold plasma. FIR heating has shown the potential to address two major concerns in the almond industry better than the conventional methods. First, current roasting processes may not ensure pasteurization of the product with a minimum 4 log reduction of *Salmonella enteritidis* PT 30 (SE PT 30) and, second, they require relatively longer processing time, which consequently increases processing costs. The almond industry has been under pressure to develop new processing methods that can produce pasteurized products in a shorter time.

Figure 12.1 shows the pilot-scale catalytic IR equipment with double-sided heating (Catalytic Industrial Group Inc., Independence, Kansas) which was built and has been used for dry roasting of almonds. The pilot-scale IR device had eight emitters, four at the top and four at the bottom, for a total heating area of 269.24×60.96 cm. Degrees of roasting of standard commercial almonds as represented by the overall color ΔE of medium-roasted and heavily roasted kernels are 11.5 and 21.4, respectively (Pan et al., 2008a; Ozdemir and Devres, 2000). Overall color changes of roasted almonds obtained under various IR treatment conditions are shown in Figure 12.2. Roasting times for producing medium-roasted and heavily roasted almonds by using different dry-roasting conditions are listed in Table 12.2. For the hot air heating alone, 34, 18, and 13 min were required to reach medium roasting at 130°C, 140°C, and 150°C, respectively. For sequential IR radiation and hot air (SIRHA), times of hot air heating reduces to 21, 11, and 5 min, excluding the additional IR preheating time of 39,

FIGURE 12.1 Pilot-scale catalytic infrared heating equipment. (From Pan et al., Infrared heating for improved safety and processing efficiency of dry-roasted almonds. Report for Almond Board of California, 2008a.)

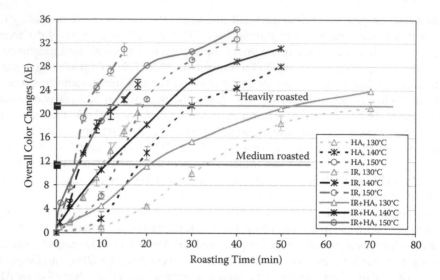

FIGURE 12.2 Overall color changes (ΔE) of almonds under different roasting conditions. (From Pan et al., Infrared heating for improved safety and processing efficiency of dry-roasted almonds. Report for Almond Board of California, 2008a.)

TABLE 12.2
Roasting Times and Time Reductions for Producing Medium-Roasted and Heavily Roasted Almonds under Different Roasting Methods and Conditions

Roasting Method		Hot Air			Infrared			Sequential IR and Hot Air		
Roasting temperature (°C)		130	140	150	130	140	150	130	140	150
Medium	Roasting time (min)	34	18	13	11	6	4	21	11	5
	Time reduction (%)[a]	—	—	—	68	67	69	38	39	62
Heavily	Roasting time (min)	72	30	19	20	14	7	52	24	12
	Time reduction (%)[a]	—	—	—	72	53	63	28	20	37

Source: Pan et al., Infrared heating for improved safety and processing efficiency of dry-roasted almonds. Report for Almond Board of California, 2008a.

[a] Time reduction (%) = (Time of hot air roasting − Time of IR or SIRHA roasting)/(Time of hot air roasting) × 100%.

44, and 53 s, respectively. Compared to hot air roasting alone and at 130°C, 140°C, and 150°C, the time reductions are 68%, 67%, 69%, and 38%, 39%, 62% with IR alone and SIRHA roasting, respectively. The time required to heat almonds to 150°C was less than 1 min using IR, compared to about 15 min with hot air heating.

Based on the dry roasting studies, it was concluded that SIRHA roasting is a superior roasting method for producing pasteurized roasted almonds with potential to

TABLE 12.3
Reductions in *Pediococcus* Population Size on Medium-Roasted Almonds under Different Conditions and Statistical Analysis Results at Different Confidence Levels

Treatment Temperature (°C)	130	140	150
Hot air treatment	3.58a A	4.62a A	5.39a AB
Infrared treatment	2.94a A	3.21a B	1.12a D
Sequential IR and hot air treatment	4.1a A	5.82b A	6.96b A

Note: The same letters in lowercase in the same row mean no significant difference at $P \le 0.01$; the same letters in uppercase in the same column mean no significant difference at $P \le 0.01$.

Treatment Temperature (°C)	130	140	150
Hot air treatment	3.58a AB	4.62a B	5.39a B
Infrared treatment	2.94a B	3.21a C	4.12b B
Sequential IR and hot air treatment	4.1a A	5.82b A	6.96c A

Note: The same letters in lowercase in the same row mean no significant difference at $P \le 0.05$; the same letters in uppercase in the same column mean no significant difference at $P \le 0.05$.

Source: Pan et al., Infrared heating for improved safety and processing efficiency of dry-roasted almonds. Report for Almond Board of California, 2008a.

reduce costs due to reduced roasting time compared to current hot air roasting under the same roasting temperature. The roasting process can be easily implemented in the industry by adding an IR pre-heating device in front of regular hot air roasters. The roasting using IR alone is recommended only for pasteurization that targets 4-log bacterial reduction. Table 12.3 shows the reductions in *Pediococcus* population size on medium-roasted almonds under different conditions. The *Pediococcus* spp. (*Enterococcus faecium*) NRRL B-2354 was used in these experiments as a surrogate of *Salmonella enteriditis* PT 30 (SE PT30). The strain has been identified and validated by the Almond Board of California Technical Expert Review Panel for quality control in almond thermal processes.

In the case of raw almond pasteurization, it is noted that for the tested range of IR heating temperature, holding at 90°C for 10 to 15 min reduced the *Pediococcus* population by over 5 log, and holding at 80°C for over 22 min provided over 4 log reduction (Table 12.4). Holding almonds at 70°C even for 60 min was not sufficient enough to meet the pasteurization requirements. IR heating of raw almonds to 100°C, 110°C, and 120°C to produce pasteurized raw almonds results in 0.320, 0.583, and 0.620 log reduction of *Pediococcus* population, respectively (Table 12.4). The studied ranges of heating times and temperatures slightly change the hue° values (around 3°) of almond's skin and its flesh (Tables 12.5 and 12.6). However, color changes may not be distinguished by observers. Therefore, the use of IR is a promising technology

TABLE 12.4

Log Reduction Value of *Pediococcus* Population

Treatment	Total Reduction	Stage Reduction	Treatment	Total Reduction	Stage Reduction
IR heating to 100°C	0.320	0.320	IR120°C-80°C-15min	3.643	2.833
IR heating to 110°C	0.583	0.583	IR110°C-80°C-15min	3.497	2.800
IR heating to 120°C	0.620	0.620	IR100°C-80°C-15min	2.928	2.318
IR120°C + Cooling to 90°C	1.350	0.730	IR120°C-80°C-22min	4.109	3.299
IR120°C + Cooling to 80°C	1.430	0.810	IR110°C-80°C-22min	3.909	3.212
IR120°C+ Cooling to 70°C	1.670	1.050	IR100°C-80°C-22min	3.665	3.055
IR110°C + Cooling to 90°C	1.140	0.557	IR120°C-80°C-30min	6.838	6.028
IR110°C + Cooling to 80°C	1.280	0.697	IR110°C-80°C-30min	6.105	5.408
IR110°C + Cooling to 70°C	1.480	0.897	IR100°C-80°C-30min	4.989	4.379
IR100°C + Cooling to 90°C	0.790	0.470			
IR100°C + Cooling to 80°C	0.930	0.610	IR120°C-90°C-5min	5.700	4.970
IR100°C + Cooling to 70°C	0.950	0.630	IR110°C-90°C-5min	4.143	3.586
			IR100°C-90°C-5min	3.810	3.340
IR120°C-70°C-15min[a]	2.429	1.379	IR120°C-90°C-10min	7.629	5.087
IR110°C-70°C-15min	2.327	1.430	IR110°C-90°C-10min	6.460	5.903
IR100°C-70°C-15min	2.050	1.420	IR100°C-90°C-10min	5.678	5.208
IR120°C-70°C-30min	3.271	2.221	IR120°C-90°C-15min	8.258	7.528
IR110°C-70°C-30min	2.999	2.102	IR110°C-90°C-15min	7.779	7.222
IR100°C-70°C-30min	2.929	2.299	IR100°C-90°C-15min	5.981	5.511
IR120°C-70°C-45min	3.692	2.642			
IR110°C-70°C-45min	3.438	2.541			
IR100°C-70°C-45min	2.908	2.278			
IR120°C-70°C-60min	3.846	2.796			
IR110°C-70°C-60min	3.692	2.795			
IR100°C-70°C-60min	3.234	2.604			

Source: Pan et al., Infrared heating for raw almond pasteurization. Final Report for Almond Board of California, 2009a.

Note: The reported total reduction value in *Pediococcus* population is the cumulative value at the end of each period (heating by IR, cooling, and holding). Precisely, a total reduction value at a holding period shows "Reduction$_{heating\ by\ IR}$ + Reduction$_{cooling}$ + Reduction$_{holding}$." Stage reduction value refers to the reduction of *Pediococcus* population corresponding to the respective stage. To be precise, a stage reduction value reported in the holding period means "Total Reduction − (Heating by IR + Cooling)."

[a] Format of the notation is IR heating temperature = Holding temperature = Holding time.

for surface pasteurization of raw almonds without significantly compromising the raw almond quality attributes. Based on bacterial reduction, preservation of sensory quality, and consumption of energy, the recommendation to almond processors is to either heat the almonds to 120°C by IR and hold it at 90°C for 5 min or to heat the almonds to 110°C by IR and hold it at 90°C for 10 min or to heat the almonds to

TABLE 12.5

Color Parameters ($L*a*b*$, ΔHue°, and ΔE) of Almond's Skin

Treatment	$L*$	$a*$	$b*$	ΔHue°	ΔE
Raw almond	48.92 ± 2.34	16.80 ± 0.88	31.91 ± 2.23	N/A	N/A
IR heating to 100°C	49.69 ± 2.12	16.60 ± 0.69	32.77 ± 1.55	1.34 ± 0.90	2.38 ± 1.46
IR heating to 110°C	49.05 ± 2.49	17.03 ± 0.58	31.94 ± 2.25	1.37 ± 1.13	2.89 ± 1.66
IR heating to 120°C	48.27 ± 1.96	17.27 ± 0.63	31.23 ± 2.03	1.80 ± 1.15	2.48 + 1.71
IR120°C 70°C-15min	49.75 ± 0.86	16.71 ± 0.29	32.65 ± 0.77	0.68 ± 0.54	1.05 ± 0.71
IR110°C-70°C-15min	49.53 ± 1.24	17.44 ± 0.51	32.57 ± 0.46	0.75 ± 0.84	1.37 ± 0.33
IR100°C-70°C-15min	49.57 ± 0.42	16.88 ± 0.29	32.19 ± 0.91	0.22 ± 0.24	0.89 ± 0.24
IR120°C-70°C-30min	48.92 ± 0.72	17.23 ± 0.21	32.36 ± 0.60	0.49 ± 0.23	0.90 ± 0.37
IR110°C-70°C-30min	48.74 ± 0.33	17.28 ± 0.08	32.67 ± 0.96	0.69 ± 0.18	1.11 ± 0.45
IR100°C-70°C-30min	49.82 ± 0.52	17.04 ± 0.05	33.28 ± 0.83	0.62 ± 0.39	1.25 ± 0.88
IR120°C-70°C-45min	48.33 ± 1.24	17.20 ± 0.29	32.46 ± 1.11	0.35 ± 0.41	1.56 ± 0.84
IR110°C-70°C-45min	49.01 ± 1.63	17.04 ± 0.24	33.30 ± 1.13	0.86 ± 0.67	1.91 ± 0.69
IR100°C-70°C-45min	48.65 ± 1.41	17.19 ± 0.54	33.80 ± 1.55	0.77 ± 0.46	2.29 ± 1.19
IR120°C-70°C-60min	49.23 ± 2.90	17.39 ± 0.95	32.60 ± 2.54	1.26 ± 0.85	2.04 ± 0.73
IR110°C-70°C-60min	48.00 ± 0.23	17.79 ± 0.19	33.06 ± 0.10	0.70 ± 0.23	1.79 ± 0.30
IR100°C-70°C-60min	48.39 ± 0.18	17.59 ± 0.11	32.63 ± 0.59	0.80 ± 0.63	1.28 ± 0.26
IR120°C-80°C-15min	47.00 ± 0.65	17.27 ± 0.51	30.34 ± 0.72	2.08 ± 0.67	2.97 ± 0.92
IR110°C-80°C-15min	47.58 ± 0.33	17.02 ± 0.28	30.16 ± 0.34	1.89 ± 0.50	2.64 ± 0.44
IR100°C-80°C-15min	47.44 ± 0.95	17.58 ± 0.46	31.20 ± 1.87	1.86 ± 0.85	2.52 ± 1.49
IR120°C-80°C-30min	49.45 ± 0.81	17.36 ± 0.15	32.90 ± 0.60	0.48 ± 0.29	1.13 ± 0.60
IR110°C-80°C-30min	49.68 ± 0.67	17.41 ± 0.46	33.31 ± 0.09	0.43 ± 0.31	1.50 ± 0.12
IR100°C 80°C-30min	49.62 ± 0.52	17.25 ± 0.07	33.07 ± 0.89	0.46 ± 0.17	1.25 ± 0.76
IR120°C-90°C-15min	47.08 ± 0.47	17.85 ± 0.20	30.60 ± 0.12	2.74 ± 0.30	2.87 ± 0.44
IR110°C-90°C-15min	47.76 ± 0.45	17.69 ± 0.22	30.99 ± 0.93	2.18 ± 0.69	2.16 ± 0.78
IR100°C-90°C-15min	46.16 ± 0.83	17.42 ± 0.23	29.82 ± 0.66	2.74 ± 0.85	3.21 ± 0.71

Source: Pan et al., Infrared heating for raw almond pasteurization. Final Report for Almond Board of California, 2009a.

100°C by IR and hold it at 90°C for 10 min. Either of these treatments provides an over 5.5 log reduction of *Pediococcus*.

12.3.4 SIMULTANEOUS IR DRYING AND DISINFESTATIONS

The development of rapid, nonchemical, safe alternative methods to eliminate insect pests, for instance, from paddy rice, while retaining high quality is an ideal alternative required to replace pesticides such as methyl bromide. The schematic diagram of the catalytic IR heating device used by Pan et al. (2004), who worked with rice processors to eliminate adult insects, larvae, and eggs and retain the milling quality, is illustrated in Figure 12.3. The unit consists of a catalytic IR emitter and vibration bed with adjustable speed for fully mixing the rice during heating. The flexible

TABLE 12.6
Color Parameters ($L*a*b*$, ΔHue°, and ΔE) Change of Almond's Flesh

Treatment	$L*$	$a*$	$b*$	ΔHue°	ΔE
Raw almond	79.68 ± 0.43	1.04 ± 0.29	19.89 ± 0.53	N/A	N/A
IR heating to 100°C	76.80 ± 0.56	0.90 ± 0.31	22.99 ± 0.48	0.82 ± 0.65	4.25 ± 0.65
IR heating to 110°C	80.75 ± 0.40	−0.07 ± 0.20	23.34 ± 0.30	3.17 ± 0.49	3.80 ± 0.31
IR heating to 120°C	80.59 ± 0.53	−0.15 ± 0.21	22.85 ± 0.79	3.37 ± 0.53	3.39 ± 0.56
IR120°C-70°C-15min	78.13 ± 1.07	−0.35 ± 0.16	23.50 ± 0.62	3.84 ± 0.38	4.27 ± 0.55
IR110°C-70°C-15min	79.50 ± 0.16	−0.53 ± 0.45	24.19 ± 0.57	4.25 ± 1.10	4.62 ± 0.39
IR100°C-70°C-15min	80.70 ± 1.30	0.61 ± 0.19	23.53 ± 1.23	1.52 ± 0.41	4.05 ± 0.64
IR120°C-70°C-30min	77.79 ± 4.64	−0.05 ± 0.40	24.13 ± 2.11	3.16 ± 0.99	6.04 ± 2.35
IR110°C-70°C-30min	76.37 ± 1.90	0.44 ± 0.21	24.71 ± 0.27	1.97 ± 0.49	6.00 ± 1.30
IR100°C-70°C-30min	80.67 ± 0.55	0.62 ± 0.34	22.93 ± 0.54	1.43 ± 0.86	3.25 ± 0.71
IR120°C-70°C-45min	79.93 ± 0.59	−0.05 ± 0.28	23.93 ± 0.44	3.10 ± 0.67	4.23 ± 0.46
IR110°C-70°C-45min	80.76 ± 0.59	0.01 ± 0.28	23.09 ± 1.41	2.95 ± 0.69	3.63 ± 1.19
IR100°C-70°C-45min	78.57 ± 1.79	0.60 ± .017	22.79 ± 1.04	1.48 ± 0.45	3.32 ± 1.60
IR120°C-70°C-60min	77.78 ± 0.38	−0.44 ± 0.12	24.14 ± 0.45	4.05 ± 0.31	4.90 ± 0.48
IR110°C-70°C-60min	78.30 ± 0.64	0.00 ± 0.25	23.75 ± 0.21	3.00 ± 0.59	4.25 ± 0.46
IR100°C-70°C-60min	78.88 ± 2.50	−0.43 ± 0.94	24.91 ± 0.37	4.00 ± 2.17	5.62 ± 0.44
IR120°C-80°C-15min	80.39 ± 0.88	−0.49 ± 0.33	23.62 ± 1.04	4.17 ± 0.79	4.20 ± 0.79
IR110°C-80°C-15min	80.49 ± 1.86	−0.38 ± 0.39	23.46 ± 0.38	3.91 ± 0.94	4.21 ± 0.59
IR100°C-80°C-15min	75.66 ± 3.71	0.30 ± 0.17	24.38 ± 0.82	2.29 ± 0.38	5.75 ± 1.32
IR120°C-80°C-30min	78.61 ± 0.53	−0.23 ± 0.11	24.07 ± 0.02	3.53 ± 0.26	4.52 ± 0.12
IR110°C-80°C-30min	79.22 ± 0.92	0.55 ± 0.10	24.03 ± 0.34	1.70 ± 0.22	4.26 ± 0.40
IR100°C-80°C-30min	78.46 ± 0.70	0.96 ± 0.10	24.18 ± 0.26	0.73 ± 0.24	4.51 ± 0.20
IR120°C-90°C-15min	80.31 ± 0.66	−0.40 ± 0.13	23.22 ± 0.58	3.96 ± 0.30	3.74 ± 0.43
IR110°C-90°C-15min	77.66 ± 1.24	−0.06 ± 0.36	25.41 ± 0.35	3.12 ± 0.81	6.07 ± 0.51
IR100°C-90°C-15min	80.83 ± 1.01	0.47 ± 0.17	23.73 ± 0.35	1.87 ± 0.41	4.15 ± 0.07

Source: Pan et al., Infrared heating for raw almond pasteurization. Final Report for Almond Board of
California, 2009a.

Teflon bottom or vibration bed (0.8 × 0.4 m) vibrates in four sections. The wavelike
vibrations are created by rotating rollers arranged under the flexible bottom at a
vibration frequency of 450 rpm. Figure 12.4 shows the emerging adult insects in
IR-treated storage rice samples. No insects were found in samples treated at 50°C or
above. The required temperature and holding time were 50°C and 1 min. Under such
treatment, rice milling quality was not affected, but IR caused about 1% moisture
loss that could be minimized by reducing the rice bed thickness and increasing the
heating rate (Table 12.7). Pan et al. (2008b) further demonstrated that for freshly harvested paddy, the rice tempering process after rapid IR heating and moisture removal
is essential to achieve high rice milling quality and improve the amount of moisture
removal during cooling (Figures 12.5 to 12.9).

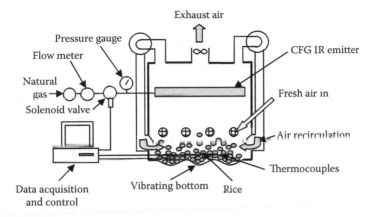

FIGURE 12.3 Schematic diagrams of catalytic infrared heating device. (From Pan et al., Rice utilization and product development. Annual comprehensive research report to the California Research Board, 2004.)

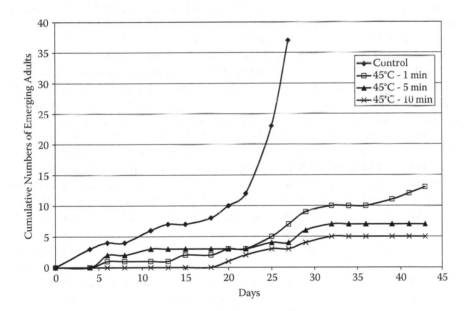

FIGURE 12.4 Emerging adult insects in infrared-treated samples (no insects found for the samples treated at 50°C or above). (From Pan et al., Rice utilization and product development. Annual comprehensive research report to the California Research Board, 2004.)

In general, for both the high and low initial MC rice samples, IR dried rice with tempering followed by natural cooling had similar and higher total rice yield (TRY) compared to the controls (Figures 12.5a and 12.5b). On average, the TRYs of low and high MC rice dried using IR followed by natural cooling were 68.0% and 68.1%, respectively, which were 0.3 and 0.7 percentage points more than the controls. In

TABLE 12.7

Moisture Change and Milling Quality of Infrared-Treated California Medium-Grain M202 Rice Samples

Treatments	Moisture Content (%)	Total Rice Yield (%)	Head Rice Yield (%)	Whiteness Index
Control	13.5 ± 0.1	68.5 ± 0.4	54.1 ± 0.7	44.4 ± 0.3
50°C-1min	12.3 ± 0.0	69.2 ± 0.4	53.8 ± 0.6	45.1 ± 0.3
50°C-5min	12.0 ± 0.0	69.5 ± 0.2	54.6 ± 0.1	44.7 ± 0.3
60°C-1min	11.9 ± 0.1	69.8 ± 0.4	52.4 ± 0.0	45.1 ± 0.4
60°C-5min	11.4 ± 0.1	70.3 ± 0.4	44.9 ± 0.4	44.5 ± 0.2

Source: Pan et al., Rice utilization and product development. Annual comprehensive research report to the California Research Board, 2004.

particular, the rice dried at about 60°C with natural cooling had the highest TRYs of 68.4% for low MC rice and 68.6% for high MC rice, compared to the controls of 67.7% and 67.4%, respectively. This meant that the TRYs of IR dried rough rice were 0.7 to 1.2 percentage points higher than the controls. However, the samples treated by other methods had much lower TRYs than the controls, especially the rice with low MC dried at high temperature. Similar trends were also observed for the HRYs (Figures 12.6a and 12.6b). The low MC rice samples dried using IR with tempering and natural cooling had significantly higher HRY (0.6 to 1.9 percentage points) than the control, and the highest HRY of 65.2% was obtained at a rice temperature of 61.2°C. For the high MC rice, the rice dried followed by tempering and natural cooling had the same HRY (63.6%) at 58.8°C as the control and slightly lower HRY at 42.8°C and 55.5°C than the control. All other post-heating treatments resulted in much lower HRYs. When the results for the whiteness index (WI) of the milled rice were examined, it could be seen that the IR dried rice generally had higher WI values than the controls, especially for the low MC rice, even though the differences between the controls and some of the treated rice samples were not significant (Figures 12.7a and 12.7b). The results indicated that most of the IR dried rice with tempering followed by natural cooling had a similar milling degree to the control. It seems that there is a trend that WI increased with an increase in the rice drying temperature for the non-tempering treatments, especially for the low MC rice. This could be due to the difference in the hardness of rice subjected to different treatments and/or the contribution of broken kernels to the color. Based on the milling quality results, it can be concluded that rough rice can be dried using IR followed by tempering and natural cooling to achieve superior rice milling quality. It is recommended that the rice temperature of IR heating be controlled at close to or below 60°C. For the current rice drying practice, the drying temperature or heated air temperature is controlled way below 60°C to avoid creating fissures, lowering the HRY. The reason the high temperature of IR heating did not damage the rice quality could be due to the relatively uniform heating in the rice kernel due to the IR penetration, resulting in a lower moisture gradient compared to conventional heated

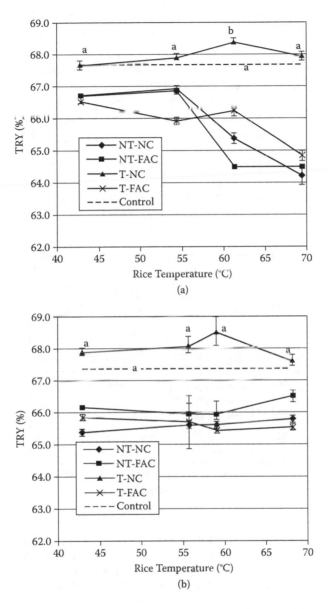

FIGURE 12.5 Total rice yields (TRY) of rough rice with initial MCs of (a) 20.6% and (b) 25.0% under different drying treatments. (T—Tempering, NT—No tempering, NC—Natural cooling, FAC—Forced air cooling; Values with letter *a* were not significantly different from the control sample at $p < 0.05$.) (Pan et al. 2008b. *J. Food Eng.*, 84(3), 469–479.)

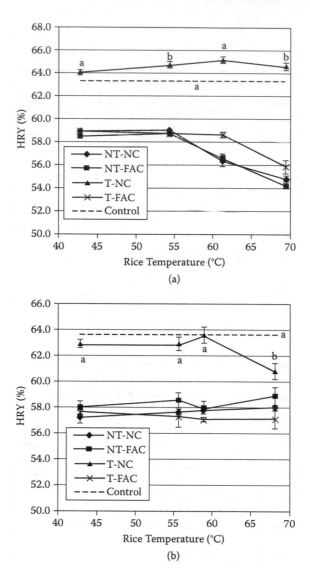

FIGURE 12.6 Head rice yields (HRY) of rice with initial MCs of (a) 20.6% and (b) 25.0% under different drying treatments. (T—Tempering, NT—No tempering, NC—Natural cooling, FAC—Forced air cooling; Values with letter *a* were not significantly different from the control sample at $p < 0.05$.) (Pan et al. 2008b. *J. Food Eng.*, 84(3), 469–479.)

air drying. The results indicate the rice milling quality may not be compromised with a relatively large amount of moisture removal in a single drying pass with a high drying rate if the rice can be heated quickly and uniformly, minimizing the moisture gradient. When a large amount of moisture is removed during IR heating, tempering is very important to re-establish the moisture equilibrium in the rice kernels. The study also showed that the cooling method following the tempering was

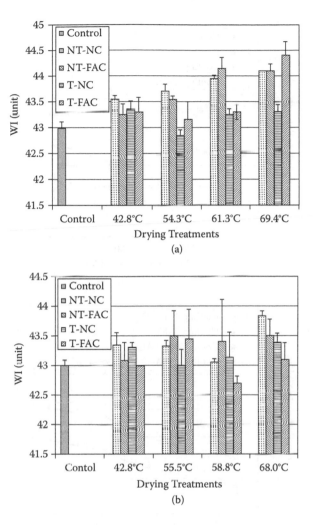

FIGURE 12.7 Whiteness (WI) of milled rice with initial MCs (a) 20.6% and (b) 25.0% under different drying treatments. (T—Tempering, NT—No tempering, NC—Natural cooling, FAC—Forced air cooling.) (Pan et al. 2008b. *J. Food Eng.*, 84(3), 469–479.)

very important. Rapid cooling using forced air can significantly lower the rice milling quality. Because a relative large amount of moisture was removed during forced air cooling, the cooling might regenerate significant moisture and temperature gradients causing fissures. Based on the glass transition hypothesis, the temperature and moisture at the rice surface were lowered first, and the starch reached a glassy state during cooling. At the same time, the temperature and moisture at the centers of the rice kernels were still relatively high, and the starch remained in a rubbery state. The differences in the thermomechanical properties of the starch at different stages would generate stresses and fissures, resulting in breakage during milling and a lower rice milling quality. Therefore, controlled slow cooling will be very important for high temperature rice drying. Since the natural cooling effectively

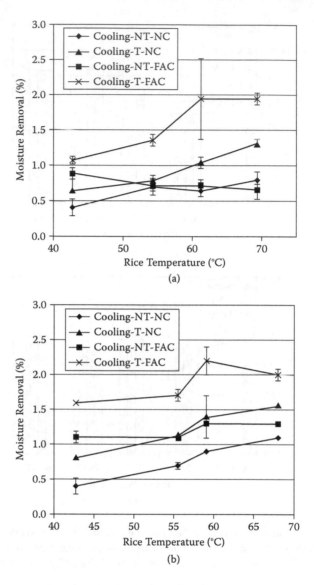

FIGURE 12.8 Moisture removal under different cooling methods with and without tempering for rough rice with initial MC of (a) 20.6% and (b) 25.0%. (T—Tempering, NT—No tempering, NC—Natural cooling, FAC—Forced air cooling.) (Pan et al. 2008b. *J. Food Eng.*, 84(3), 469–479.)

preserved the quality, controlled slow cooling could be accomplished by low rates of air flow through a bin of rice. The clear trends of tempering vs. non-tempering and natural cooling vs. forced air cooling are seen in Figures 12.8a and 12.8b. For low MC rice, the moisture removals from the tempered rice samples under natural cooling and forced air cooling were 0.6 to 1.3 and 1.1 to 1.9 percentage points, respectively, in the tested temperature range from 42.8°C to 69.4°C. In contrast, the

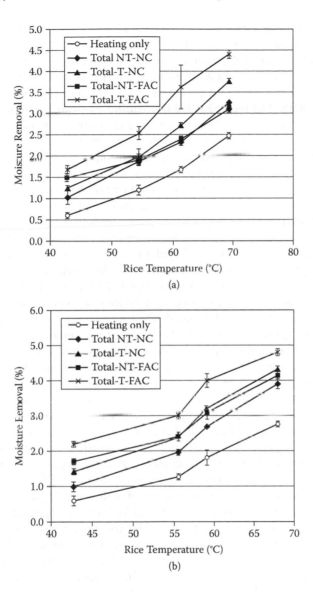

FIGURE 12.9 Total moisture removal under different cooling methods with and without tempering for rough rice with initial MC of (a) 20.6% and (b) 25.0%. (T—Tempering, NT—No tempering, NC—Natural cooling, FAC—Forced air cooling.) (Pan et al. 2008b. *J. Food Eng.*, 84(3), 469–479.)

non-tempered rice had 0.4 to 0.8 and 0.7 to 0.9 percentage point moisture removals under natural cooling and forced air cooling, respectively. Tempering resulted in 0.2 to 0.5 percentage point higher MC removal than non-tempering, which showed that the tempering treatment significantly improved the moisture removal during cooling compared to non-tempered samples. The forced air cooling also removed up to 0.9 percentage points more moisture than natural cooling in the tested temperature range. However, at the high heating temperature of 69.4°C without tempering, similar moisture removals were achieved with both natural cooling and forced air cooling. This was due to the high moisture gradients after more than 2.5 percentage point moisture removal, and moisture diffusion in the rice kernels became the limiting factor for further improving the drying rate by using an increased drying force of forced air cooling. The high MC rice had similar moisture removal trends to the low MC rice during cooling, even though more moisture was removed compared to the low MC rice. The tempered rice had moisture removals of 1.6 to 2.2 percentage points for forced air cooling and 0.8 to 1.5 percentage points for natural cooling, compared to 1.1 to 1.3 percentage points for forced air cooling and 0.4 to 1.1 percentage points for natural cooling of non-tempered rice in the tested temperature range. The tempering treatment resulted in 0.4 to 0.5 and 0.5 to 0.9 percentage points higher moisture removals than the non-tempering treatment with natural cooling and forced air cooling, respectively. The results indicate that tempering is even more important for high MC rice than low MC rice in order to have high MC removal during cooling. Accordingly, the tempering process reduced the moisture gradient in the rice kernels and allowed the moisture to equilibrate before the rice kernels were cooled. Without tempering, there was a significant moisture gradient in the rice kernels and a low MC near the surface, which resulted in less moisture removal during cooling. In general, reduced moisture gradient in the tempered rice kernels and forced air cooling increased moisture removal during the cooling process. Therefore, the tempering process is a critical step in increasing moisture removal during cooling. In order to achieve high moisture removal during cooling, a combination of tempering and forced air cooling could be used, even though excess moisture removal could cause rice fissures, lowering the rice milling quality, which would need to be considered.

The trend of total moisture removal at different temperatures with different tempering and cooling treatments was more or less parallel to the moisture removal caused by heating only (Figures 12.9a and 12.9b). The highest total MC removals from the rice were 1.7 to 4.4 and 2.2 to 4.8 percentage points for low and high MC rice samples, respectively, which were achieved with tempering and forced air cooling among the treatments. But the lowest total MC removals generally occurred for rice experiencing no tempering and a natural cooling treatment. For rice treated with tempering and natural cooling, the total moisture removals were 1.4, 2.4, 3.2 and 4.3 percentage points for the high MC rice and 1.3, 2.0, 2.7 and 3.8 percentage points for the low MC rice over the tested temperature range. The moisture removals were the second highest among the treatments when the temperatures were above 55°C. These numbers indicated that 2.7 to 3.2 percentage points of moisture were removed with 1 min heating followed by tempering and natural cooling. The drying rates were much higher than the 2 to 3 percentage point moisture removal with 15 to 20 min heating of the current conventional heated air drying. For total moisture removal, the moisture removed due to sensible heat during cooling was a very significant portion. For example, 37% and 44% of total moisture removals occurred during cooling when the low and high MC rice samples, respectively, were heated for 60 s (to about 60°C), followed by tempering and natural cooling. Because no additional heating energy is needed during the cooling, the high moisture removal could further improve the energy efficiency of the IR drying process.

12.3.5 IR RADIATION DRY PEELING

Hot lye peeling and steam peeling are so far the commercialized techniques adopted by the fruit and vegetable processing industry. Lye peeling has been the more widely industrialized method of the two, producing high-quality peeled fruit and vegetable products. However, lye peeling has resulted in significantly negative impacts on the environment and related industries. The associated burden of lye peeling related to salinity management of wasterwater, escalated cost, and threat to long-term supply of water have made this technology unattractive to the processors. On the other hand, steam peeling results in inferior products such as losses in product appearance, firmness, and yield (Pan et al., 2009b).

Due to higher efficiency of IR heating and lower penetration depth, IR heating has found an application in this area, albeit in the infant stages of research, as a suitable method for peeling fruits and vegetables. Hart et al. (1970) pioneered the term *dry peeling* process because no water is used in the IR peeling process. Sproul et al. (1975) had earlier studied the application of IR peeling on white potatoes and peaches, and noted significant reductions in peeling loss, wastewater, and use of caustic lye. The most recent advance (Pan et al., 2009b) describes the development of a novel IR peeling method for tomatoes as an alternative to lye and steam peeling. IR peeling is emerging as a cost-effective and environmentally friendlier method compared to conventional methods. The new IR peeling system described by Pan et al. (2009b) is equipped with catalytic IR emitters powered by natural gas that could make the heating process energy efficient. Pan et al. (2009b) have reported on the superiority of IR dry peeling over lye, sequential lye–IR, or sequential enzyme–IR peeling method. Table 12.8 contrasts IR dry peeling and lye peeling performance on tomatoes. IR dry peeling reduces tomato peeling loss to significantly lower levels than hot lye peeling. It has also been observed that treatment using lye and enzyme as a pretreatment of IR peeling does not provide advantageous synergistic effects over IR peeling alone (Pan et al., 2009b). Whereas sequential enzyme–IR peeling gives better peeling ease, it results in much higher peeling loss and longer treatment times compared to IR peeling alone. Sequential lye–IR peeling causes a significantly higher peeling loss compared to IR dry peeling alone. Presently, IR dry peeling has potential to solve the crises of water and lye usage in the fruit and vegetable peeling industry.

12.4 ECONOMIC BENEFITS

IR heating applications are expected to increase as a result of higher demands for safe products of high nutritional value and organoleptic quality (Sakai and Mao 2005). Mujumdar (2002) highlighted product quality and energy savings to be very important aspects in any drying process. Innovations in designing equipment using IR energy are therefore expected to lead to the advancement of energy efficiency

TABLE 12.8

Effects of Heating Time on Tomato Peeling with Lye and Infrared Heating for Tomato Sun6366

Methods and Conditions	Peelability (cm²/g)	Ease of Peeling	Peeling Loss (%)	Peeled Firmness (kg)	Surface Temperature (°C)
Lye$_{10}$-30s	0.004	3.5a	11.46	1.5a	95
Lye$_{10}$-45s	0.008	4.1a,b	11.68	1.3b	95
Lye$_{10}$-60s	0.004	4.7b	13.37	1.4b	95
Lye$_{10}$-75s	0.003	4.9c	13.55	1.3b	95
IR$_{12}$-30s	0.020a	1.6a	7.64	1.8a	57.9a
IR$_{12}$-45s	0.004b	3.0b	6.11	1.8a	66.2b
IR$_{12}$-60s	0.002b	4.1c	7.74	1.5a,b	70.1c
IR$_{12}$-75s	0.002b	4.6d	9.41	1.6b	76.8d

Source: Pan et al. 2009b. Development of infrared radiation heating method for sustainable tomato peeling, *Applied Engineering in Agriculture*, in press.

Note: Subscript 10 of Lye$_{10}$ stands for the concentration of the lye solution. Subscript 12 of IR$_{12}$ stands for the gap of the two IR emitters. Mean separation was via Duncan's multiple range test. Means with a different letter (a, b, c, d) in each section are significantly different at the 0.05 level.

with faster processing and lower energy costs. Dudkiewicz and Jezowiecki (2009) reported that product temperature rise of 1°C is equivalent to a 6% increase in energy consumption compared to a classical heating method.

Table 12.9 compares the performance of FIR and conventional ovens, and reveals that FIR ovens are superior in terms of energy cost and compactness of size (Sakai and Hanzawa, 1994). IR heating systems have satisfactory energy efficiency and energy conservation compared to the conventional heating systems. Xin et al. (1997) compared the energy use between conventional and energy-efficient radiant heat lamps. A possible benefit of using the energy-efficient heat lamp was an annual energy saving of $5400. Pan et al. (2005) demonstrated the feasibility of simultaneous IR blanching and drying with significant reductions in processing time and high-quality products compared to the current processing technologies. The application of their new industrial-scale mobile IR dry-blanching equipment is described in detail in Chapter 9. The new mobile IR dry-blanching equipment can be controlled to achieve optimized processing conditions. In order to avoid unnecessary energy use and to obtain high quality of either blanched, partially, or fully dehydrated products, specific adjustment of the equipment setting based on each product is necessary. A summary of energy analysis of a selected condition of the IDB unit (Figure 12.10) for blanching of potato is presented in Table 12.10. The selected test conditions for potato processing are thus as follows: gas supply at section 1 equals 100% and at section 2 is off; conveyor speed is set at 3.175 ft/min; fan is set at off mode; the exposure time to IR is 141.5 s; the total product residence time in unit 283 s; on high fire, 168.809 kW

TABLE 12.9
Comparative Performance Characteristics of (A) FIR and (B) Conventional Heating Ovens (LPB Implies Liquefied Petroleum Gas)

a) Oven for Baking Rice Crackers:

	Type of Oven		FIR Performance
Criterion for Comparison	FIR (A)	LPG(B)	(A/B) × 100%
Caloric consumption	223,200 kJ/h	836,000 kJ/h	26.7%
Production rate	10,000 pieces/h	10,000 pieces/h	100%
Baking time	10 min	15 min	66.7%
Energy cost	$13.70/h	$25.10/h	54.5%

b) Oven for Roasting Fish Paste

	Type of Oven		FIR Performance
Criterion for Comparison	FIR (A)	Nichrome(B)	(A/B) × 100%
Length of furnace	9 m	19 m	47.4%
Production rate	1470 kg/day	500 kg/day	294%
Electric power consumption	0.06 kW/kg	0.23 kW/kg	26.1%

Source: Sakai and Hanzawa. 1994. *Trends Food Sci. Tech.*, 5, 357.

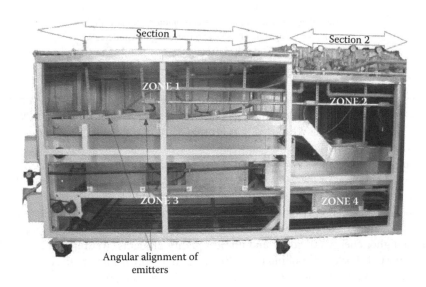

FIGURE 12.10 Side view of part of the IR equipment with imaginary zones and imaginary sections. (From Pan et al., Test results and performance of mobile infrared heating equipment for processing fruits and vegetables. USDA-ARS-WRRC Report, 2009c.)

TABLE 12.10

Summarized Data of Energy Analysis for a Selected Condition of the New Mobile IR Dry Blancher during the Dry Blanching of Russet Potatoes Slices

Parameter	Amount[a]
Power consumption	42.2 kW
Energy losses	
Heat loss by natural gas to IR	4777.3 kJ
Heat loss by natural convection	1074.4 kJ
Heat loss by radiation	140 kJ
Total heat loss	5986.7 kJ
Operation efficiency	74.9%
Total natural gas consumption	0.64 m^3
Cost per lb of tomato	0.9 c/lb

Source: Pan et al., Test results and performance of mobile infrared heating equipment for processing fruits and vegetables. USDA-ARS-WRRC Report, 2009c.

Note: Selected test conditions for potato processing are thus: gas supply at section 1 equals 100% and at section 2 is off; conveyor speed is set at 3.175 ft/min; fan is set at off mode; the exposure time to IR is 141.5 s and total product residence time in unit 283 s.

[a] The calculations are based on a continuous process with the total processing time of 283 s and four running emitters. See Appendix 12.1 for detailed explanations on calculations.

of energy is supplied to eight emitters by natural gas; and four emitters are running. Given the conversion rate for natural gas to IR radiation as approximately 80%, the summarized calculations (Table 12.10) are based on the assumptions that (1) heat losses are by natural convection and radiation, (2) the face-material of the equipment is polished stainless steel ($\varepsilon = 0.075$), (3) the ambient temperature is 20°C, and (4) the temperatures of the walls surrounding the equipment are same as the ambient temperature. (See Appendix 12.1 for a detailed explanation.)

Innovations and improvements in IR heating systems combined with microwave, radio-frequency or ohmic heating, and impingement systems are also expected to lead to higher energy efficiency and higher-quality products. Andrejko et al. (2008) have reported shortened cooking times (from 12 to 5 min), leading to the economical advantages, via the use of IR heating for pea seeds cooking process.

12.5 INDUSTRIALIZATION NEEDS

The commercialization of any new technology and foods produced using newer technologies often involves many unknowns. Although IR heating technology has demonstrated many advantages, the major obstacles to the adoption of most IR technologies in the agricultural and food industries are the lack of understanding of IR technology and lack of its extension beyond the conventional uses in other manufacturing industries such as textile, automotive, etc. To date, IR heating has demonstrated the advantage of efficient heat transfer to the food and agricultural products by reducing processing time and energy costs. In order to achieve optimum industrial systems, application of the available knowledge base on IR heating systems and the related process modeling to advanced process control is crucial. Research in control and optimization of IR drying processes, as well as in the use of different IR heating systems for different purposes, is quite limited (Dhib, 2007), and this is especially obvious in regard to food-processing applications. Particularly, optimization of IR processes with respect to sensory evaluation, quality changes, energy savings, and economical analysis compared to the conventional processes will put the spotlight on IR technology. Industrial implementations of IR processes for food and agricultural process application need careful design that integrates variable heat fluxes and wavelengths to avoid product overheating and overprocessing on the surface in order not to compromise product quality.

Already, the effort in the research fields to utilize the existing knowledge base to improve equipment design is demonstrated by researchers across the food processing fields. For instance, in an industry-partnered project, a new IR heating unit has been designed in the Department of Food Engineering at the University of Mersin that utilizes flat, electrically operated ceramic emitters. The designers introduced an IR tunnel using the ceramic IR heaters to reduce the microbial load of spices with minimal damage to economical and quality parameters (Figure 12.11). Other avenues of application continue to be featured, such as FIR heat sterilization units to sterilize floors of food processing plants at lower costs (Takenaka Corporation, Japan, http://www.takenaka.co.jp/takenaka_e/news_e/pr0105/m0105_01.htm).

12.6 CONCLUSION

The development and implementation of IR technologies in the food and agricultural sectors as alternative and sustainable methods will benefit the environment and reduce energy and water use. Overall, efficient IR processes having environmental and economic benefits as compared to conventional thermal processes can be achieved by precise process control and a well-designed equipment and processing system. The need for improved product quality, safety, and energy and processing efficiency are expected to drive the industrialization of IR technologies for food and agricultural processing.

FIGURE 12.11 Infrared (IR) heating unit with electrical powered heating elements.

REFERENCES

Afzal, T.M. and Abe, T. 1997. Modeling far infrared drying of rough rice. *J. Microw. Power Electromagn. Energ.* 32, 80.

Afzal, T.M. and Abe, T. 1998. Diffusion in potato during far infrared radiation drying. *J. Food Eng.* 37, 353.

Andrejko, D., Rydzak, L., Slaska-Grzywna, B., Gozdziewska, M., and Kobus, Z. 2008. Influence of preliminary thermal processing applying infra-red radiation on pea seeds cooking process. *Int. Agrophysics*, 22, 17.

Cenkowski, S. and Sosulski, F.W. 1997. Physical and cooking properties of micronized lentils. *J. Food Proc. Eng.*, 20, 249.

Cenkowski, S. and Sosulski, F.W. 1998. Cooking characteristics of split peas treated with infrared heat. *Trans. ASAE*, 41, 715.

Dagerskog, M. 1979. Infra-red radiation for food processing II. Calculation of heat penetration during infra-red frying of meat products. *Lebensm. Wiss. u. Tech.*, 12, 252.

Das, I., Bal, S., and Das, S.K. 2003. Optimization of the vibration-aided infrared heating process using response surface methodology: An application to the drying and parboiling of high moisture paddy. Paper no 036040, ASAE Annual Meeting.

Dhib, R. 2007. Infrared drying: From process modeling to advanced process control. *Drying Tech.*, 25, 97.

Dudkiewicz, E. and Jezowiecki, J. 2009. Measured radiant thermal fields in industrial spaces served by high intensity infrared heater. *Energy Buildings*, 41, 27.

Fasina, O.O. and Tyler, R.T. 2001. Infrared heating of biological materials. In *Food Processing Operations Modeling—Design and Analysis*, Irudayaraj, J. Ed., New York: Marcel Dekker, chap. 7.

Fasina, O., Tyler, B., Pickard, M., Zheng, G-H. and Wang, N. 2001. Effect of infrared heating on the properties of legume seeds. *Int. J. Food Sci. Tech.*, 36, 79.

Fu, W-R. and Lien, W-R. 1998. Optimization of far infrared heat dehydration of shrimp using RSM. *J. Food Sci.*, 63, 80.

Gabel, M.M., Pan, Z., Amaratunga, K.S.P., Harris, L.J., and Thompson, J.F. 2006. Catalytic infrared dehydration of onions. *J. Food Sci.*, 71, E351.

Hamanaka, D., Dokan, S., Yasunaga, E., Kuroki, S., Uchino, T., and Akimoto, K. 2000. The sterilization effects of infrared ray on the agricultural products spoilage microorganisms (part I). Paper no. 006090, ASAE Annual Meeting.

Hamanaka, D., Uchino, T., Furuse, N., Han, W. and Tanaka, S. 2006a. Effect of wavelength of infrared heaters on the inactivation of bacterial spores at various water activities. *Int. J. Food Microbiology*, 108, 281.

Hamanaka, D., Uchino, T., Inoue, A., Kawasaki, K. and Hori, Y. 2006b. Development of the rotating type grain sterilizer using infrared radiation heating, presented in *IUFOST 13th World Congress Food Sci. Tech.*, pp. 1615–1616.

Hart, M.R., Graham, R.P., Huxsoll, C.C., and Williams, G.S. 1970. An experimental dry caustic peeler for cling peaches and other fruits. *J. Food Sci.*, 35, 839.

Hebbar, H.U., Vishwanathan, K.H., and Ramesh, M.N. 2004. Development of combined infrared and hot air dryer for vegetables. *J. Food Eng.*, 65, 557.

Hidaka, Y., Kubota, K., and Ichikawa, T. 2004. Development of a far-infrared radiation dryer for grain, Paper no: 046154, ASABE Annual Meeting.

Huang, L. 2004. Infrared surface pasteurization of turkey frankfurters. *Innovation Food Sci. Emerging Tech.*, 5, 345.

Huang, L. and Sites, J. 2008. Elimination of *Listeria monocytogenes* on hotdogs by infrared surface treatment. *J. Food Sci.*, 73, M27.

Jun, S. and Irudayaraj, J. 2004. Explore the mechanism of selective infrared heating on disinfection of fungal spores. *App. Eng. Agriculture*, 20, 481.

Khan, M.K. and Vandermey, P.A. 1985. Quality assessment of ground beef patties after infrared heat processing in a conveyorized tube broiler for foodservice use. *J. Food Sci.*, 50, 101.

Khir, R., Pan, Z., and Salim, A., Drying rates of thin layer rough rice drying using infrared radiation. Paper no. 066011, ASABE Annual Meeting, 2006.

Khir, R., Pan, Z., Salim, A., and Thompson, J.F., Drying characteristics and quality of rough rice under infrared radiation heating, Paper no: 076261, ASABE Annual Meeting, 2007.

Krishnamurthy, K., Jun, S., Irudayaraj, J., and Demirci, A., Efficacy of infrared heat treatment for inactivation of *Staphylococcus aureus* in milk, *J. Food Process Eng.*, 31, 798, 2008.

Li, X., Pan, Z., Bingol, G., McHugh, T.H., and Atungulu, G.G. 2009. Feasibility study of using infrared radiation heating as a sustainable tomato peeling method. Paper no. 095689, ASABE Annual Meeting, 2009.

Liu, C.M., Sakai, N. and Hanzawa, T. 1999. Three dimensional analysis of heat transfer during food thawing by far-infrared radiation. *Food Sci. Tech. Res.*, 5, 294.

Lloyd, B.J., Farkas, B.E., and Keener, K.M. 2003. Characterization of radiant emitters used in food processing. *J. Micro. Power Electromagn. Energ.*, 38, 213.

Maria, B.T., Pan, Z., Steven, H., Zhu, Y., and Tara, M. 2008. Reduction of *Salmonella enteritidis* population sizes on almond kernels with infrared heat. *J. Food Protection*, 71, 897.

Meeso, N., Nathakaranakule, A., Madhiyanon, T., and Soponronnarit, S. 2008. Different strategies of far-infrared application in paddy drying. *Int. J. Food Eng.*, 4, Article 10.

Metussin, R., Alli, I., and Kermasha, S. 1992. Micronization effects of composition and properties of tofu. *J. Food Sci.*, 57, 418.

Mujumdar, A.S. 2002. Drying research-current state and future trends. *Developments Chem. Eng. Mineral Process.*, 10, 225.

Ozdemir, M. and Devres, O. 2000. Analysis of color development during roasting of hazelnuts using response surface methodology. *J. Food Eng.*, 45, 17–24.

Pan Z., Bingol, G., and McHugh, T. 2009c. Test results and performance of mobile infrared heating equipment for processing fruits and vegetables. USDA-ARS-WRRC Report.

Pan, Z., Khir, R., Godfrey, L.D., Lewis, R., Thompson, J.F., and Salim, A. 2008b. Feasibility of simultaneous rough rice drying and disinfestations by infrared radiation heating and rice milling quality. *J. Food Eng.*, 84(3), 469–479.

Pan, Z., Li, X., Bingol, G., McHugh, T., Atungulu, G.G. 2009b. Development of infrared radiation heating method for sustainable tomato peeling. Applied Engineering. 25(6), 935–941.

Pan, Z., and McHugh, T.H. 2004. Novel infrared dry-blanching (IDB), infrared blanching, and infrared drying technologies for food processing. U.S. Patent Application 20060034981. Filed 8/13/2004, published 2/16/2006.

Pan, Z., Olson, D.A., Amaratunga, K.S.P., Olsen, C.W., Zhu, Y., and McHugh, T.H. 2005. Feasibility of using infrared heating for blanching and dehydration of fruits and vegetables. Paper no. 056086, ASAE Annual Meeting.

Pan, Z., Solar, M. L., and Yokohama, W.H. 2004. Rice utilization and product development. Annual comprehensive research report to the California Research Board.

Pan Z., Yang, J., Bingol, G., Brandl, M., McHugh, T. 2009a. Infrared heating for raw almond pasteurization. Final Report for Almond Board of California.

Pan Z., Yang, J., Brandl, M., McHugh, T., and Bingol, G. 2008a. Infrared heating for improved safety and processing efficiency of dry-roasted almonds. Report for Almond Board of California.

Ranjan, R., Irudayaraj, J., and Jun, S. 2002. Simulation of infrared drying process. *Drying Tech.*, 20, 363.

Ratti, C. and Mujumdar, A.S. 1995. Infrared drying. In *Handbook of Industrial Drying*. Vol. I, Mujumdar, A.S., Ed., New York: Marcel Dekker, pp. 567–588.

Rosenthal, I., Rosen, B., and Berstein, B. 1996. Surface pasteurization of cottage cheese. *Milchwissenchaft*, 51, 198.

Sakai, N. and Hanzawa, T. 1994. Applications and advances in far-infrared heating in Japan. *Trends Food Sci. Tech.*, 5, 357.

Sakai, N. and Mao, W. 2005. Infrared heating, in *Thermal Food Processing*, Sun, D-W., Ed., Boca Raton, FL: CRC Press, chap. 16.

Sandu, C. 1986. Infrared radiative drying in food engineering: A process analysis. *Biotech. Progress*, 2, 209.

Sawai, J., Sagara, K., Kasai, S., Igarashi, H., Hashimoto, A., Kokugan, T., Shimizu, M., and Kojima, H. 2000. Far-infrared irradiation-induced injuries to *Escherichia coli* at below the lethal temperature. *J. Ind. Microbiol. Biotech.*, 24, 19.

Sheridan, P. and Shilton, N. 1999. Application of far infra-red radiation to cooking of meat products. *J. Food Eng.*, 41, 203.

Shi, J., Pan, Z., McHugh, T.H., Wood, T., Zhu, Y., Avena-Bustillos, E.J., and Hirschberg, E. 2008. Effect of berry size and sodium hydroxide pretreatment on the drying characteristics of blue berries under infrared radiation heating. *J. Food Sci.*, 73, E259.

Skjoldebrand, C. and Anderson, C. 1989. A comparison of infrared bread baking and conventional baking. *J. Microwave Power Electromagn. Energ.*, 24, 91.

Sproul, O.J., Vennes. W., Knudson, W., and Cyr, J.W. 1975. Infrared dry caustic vs. wet caustic peeling of potatoes. Environmental Protection Technology Series, Corvallis, Oregon.

Staack, N., Ahrne, L., Borch, E., and Knorr, D. 2008a. Effect of infrared heating on quality and microbial decontamination in paprika powder. *J. Food Eng.*, 86, 17.

Staack, N., Ahrne, L., Borch, E., and Knorr, D. 2008b. Effect of temperature, pH and controlled water activity on inactivation of spore of *Bacillus cereus* in paprika powder by near-IR radiation. *J. Food Eng.*, 89, 319.

Tanaka, F., Verboven, P., Scheerlinck, N., Morita, K., Iwasaki, K., and Nicolai, B. 2007. Investigation of far infrared radiation as an alternative technique for surface decontamination of strawberry. *J. Food Eng.*, 79, 445.

Wade, P. 1987. Biscuit baking by near-infrared radiation. *J. Food Eng.*, 6, 165.

Wang, J. and Sheng, K.C. 2004. Modeling of multi-layer far-infrared dryer. *Drying Tech.*, 22, 809.

Xin, H., Zhou, H., and Bundy, D.S. 1997. Comparison of energy use and piglet performance between conventional and energy-efficient heat lamps. *App. Eng. Agric.*, 13, 95.

Zhu, Y. and Pan, Z. 2009. Processing and quality characteristics of apple slices under simultaneous infrared dry-blanching and dehydration with continuous heating. *J. Food Eng.*, 90, 441.

APPENDIX 12.1

ENERGY EFFICIENCY AND COST ANALYSIS OF A SMALL COMMERCIAL IR UNIT FOR DRY BLANCHING

ENERGY LOSS DUE TO CONVERSION OF NATURAL GAS TO INFRARED ENERGY

The conversion rate for natural gas to infrared radiation is approximately 80%.

Energy loss due to conversion = 168.809 × 0.20 = 33.7618 kW

ASSUMPTIONS

1. Heat losses are by natural convection and radiation.
2. The face material of the equipment is polished stainless steel ($\varepsilon = 0.075$).
3. The ambient temperature is 20°C.
4. The temperature of the walls of the building surrounding the equipment is the same as the ambient temperature.

NOMENCLATURE

A: Area (m²)
h: Convection heat transfer coefficient (W/m²·K)
k: Thermal conductivity (W/m·K)
L: Length (m)
Nu: Nusselt number
P: Wetted perimeter (m)

Pr: Prandtl number
Ra: Rayleigh number
T: Temperature

GREEK SYMBOLS

ρ: Density (kg/m^3)
σ: Stefan-Boltzmann constant (5.67 × 10^{-8} W/m^2·K^4)
μ: Viscosity (kg/m·s)
ε: Emissivity
β: Volume expansion (1/K)

SUBSCRIPTS

c: Critical
f: film
s: Surface
∞: Ambient

HEAT LOSSES BY NATURAL CONVECTION AND RADIATION FROM FACES OF THE EQUIPMENT

A1. Natural Convection Heat Losses from Sides

$T_s = 63°C, \ T_\infty = 20°C$

Total area $= 180 \times 66 \, \text{in.}^2 = 7.664 \, \text{m}^2$

$$Tf = \frac{63+20}{2} = 41.5°C$$

$$\beta = \frac{1}{T_f}$$

Air at 41.5°C

$\rho = 1.1248 \, \text{kg/m}^3, \ \mu = 1.9131 \times 10^{-5} \, \text{kg/ms}, \ Pr = 0.7047$

$k = 0.02735 \, \text{W/m°C}$

$$Ra = Gr \cdot Pr = \frac{g \cdot \beta \cdot \left(T_s - T_\infty\right) \cdot L_c^3}{\left(\dfrac{\mu}{\rho}\right)^2} = \frac{9.81 \times \dfrac{1}{314.65} \times \left(63 - 20\right) \times 1.6764^3}{\left(\dfrac{1.9132 \times 10^{-5}}{1.1248}\right)^2} \times 0.707$$

$$Ra = 2.1709 \times 10^{10}$$

$$Nu = 0.1 \cdot Ra^{1/3}$$

$$Nu = 278.96 = \frac{h \cdot L_c}{k}$$

$$h = \frac{278.96 \times 0.02735}{1.6764} = 4.55 \frac{W}{m^2 \, ^\circ C}$$

$$Q_{loss} = h \cdot A \cdot \Delta T = 4.55 \times 7.664 \times \left(63 - 20\right)$$

$$Q_{loss} = 1499.854 \, W = 1.5 \, kW$$

Losses from two sides = $2 \times 1.5 = 3$ kW

A2. Radiation Heat Loss from Sides

$$Q = \varepsilon \sigma A \left(T_s^4 - T_{surroundings}^4\right)$$

$$Q = 0.075 \times \left(5.67 \times 10^{-8} \frac{W}{m^2 K^4}\right) \times \left(7.664 \, m^2\right) \left(\left(63 + 273.15\right)^4 - \left(20 + 273.15\right)^4 \, K\right)$$

$$Q = 175.44 \, W = 0.175 \, kW$$

Losses from two sides = $2 \times 0.175 = 0.35$ kW

B1. Heat Losses from Front Side

$T_s = 99^\circ C$, $T_\infty = 20^\circ C$

Total area = $28.75 \times 77 \, in.^2 = 1.428 \, m^2$

$$Tf = \frac{99 + 20}{2} = 59.5^\circ C$$

$$\beta = \frac{1}{T_f} = \frac{1}{332.65} \left(\frac{1}{K}\right)$$

Air at 59.5°C

$\rho = 1.0602 \text{ kg/m}^3$, $\mu = 1.995 \times 10^{-5} \text{ kg/ms}$, $Pr = 0.700817$

$k = 0.02871 \text{ W/m°C}$

$$Ra = Gr \cdot Pr = \frac{g \cdot \beta \cdot (T_s - T_\infty) \cdot L_c^3}{\left(\dfrac{\mu}{\rho}\right)^2} = \frac{9.81 \times \dfrac{1}{332.65} \times (99 - 20) \times 0.730^3}{\left(\dfrac{1.995 \times 10^{-5}}{1.0602}\right)^2} \times 0.70817$$

$Ra = 1.7938 \times 10^9$

$Nu = 0.1 \cdot Ra^{1/3}$

$Nu = 121.503 = \dfrac{h \cdot L_c}{k}$

$h = \dfrac{121.503 \times 0.02871}{0.730} = 4.778 \dfrac{W}{m^2 °C}$

$Q_{loss} = h \cdot A \cdot \Delta T = 4.778 \times 1.428 \times (99 - 20)$

$Q_{loss} = 539.08 \text{ W} = 0.539 \text{ kW}$

B2. Radiation Heat Loss from Front Side

$$Q = \varepsilon \sigma A \left(T_s^4 - T_{surroundings}^4 \right)$$

$$Q = 0.075 \times \left(5.67 \times 10^{-8} \frac{W}{m^2 K^4} \right) \times (1.428 \, m^2) \left((99 + 273.15)^4 - (20 + 273.15)^4 \, K \right)$$

$Q = 71.63 \text{ W} = 0.071 \text{ kW}$

C1. Heat Losses from Rear Side

$T_s = 67°C$, $T_\infty = 20°C$

Total area $= 32.75 \times 77 \text{ in.}^2 = 1.627 \text{ m}^2$

$Tf = \dfrac{67 + 20}{2} = 43.5°C$

$\beta = \dfrac{1}{T_f} = \dfrac{1}{316.65} \left(\dfrac{1}{K} \right)$

Air at 43.5°C

$\rho = 1.1176 \text{ kg/m}^3$, $\mu = 1.9223 \times 10^{-5} \text{ kg/ms}$, $Pr = 0.7043$

$k = 0.027502 \text{ W/m°C}$

$$Ra = Gr \cdot Pr = \frac{g \cdot \beta \cdot (T_s - T_\infty) \cdot L_c^3}{\left(\dfrac{\mu}{\rho}\right)^2} = \frac{9.81 \times \dfrac{1}{316.65} \times (67-20) \times 0.8318^3}{\left(\dfrac{1.9223 \times 10^{-5}}{1.1176}\right)^2} \times 0.7043$$

$Ra = 1.995 \times 10^9$

$Nu = 0.1 \cdot Ra^{1/3}$

$Nu = 125.885 = \dfrac{h \cdot L_c}{k}$

$h = \dfrac{125.885 \times 0.027502}{0.8318} = 4.1621 \dfrac{\text{W}}{\text{m}^2\text{°C}}$

$Q_{loss} = h \cdot A \cdot \Delta T = 4.1621 \times 1.627 \times (67-20)$

$Q_{loss} = 318.279 \text{ W} = 0.318 \text{ kW}$

C2. Radiation Heat Loss from Rear Side

$$Q = \varepsilon \sigma A \left(T_s^4 - T_{surroundings}^4\right)$$

$$Q = 0.075 \times \left(5.67 \times 10^{-8} \frac{\text{W}}{\text{m}^2\text{K}^4}\right) \times \left(1.627 \text{ m}^2\right) \left((67+273.15)^4 - (20+273.15)^4 \text{ K}\right)$$

$Q = 41.525 \text{ W} = 0.041 \text{ kW}$

D1. Heat Losses from Top and Bottom Sides

$T_s = 70\text{°C}$, $T_\infty = 20\text{°C}$

Total area $= 77 \cdot 180 \text{ in.}^2 = 8.941 \text{ m}^2$

$Tf = \dfrac{70+20}{2} = 45\text{°C}$

$$\beta = \frac{1}{T_f} = \frac{1}{318.15}\left(\frac{1}{K}\right)$$

$$L_c = \frac{A}{P} = \frac{8.941\ \text{m}^2}{13.055\ \text{m}} = 0.6849\ \text{m}$$

Air at 45°C

$$\rho = 1.1122\ \text{kg/m}^3,\ \mu = 1.9292\times10^{-5}\ \text{kg/ms},\ Pr = 0.704$$

$$k = 0.02761\ \text{W/m}°\text{C}$$

$$Ra = Gr \cdot Pr = \frac{g \cdot \beta \cdot (T_s - T_\infty) \cdot L_c^3}{\left(\dfrac{\mu}{\rho}\right)^2} = \frac{9.81 \times \dfrac{1}{318.15} \times (70 - 20) \times 0.6849^3}{\left(\dfrac{1.9223\times10^{-5}}{1.1176}\right)^2} \times 0.704$$

$$Ra = 1.159 \times 10^9$$

$Nu = 0.1 \cdot Ra^{1/3}$ For top side

$$Nu = 157.56 = \frac{h \cdot L_c}{k}$$

$$h = \frac{157.56 \times 0.02761}{0.6849} = 6.3516\ \frac{\text{W}}{\text{m}^2}$$

$$Q_{loss,top} = h \cdot A \cdot \Delta T = 6.3516 \times 8.941 \times (70 - 20)$$

$$Q_{loss,top} = 2839.51\ \text{W} = 2.839\ \text{kW}$$

$Nu = 0.27 \cdot Ra^{1/4}$ For bottom side

$$Nu = 49.817 = \frac{h \cdot L_c}{k}$$

$$h = \frac{49.817 \times 0.02761}{0.6849} = 2.0082\ \frac{\text{W}}{\text{m}^2°\text{K}}$$

$$Q_{loss,bottom} = h \cdot A \cdot \Delta T = 2.0082 \times 8.941 \times (70 - 20)$$

$$Q_{loss,bottom} = 897.79\ \text{W} = 0.897\ \text{kW}$$

D2. Radiation Heat Loss from Top and Bottom Sides

$$Q = \varepsilon \sigma A \left(T_s^4 - T_{surroundings}^4 \right)$$

$$Q = 0.075 \cdot \left(5.67 \cdot 10^{-8} \frac{W}{m^2 K^4} \right) \cdot \left(2 \times 8.941 \ m^2 \right) \left(\left(70 + 273.15 \right)^4 - \left(20 + 273.15 \right)^4 \ K \right)$$

$$Q = 492.787 \ W = 0.492 \ kW$$

Total Heat Loss by Natural Convection

$$\sum Q_{loss} = Q_{sides} + Q_{rear} + Q_{front} + Q_{top} + Q_{bottom}$$

$$\sum Q_{loss} = \left(3 + 0.318 + 0.539 + 2.839 + 0.897 \right) \ kW$$

$$\sum Q_{loss} = 7.593 \ kW$$

Total Heat Loss by Radiation

$$\sum Q_{loss} = Q_{sides} + Q_{rear} + Q_{front} + Q_{top \wedge bottom}$$

$$\sum Q_{loss} = \left(0.35 + 0.0.071 + 0.0.041 + 0.0.492 \right) \ kW$$

$$\sum Q_{loss} = 0.954 \ kW$$

IN SUMMARY

The following calculations are based on a continuous process with the total processing time of 283 s and four running emitters (only in section 1). Thus:

a. On high natural gas supply, 168.809 kW of energy is supplied to eight emitters by natural gas.
b. Electric power consumption per emitter = 168.809/8 = 42.205 kW
c. Energy consumption by four emitters during continuous process = 4 × 42.205 × 141.5 = 23,886.47 kJ
d. Energy loss due to natural gas to IR conversion = 23,886.47 × 0.20 = 4777.29 kJ

e. Energy loss due to heat loss by natural convection = 7.593 × 141.5 = 1074.41 kJ

f. Energy loss due to heat loss by radiation = 0.954 × 141.5 = 134.991 kJ

g. Total heat loss = 4777.29 + 1074.41 + 134.991 = 5986.691 kJ

h. Operation efficiency (O_{eff})

$$O_{eff} = \frac{\text{Energy delivered to potato slices}}{\text{Input energy}} \times 100 = \frac{17{,}889.78}{23{,}886.47} \times 100 = 74.93\%$$

i. 1 m³ of natural gas provides 37,290 kJ energy (1000 Btu/ft³), and its price is 28 cents ($8/1000 ft³) (data provided by equipment manufacturer)

Hence, total natural gas consumption during operation =

$$\frac{23{,}886.47\,\text{kJ}}{\dfrac{37{,}290\,\text{kJ}}{1\,\text{m}^3}} = 0.640\,\text{m}^3 \cdot$$

j. Cost of the operation $= 0.640\,\text{m}^3 \times 28\dfrac{\text{cents}}{\text{m}^3} = 17.935$ cents

k. Test condition was found to be suitable for 2.88-mm-thick potato slices, which have a load density of 2.106 kg/m²

l. Amount of potato slice that can be processed =

$$0.0161\left(\frac{\text{m}}{\text{s}}\right) \cdot 2.106\left(\frac{\text{kg}}{\text{m}^2}\right) \cdot \left(\frac{77 \times 2.54}{100}\,\text{m}\right) \cdot (141.5\,\text{s}) = 9.383\,\text{kg}$$

m. Cost per pound of potato $= \dfrac{0.453\left(\text{kg} = 1\,\text{lb}\right) \cdot 17.935\,\text{cents}}{9.383\,\text{kg}} = 0.86\dfrac{\text{cent}}{\text{lb}}$

Index

A

Absorption phenomena, 5
Acid-dipping treatment, 128, 130
Airflow model, 45
Air jet impingement, 212
Almond
    color, 249, 250
    roasting, 215
American National Standards Institute (ANSI),
        61
Amino acids, 215
ANSI, *see* American National Standards Institute
Anthocyanin, 131
Antioxidants, 217
Apple, 107
    color change, 190
    IR drying, 97
    NIR energy penetration depth, 15
    penetration depth, 28
    PPO extract, 179
    raw, infrared penetration depth, 75
    slices, thin-layer drying, 103
Ascorbic acid
    loss, 108
    oxidation of, 215
Asparagine, 215
Aspartic acid, 215
Aspergillus niger, 227, 229, 230

B

Baby carrots, 193
Bacillus subtilis, 226, 227, 228, 230
Bacteria
    inactivation, 226
    resistance to IR heating, 228
Baking and roasting, 203–223
    air jet impingement, 212
    amino acids, 215
    antioxidants, 217
    ascorbate, 217
    ascorbic acid, oxidation of, 215
    baking, 204–214
        IR-assisted baking, 207–214
        IR baking, 205–206
    bread samples, firmness of, 210
    caramelization products, 207
    carotene, 217

coffee roasting, 217
crispness, patents, 207
crumb structure, 211
denatured enzymes, 217
enzymes responsible for browning, 215
fat replacer, protein-based, 211
flavonoids, 217
flavoprotein, 217
Fourier transform infrared spectroscopy, 212
gluten-free cakes, 208
lipid oxidation, 216
Maillard reaction, 215
main flavor-determining process, 216
microwave baking, disadvantages, 207
monosaccharides, 215
NIR source, 205
nonenzymatic browning, 215
pasteurization, 216
polyphenols, 217
protein-based fat replacer, 211
protein denaturation, 204
protein digestibility, 215
roasting, 215–218
    IR-assisted roasting, 218
    IR roasting, 216–218
starch gelatinization, bakery products, 212
sugar caramelization, 215
tannin, 217
Banana
    Anamur, 179
    color changes, 159
    color degradation, 113
    cross section SEM, 124, 125
    dried, crispness, 163, 164
    drying curves, 150
    drying experiments, 148
    final moisture content, 130
    FIR-LPSSD drying, 161
    FIR-VACUUM drying, 162, 163
    porosity, 164
    rate of moisture reduction, 146
    SEM photographs, 162
    slices, thin-layer drying, 103
    yellowness, 160
Barley, 13, 107
Beer–Lambert's law, 27
Beer's law, 6
Beet, 177
β-Carotene, 12, 141

Biphasic model, 186, 188
Biscuit
    infrared penetration depth, 75
    products, 241
Blackbody, 21
    emissive power spectrum, 4
    material absorption, 90
    perfect, 78
    view factor equation, 78
Blanching, *see* Dry blanching
Blueberry, drying characteristics, 240
Boltzmann's constant, 3
Bouguer's law, 75
Bread
    baking, 239
    biscuit, NIR energy penetration depth, 15
    dried, penetration depth, 28
    penetration depth, 28
    rye, penetration depth, 28
    samples, firmness of, 210
    wheat
        infrared penetration depth, 75
        NIR energy penetration depth, 15
        penetration depth, 28

C

Campylobacter jejuni, 225
Capillary-porous colloidal systems, 29
Carbohydrates
    absorption wavelength, 171
    differential absorption, 91
    physicochemical properties, 29
Carbon twin IR emitter, 69
Carrot
    blanching, 193, 239
    case hardening, 108
    color degradation, 113
    drying rate versus moisture content, 106
    energy consumption during drying of, 107
    infrared penetration depth, 75
    IR drying studies, 97, 103
    NIR energy penetration depth, 15
    penetration depth, 28
    temperature, 143
    thin-layer drying, 103
Case hardening, 92, 108, 160
Cashew, 96
Catalytic gas-fired IR emitter, 68
Catalytic IR (CIR) energy, 133
Ceramic burner, 68
Ceramic IR emitters, 65
Cereal surface, contamination, 228
CFD, *see* Computational fluid dynamics
Chemical groups, IR absorption bands, 12
Chestnut, 239

Chloramphenicol (CP), 231
CIR energy, *see* Catalytic IR energy
Clostridium perfringens, 225
Cocoa, 215
Coffee, 217, 239
Color
    almond, 249, 250
    baked products, 205, 207
    banana, 146, 159
    carrot, 145
    cashew, 97
    change
        almond, 246, 247
        apple, 180, 183, 190
        banana, 159
        onion, 109
        potato, 190
    degradation, 113
    emitters, 58, 66
    hazelnuts, 218
    as indicator of heat output, 58
    loss, 108
    rice, 252
    stabilization, 169
    strawberry, 131
    temperature, 63
    values, strawberry, 131
    wheat, 228
Computational fluid dynamics (CFD), 48
Convective hot air drying, 102
Cookie dough, multiphase porous media model,
        46
Corn meal, inactivation of pathogenic
        microorganisms, 229
CP, *see* Chloramphenicol
Crispness
    comparisons, strawberry slices, 126, 127
    dried bananas, 163, 164
    effect of drying methods on, 124
    patents, 207

D

Dehydrofrozen products, 192
Differential scanning calorimetry (DSC), 212
Diffuse reflection, 23
Direct flame IR radiator, 68
Dough
    macaroni, penetration depth, 28
    wheat
        infrared penetration depth, 75
        NIR energy penetration depth, 15
        penetration depth, 28
Dry blanching (infrared), 169–201
    activation energy, 177, 182
    advances, 192–197

equipment description, 192–193
performance of pilot-scale infrared dry
    blancher, 193–197
characterization of processing conditions,
    172–192
    apple slice processing (case study), 172–192
    comparison between continuous and
        intermittent heating modes, 191–192
    continuous heating mode, 172–181
    intermittent heating mode, 182–191
chemical groups, infrared absorption band
    characteristics of, 171
dehydrofrozen products, 192
effective moisture diffusivity, 183
electrolyte leakage, 171
empirical models, 174
enzyme activity, 177
enzyme inactivation curves, 188
high-pressure blanching, 170
high-temperature short-time blanching, 170
hot gas blanching, 170
incident radiation energy, 171
infrared dry blanching, 171–172
    absorption band characteristics of
        chemical groups of foods, 171
    process parameters, 171–172
low-temperature long-time blanching, 170
moisture diffusivity, 183
moisture loss, 195
moisture ratio, 175
nomenclature, 198
ohmic heating, 170
overview of blanching methods, 170–171
protein, chemical groups, 171
semi-empirical models, 174
surface methodology, 170
two-fraction model, 186
Drying (infrared), 89–99
advantages of IR radiation, 91–92
application of IR radiation in industrial
    drying, 93–95
case hardening, 92
characteristics of IR radiation, 92–93
far-IR, 93
future directions, 97
IR radiation and its propagation during
    drying, 89–90
mechanism, 91
near-IR, 93
propagation of IR radiation, 90
rotational energy, 90
solar lamps, 96
specific applications, 95–97
thermal radiation drying, 91
Dry peeling, 259
DSC, see Differential scanning calorimetry

E

Economic benefits, see Industrial applications
    and economic benefits
Effective inversion temperature, 151
Electrically heated emitters, 94
Emitters and infrared heating system design,
    57–88
carbon twin IR emitter, 69–70
classification of IR emitters, 58–70
    carbon twin IR emitter, 69–70
    electric IR emitter, 60–66
    gas-fired IR emitters, 66–69
    mechanism of infrared absorption by
        foods, 60
comparison of different forms of IR heat,
    72–73
comparison of emitter forms, 59
design variables of IR heating systems, 73–86
    design arrangement for components,
        82–86
    distance between emitter and object, 82–84
    distance between emitters, 85–86
    efficiency of IR heating system, 80–82
    power required for heating, 73–80
drying chamber, power required for heating
    in, 80
electric IR emitter, 60–66
    ceramic IR emitters, 65–66
    quartz tube IR emitter, 64–65
    radiant panels, 66
    reflector-type IR incandescent lamps,
        61–63
    tubular metal-sheathed elements, 66
gas-fired IR emitters, 66–69
    catalytic gas-fired IR emitter, 68–69
    ceramic burner, 68
    direct flame IR radiator, 68
    high-intensity porous burner, 68
    metal fiber burner, 68
heat utilization efficiency, 82
infrared incandescent lamp, 62, 64
longwave IR emitters, 60
nomenclature, 86–87
perfect blackbody, 78
protein, FIR energy absorption, 60
radiation efficiency, definition, 71
relation between wavelength and temperature,
    59
selection criteria of IR emitters, 70–72
    coefficient of radiation efficiency, 71
    response time or speed of heating, 71
    stability of distributed radiation, 72
    temperature of emitter, 70–71
    working life span, 71
sensible heat gain, 79

shortwave emitters, 58
specification of IR emitters, 72
    filament temperature, 72
    maximum power density, 72
    operating voltage, 72
Energy
    absorption mechanisms, 10
    balance, definition, 6
    catalytic IR, 133
    consumption, drying of potato and carrot, 107
    drying, 106
    electromagnetic, 26
    FIR, absorption, 10–11, 60
    rotational, 90
    -scattering polydispersed media, 29
Enterococcus faecium, 247
Enzymes, browning, 215
Equipment, heat losses by natural convection and
        radiation from faces of, 268–273
    heat losses from front side, 269–270
    heat losses from rear side, 270–271
    heat losses from top and bottom side, 271–272
    natural convection heat losses from sides,
        268–269
    radiation heat loss from front side, 270, 271,
        273
    radiation heat loss from sides, 269
Escherichia coli, 230, 231
Escherichia coli O157:H7, 225
Evaporative cooling, 108
Extinction
    definition, 6
    expression of, 27
    spectral extinction coefficient, 6

F

Far-IR radiation (FIR), 2, 93, 118, 238
Fat(s)
    absorption wavelength, 171
    differential absorption, 91
    physicochemical properties, 29
    replacer, protein-based, 211
Fatty acids, 12, 215
Ficus carica L., 49
Fig
    origination, 49
    thermal treatment of for surface
        decontamination, 49–50
FIR, *see* Far-IR radiation
Fish paste, 239
Food(s)
    absorptivity, 9
    applications, heating equipment, 239
    -borne diseases, statistics, 225
    components, absorption bands, 31
    emissivity, definition, 43

infrared penetration depth, 75
properties, wavelength dependence of, 24
texture modification, 124
thermal conductivity, 240
Food materials, infrared radiative properties of,
        19–39
    attenuation of radiation, 29
    Beer–Lambert's law, 27
    blackbody, 21
    capillary-porous colloidal systems, 29
    characteristic dependence of absorptance, 30
    diffuse reflection, 23
    discretized food domain with food holder, 34
    electromagnetic energy, 26
    energy-scattering polydispersed media, 29
    extinction, 27
    food materials, penetration depths, 28
    heating chamber, 35
    incident energy flux, 27
    interaction of radiation with foods, 27–32
        effect of size and form of food particles,
            27–29
        effect of water content, 29–32
        food processing effects, 32
    radiative property control for optimal
        processing, 32–37
    scattering of radiation, 27–29
    spectral emissivity, 21
    spectral properties, 24
    specular reflection, 23
    starch
        gelatinization, 32
        potato, 30
    surface emissivity, definition, 21
    thermal radiation, characteristics of, 20–27
        absorptivity, 22, 25–26
        attenuation, 26–27
        emissivity, 21–22
        fundamental properties, 20–27
        reflectivity, 22–25
        transmissivity, 22–25
Food safety improvement, 225–236
    cereal surface, contamination, 228
    convection–diffusion airflow, 234
    food-borne diseases, statistics, 225
    future research potential, 234
    inactivation mechanism, 228–231
    pathogen inactivation, 226–228
        effect of peak wavelength and bandwidth,
            227
        effect of power and sample temperature,
            226
        effect of sample depth, 227–228
        types of food materials, 228
        types of microorganisms, 228
    selective heating by IR radiation, 231
    thermal death kinetics model, 234

Fourier transform infrared spectroscopy (FTIR),
   5, 212, 231
Freeze-drying, infrared radiation and
      (combined), 117–140
   acid-dipping treatment, 128, 130
   anthocyanin, 131
   catalytic IR energy, 133
   characteristics, 119–133
      dehydration cost, 133
      drying rate, 119–122
      quality attributes of products, 122–132
      rehydration potential, 122–124
      retention of color and chemical
         components, 130–132
      texture modification, 124–130
   commercial potential, 136
   crispness
      comparisons, strawberry slices, 126, 127
      effect of drying methods on, 124
   equipment, 133–134
      commercial-scale dryers, 133
      laboratory-scale dryers, 133–134
   modeling, 134–136
   nomenclature, 138
   oxidative degradation, 130
   prehydration, 117
   principles, 118–119
   protein denaturation, 127
   rehydration capability, 122
   rehydration ratio, 122, 123
   SHAFD products, 123
   starch, crystallinity, 127
   sublimation interface, 119
   sweet potato, experimental drying time, 121
   weight reduction factor, 131
Fruits, see also specific fruit
   dry-blanching of, 242
   IR drying studies, 97
   vitamin requirements, 102
FTIR, see Fourier transform infrared
      spectroscopy
Fundamentals and theory of infrared radiation,
      1–18
   absorption phenomena, 5
   blackbody emissive power spectrum, 4
   body reflection, 12
   Boltzmann's constant, 3
   capillary-porous bodies, 5
   chemical groups, IR absorption bands, 12
   classes of infrared radiation, 2
   energy absorption mechanisms, 10
   energy balance, 6
   extinction of radiation, 5–6
   food absorptivity, 9
   group frequencies, 10
   heating chamber, 7
   hydrogen bonds, 10

light scattering, 11
lipids, 11
nomenclature, 16–17
nutrition retention, 12
physiological effects of IR radiation, 10–13
Planck's law, 3
point sources, 3
protein
   chemical groups, 12
   differential energy absorption, 1
   FIR energy absorption, 10–11
quad flat packages, 14
radiative heat transfer, basic laws of, 2–5
   Planck's law, 3
   Stefan–Boltzmann's law, 5
   Wien's displacement law, 3–5
regular reflection, 12
spectral extinction coefficient, 6
spectral selectivity for IR heating, 13–16
starch
   FIR energy absorption, 10–11
   gelatinization, 13
   shortwave IR radiation treatment, 12
surface-mounted devices, 13
view factor, 3, 7
waveguide, 3, 8
Fusarium proliferatum, 227, 229

G

Garlic, thin-layer drying, 103
Gas-filled lamps, 62
Gas-fired emitters, 66, 67, 94
Glutamic acid, 215
Glutamine, 215
Gluten-free cakes, 208
Grain, wheat
   infrared penetration depth, 75
   NIR energy penetration depth, 15
Grape, IR drying, 97
Green bell pepper, blanching, 193
Green tea, 217, 239
Group frequencies, 10

H

Halogen lamp, 63, 208
Hamburger
   nutrient analysis, 12
   patties, 241
Hazelnuts, 215, 218
Heat and mass transfer modeling, 41–55
   food emissivity, definition, 43
   geometry of materials, 41
   IR heating model, 42–48
      airflow model, 45
      conductive heat transfer model, 45–47

numerical simulation, 47–48
    radiation model, 42–45
model prediction, 48–53
mold ascospore inactivation, 52
Monte Carlo ray-tracing technique, 42
multiphase porous media model, 46
photon tracking, 44
pseudorandom variable, 43
thermal treatment of fresh fruit for surface
    decontamination, 49–53
    figs, 49–50
    peach, 51–53
    strawberry, 50–51
wood heating, 48–49
Heat output, indicator of, 58
High-intensity porous burner, 68
High-temperature short-time (HTST) blanching,
    170
Histidine, 215
Honey
    inactivation of pathogenic microorganisms,
    229
    yeasts in, 228
Hot air drying, infrared and (combined), 101–116
    ascorbic acid loss, 108
    case hardening, 108
    convective hot air drying, 102
    dryer heat utilization efficiency, 107
    drying conditions versus drying times, 104
    drying energy, 106–108
    drying time, drying curve, and drying rate,
    103–106
    dry product, pore structure of, 108
    energy consumption, drying of potato and
    carrot, 107
    evaporative cooling, 108
    heating up period, 105
    intermittent and continuous uses of IR and
    convective heating, 109–110
    intermittent infrared drying, 111
    modeling studies, 110
    modeling works, 112
    need for products in powder form, 102
    oldest preservation method, 102
    optimization, 110–113
    process optimization, 113
    product quality, 108–109
    quality losses, 108
    quality parameters, 108
    rehydration capacity of dried product, 108
    rising period, 105
    second falling rate period, 106
    sun drying, 102
HTST blanching, see High-temperature short-
    time blanching

I

IDB technology, see IR dry-blanching technology
Incandescent lamp, 62, 64
Industrial applications and economic benefits,
    237–271
    development of IR food and agricultural
    processing technologies, 238–243
    baking, cooking, blanching, and roasting,
    241–243
    drying, 239–240
    equipment, 238–239
    pasteurization, 240–241
    peeling, 243
    thawing, 243
    economic benefits, 259–262
    emerging novel industrial applications,
    244–259
    combined infrared and freeze-drying,
    244
    IR dry roasting and pasteurization,
    244–249
    IR radiation dry peeling, 259
    simultaneous IR dry blanching and
    dehydration, 244
    simultaneous IR drying and
    disinfestations, 249–259
    energy efficiency and cost analysis of small
    commercial IR unit for dry blanching,
    267–274
    assumptions, 267
    energy loss due to conversion of natural
    gas to infrared energy, 267
    Greek symbols, 268
    heat losses by natural convection and
    radiation from faces of equipment,
    268–273
    nomenclature, 267–268
    subscripts, 268
    summary, 273–274
    grain sterilizer, 240
    heating equipment, food applications, 239
    industrialization needs, 263
    insects, 250
    microwave thawing, 243
    starch, thermomechanical properties, 255
    water hydration rate, 242
Infrared heating system design, see Emitters and
    infrared heating system design
Infrared (IR) radiation, 20
    advantages, 1
    artificial radiation drying, 91
    blackbody source, 3
    cell structure change by, 13
    characteristics, 92–93

conversion rate for natural gas to, 262
definition of, 20
drying of onions, 97
dry peeling, 259
energy balance for food receiving, 74
high-intensity porous burners, 68
local radiative heat flux, 35
metal fiber burner, 68
microbial inactivation, 227
penetration capacity, 228
penetration depths, 70, 203
physiological effects, 10–13
propagation during drying, 89–90
rehydration capacity and, 109
rough rice drying, 93
selective heating by, 231
IR dry-blanching (IDB) technology, 198, 244
IR radiation, *see* Infrared radiation

L

Lamp(s)
gas-filled, 62
halogen, 63, 208
incandescent, 62, 64
quarts, 73
radiation energy distribution, 83
solar, 96
tungsten halogen, 62
Law
Beer–Lambert's, 27
Beer's, 6
Bouguer's, 75
conservation of energy, 20
cosine, 82
Lambert's, 78
optics, 91
Planck's law, 3
Stefan–Boltzmann, 5, 58
Wien's displacement, 3, 21
LEDs, *see* Light-emitting diodes
Leek, 107
Light-emitting diodes (LEDs), 36
Lignans, 215
Lipids, 11, 216
Liquid foods, 239
Listeria monocytogenes, 225, 229, 240, 241
Liver, 239
Low-pressure superheated steam drying
(LPSSD), 143, 148
Low-temperature long-time (LTLT) blanching,
170
LPSSD, *see* Low-pressure superheated steam
drying

LTLT blanching, *see* Low-temperature long-time
blanching
Lysine, 215

M

Maillard reaction, 208, 215
Mass transfer modeling, *see* Heat and mass
transfer modeling
Meat, multiphase porous media model, 46
Metal fiber burner, 68
Microwave baking, disadvantages, 207
Microwave thawing, 243
Mid-IR radiation (MIR), 2, 118, 169, 238
Minerals
figs, 49
retention of, 171
MIR, *see* Mid-IR radiation
Model
airflow, 45
biphasic, 186, 188
diffusion, 134
exponential, 134, 135
first-order kinetics, 183
Henderson and Pabis, 138
internal heat transfer, 44
IR heating, 42–48
airflow model, 45
conductive heat transfer model, 45–47
numerical simulation, 47–48
radiation model, 42–45
kinetic, enzyme inactivation, 177
microbial kinetics, 50
Monte Carlo IR model, 42
multiphase porous media, 46
Page, 112, 136
peach IR heating, 53
POD inactivation, 182
PPO inactivation, 179
thermal death kinetics, 234
two-fraction, 186
Mold
inactivation, 52, 226, 229
resistance to IR heating, 228
Monilinia fructigena, 51, 52, 229
Monosaccharides, 215
Monte Carlo ray-tracing technique, 42
Multiphase porous media model, 46

N

NA, *see* Nalidixic acid
Nalidixic acid (NA), 231
Natural gas, conversion rate, 262
Near-IR radiation (NIR), 2, 93, 118, 227, 238

NIR, *see* Near-IR radiation
Nonenzymatic browning, 215
Nutrition retention, 12

O

Onion
    blanching, 193
    dehydration, 240
    drying of, 97
    inactivation of pathogenic microorganisms,
        229
    thin-layer drying, 103
Orange juice, 33
Oyster, 239

P

Packed foods, 239
Paddy rice, 103, 249
Page model, 112, 136
Particulate foods, 239
Pasteurization, 216, 239, 240–241
Pathogen inactivation, 226
PCG, *see* Penicillin
Peach, 51
    energy consumption rate, 113
    IR drying, 97
    peeling, 259
    surface decontamination, 53
    thermal treatment of for surface
        decontamination, 51–53
    thin-layer drying, 103
Peanuts, roasting, 215
Pear, thin-layer drying, 103
Pediococcus, 247, 249
Penicillin (PCG), 231
Penicillium, 228
Peroxidase (POD)
    inactivation, 186, 191
    intermittent heating mode, 242
    thermal stability of, 180
Phenylalanine, 215
Phomopsis sp., 52
Photon tracking, 44
Phytosterols, 215
Planck's law, 3
POD, *see* Peroxidase
Point sources, 3
Polyphenol oxidase (PPO), 172, 178, 183
Potato
    blanching, 193, 260
    case hardening, 108
    color
        change, 190
        degradation, 113

dry
    infrared penetration depth, 75
    penetration depth, 28
energy consumption during drying of, 107
IR drying studies, 97, 103
multiphase porous media model, 46
NIR energy penetration depth, 15
peeling, 239, 259
penetration depth, 28
raw
    infrared penetration depth, 75
    penetration depth, 28
rehydration, 124
russet, blanching, 193, 262
slice, spectral transmissivity, 25
spectral reflectance, 26
starch, 30, 32
thin-layer drying, 103
PPO, *see* Polyphenol oxidase
Protein
    -based fat replacer, 211
    chemical groups, 12, 171
    denaturation, 127, 204
    differential absorption, 91
    digestibility, 215
    FIR energy absorption, 10–11, 60
    physicochemical properties, 29
    soy, 15
Purawave™, 211

Q

QFPs, *see* Quad flat packages
Quad flat packages (QFPs), 14
Quartz lamps, 73
Quartz tube IR emitter, 64

R

Radiation
    attenuation of, 29
    efficiency, definition, 71
    -emitting diodes (REDs), 36
    extinction, 5, 6
    heat transfer
        advantages, 91
        coefficient, 86
        diffuse approximation, 22
        efficiency, 58
        far-IR radiator, 155
        gray approximation, 22
        IR wavelength, 118
        large-size IR ovens, 83
        model, 50
        photon events, 45
        thermodynamic equilibrium, 74
    scattering of, 27–29

Radiative heat transfer, basic laws of, 2–5
  Planck's law, 3
  Stefan–Boltzmann's law, 5
  Wien's displacement law, 3–5
Rapid viscoanalyser (RVA), 212
Red pepper, 97, 103, 107
REDs, see Radiation-emitting diodes
Rehydration ratio, 122, 123
RFP, see Rifampicin
Riboflavin, 12
Rice
  paddy, 249
  whiteness index, 252
Rifampicin (RFP), 231
Rising period, 105
RMSE, see Root mean squared error
Roasting, see Baking and roasting
Root mean squared error (RMSE), 136
Rough rice, thin-layer drying, 103
Russet potatoes, 193
RVA, see Rapid viscoanalyser

S

Salmonella
  enteritidis, 241, 245
  enteriditis PT 30, 216, 247
  spp., 225
Scanning electron microscopy (SEM), 13, 124, 211
Second falling rate period, 106
SEM, see Scanning electron microscopy
Sequential hot air– and freeze-dried (SHAFD) products, 123
Sequential IR radiation and freeze-drying (SIRFD), 117, 244
Sequential IR radiation and hot air (SIRHA), 245
SHAFD products, see Sequential hot air– and freeze-dried products
Simplesse™, 211
Simultaneous infrared dry blanching and dehydration (SIRDBD), 169
SIRDBD, see Simultaneous infrared dry blanching and dehydration
SIRFD, see Sequential IR radiation and freeze-drying
SIRHA, see Sequential IR radiation and hot air
SMDs, see Surface-mounted devices
Software
  CFX 11.0, 50
  computational fluid dynamics, 48
  Navier–Stokes equations, 50
  SPSS, 134
Solar lamps, 96
Soybean, inactivation of pathogenic microorganisms, 229
Soy protein, 15

Spectral emissivity, 21
Spectral properties, 24
Specular reflection, 23
Spores, inactivation, 226
SSD, see Superheated stream drying
Staphylococcus aureus, 225, 227, 241
Starch
  crystallinity, 127
  FIR energy absorption, 10–11, 60
  gelatinization, 13, 32, 212
  potato, 30
  shortwave IR radiation treatment, 12
  thermomechanical properties, 255
Stefan–Boltzmann law, 5, 58
Strawberry
  blanching, 239
  brightness/whiteness, 130
  color values, 131
  crispness comparisons, 126, 127
  cross section SEM, 124, 125
  dehydration curves, 120
  disinfection, 51
  drying rate, 119
  effect of drying method on appearance, 132
  firmness comparisons, 128
  inactivation of pathogenic microorganisms, 229
  microbial risk assessment, 50
  rehydration ratios, 123
  SIRFD process, 136
  surface decontamination, 234, 241
  thermal treatment of for surface decontamination, 50–51
Sugar caramelization, 215
Sun drying, 102
Superheated stream drying (SSD), 148
Surface emissivity, definition, 21
Surface-mounted devices (SMDs), 13
Sushi, 239
Sweet corn, thin-layer drying, 103
Sweet potato, 239
  drying characteristics, 120
  drying time, 121
  thin-layer drying, 103
System design, see Emitters and infrared heating system design

T

Temperature
  carrot, 143
  color, 63
  effective inversion, 151
  emitter, 70
  relationship between wavelength and, 59
Thawing process, main goal, 243

Theory, *see* Fundamentals and theory of infrared
    radiation
Thermal death kinetics model, 234
Thermal radiation, 20–27
    absorptivity, 25–26
    attenuation, 26–27
    drying, 91
    emissivity, 21–22
    fundamental properties, 20–27
    reflectivity, 22–25
    transmissivity, 22–25
Thiamin, 12
Threonine, 215
Tomato
    nutrient analysis, 12
    paste
        infrared penetration depth, 75
        NIR energy penetration depth, 15
        penetration depth, 28
    peeling, 239
Total rice yield (TRY), 251
TRY, *see* Total rice yield
Tuna, thawing, 243
Tungsten halogen lamps, 62
Turkey
    frankfurters, 229
    oil-browned deli, 229
Two-fraction model, 186
Tyrosine, 215

V

Vacuum infrared drying, 141–167
    advances, 148–165
        energy consumption of process, 158
        quality of dried products, 159–165
    banana slices, 146
    case hardening, 160
    crispness, dried bananas, 163, 164
    effective inversion temperature, 151
    FIR-VACUUM, 146
    foods and agricultural materials, 143–148
    internal vapor pressure, 141
    low-pressure superheated steam drying, 143,
        148
    moisture migration, 141
    rehydration behavior, 161
    schematic diagram, 142, 149
    specific energy consumption, 151
    system, 142
Vegetable(s), *see also specific vegetable*
    dry-blanching of, 242
    drying, 239
    IR drying studies, 97
    moisture loss during dry blanching, 195
    vitamin requirements, 102
View factor, 3, 7
Vitamin(s)
    losses, 106
    requirements, vegetables, 102
    retention of, 171
Vitamin A, 12
Vitamin C, 12
Vitamin E, 215

W

Water
    heat capacity, 79
    latent heat for vaporization, 47
    water hydration rate, 242
Waveguide, 3, 8
Welsh onion, 145, 160
Wheat
    discoloration, 228
    grains, penetration depth, 28
    inactivation of pathogenic microorganisms,
        229
    infrared penetration depth, 75
    NIR energy penetration depth, 15
Whiteness index (WI), 252
WI, *see* Whiteness index
Wien's displacement law, 3–5, 21

Y

Yeast
    honey, 228
    inactivation, 226
    resistance to IR heating, 228

Milton Keynes UK
Ingram Content Group UK Ltd.
UKHW040447071024
449327UK00020B/1059